Lecture Notes in Computer Science　　13766

More information about this series at https://link.springer.com/bookseries/558

Vladimir M. Vishnevskiy ·
Konstantin E. Samouylov ·
Dmitry V. Kozyrev (Eds.)

Distributed Computer and Communication Networks

Control, Computation, Communications

25th International Conference, DCCN 2022
Moscow, Russia, September 26–29, 2022
Revised Selected Papers

 Springer

Editors
Vladimir M. Vishnevskiy [iD]
V. A. Trapeznikov Institute of Control
Sciences of Russian Academy of Sciences
Moscow, Russia

Konstantin E. Samouylov [iD]
Peoples' Friendship University of Russia
Moscow, Russia

Dmitry V. Kozyrev [iD]
V. A. Trapeznikov Institute of Control
Sciences of Russian Academy of Sciences
Moscow, Russia

ISSN 0302-9743   ISSN 1611-3349 (electronic)
Lecture Notes in Computer Science
ISBN 978-3-031-23206-0   ISBN 978-3-031-23207-7 (eBook)
https://doi.org/10.1007/978-3-031-23207-7

This Springer imprint is published by the registered company Springer Nature Switzerland AG
The registered company address is: Gewerbestrasse 11, 6330 Cham, Switzerland

Preface

This volume contains a collection of revised selected full-text papers presented at the 25th International Conference on Distributed Computer and Communication Networks (DCCN 2022), held in Moscow, Russia, during September 26–29, 2022.

The conference is a continuation of the traditional international conferences of the DCCN series, which have taken place in Sofia, Bulgaria (1995, 2005, 2006, 2008, 2009, 2014); Tel Aviv, Israel (1996, 1997, 1999, 2001); and Moscow, Russia (1998, 2000, 2003, 2007, 2010, 2011, 2013, 2015, 2016, 2017, 2018, 2019, 2020, 2021) in the last 25 years. The main idea of the conference is to provide a platform and forum for researchers and developers from academia and industry from various countries working in the area of theory and applications of distributed computer and communication networks, mathematical modeling, and methods of control and optimization of distributed systems, by offering them a unique opportunity to share their views, as well as discuss the prospective developments, and pursue collaboration in this area. The content of this volume is related to the following subjects:

- Communication networks, algorithms, and protocols
- Wireless and mobile networks
- Computer and telecommunication networks control and management
- Performance analysis, QoS/QoE evaluation, and network efficiency
- Analytical modeling and simulation of communication systems
- Evolution of wireless networks toward 5G
- Centimeter- and millimeter-wave radio technologies
- Internet of Things and fog computing
- Machine learning, big data, and artificial intelligence
- Probabilistic and statistical models in information systems
- Queuing theory and reliability theory applications
- High-altitude telecommunications platforms

The DCCN 2022 conference gathered 130 submissions from authors from 18 different countries. From these, 96 high-quality papers in English were accepted and presented during the conference. The current volume contains 33 extended papers which were recommended by session chairs and selected by the Program Committee for the Springer post-proceedings. Thus, the acceptance rate is 34.4%.

All the papers selected for the post-proceedings volume are given in the form presented by the authors. These papers are of interest to everyone working in the field of computer and communication networks.

We thank all the authors for their interest in DCCN, the members of the Program Committee for their contributions, and the reviewers for their peer-reviewing efforts.

September 2022

Vladimir M. Vishnevskiy
Konstantin E. Samouylov
Dmitry V. Kozyrev

Organization

DCCN 2022 was jointly organized by the Russian Academy of Sciences (RAS), the V.A. Trapeznikov Institute of Control Sciences of RAS (ICS RAS), the Peoples Friendship University of Russia (RUDN University), the National Research Tomsk State University, and the Institute of Information and Communication Technologies of the Bulgarian Academy of Sciences (IICT BAS).

Program Committee Chairs

V. M. Vishnevskiy (Chair)	ICS RAS, Russia
K. E. Samouylov (Co-chair)	RUDN University, Russia

Publication and Publicity Chair

D. V. Kozyrev	ICS RAS and RUDN University, Russia

International Program Committee

S. M. Abramov	Program Systems Institute of RAS, Russia
A. M. Andronov	Transport and Telecommunication Institute, Latvia
T. Atanasova	IICT BAS, Bulgaria
S. E. Bankov	Kotelnikov Institute of Radio Engineering and Electronics of RAS, Russia
A. S. Bugaev	Moscow Institute of Physics and Technology, Russia
S. R. Chakravarthy	Kettering University, USA
D. Deng	National Changhua University of Education, Taiwan
S. Dharmaraja	Indian Institute of Technology, Delhi, India
A. N. Dudin	Belarusian State University, Belarus
A. V. Dvorkovich	Moscow Institute of Physics and Technology, Russia
D.V. Efrosinin	Johannes Kepler University Linz, Austria
Yu. V. Gaidamaka	RUDN University, Russia
Yu. V. Gulyaev	Kotelnikov Institute of Radio-Engineering and Electronics of RAS, Russia
V. C. Joshua	CMS College Kottayam, India
H. Karatza	Aristotle University of Thessaloniki, Greece

N. Kolev	University of São Paulo, Brazil
G. Kotsis	Johannes Kepler University Linz, Austria
A. E. Koucheryavy	Bonch-Bruevich Saint Petersburg State University of Telecommunications, Russia
A. Krishnamoorthy	Cochin University of Science and Technology, India
N. A. Kuznetsov	Moscow Institute of Physics and Technology, Russia
L. Lakatos	Budapest University, Hungary
E. Levner	Holon Institute of Technology, Israel
S. D. Margenov	IICT BAS, Bulgaria
N. Markovich	ICS RAS, Russia
A. Melikov	Institute of Cybernetics of the Azerbaijan National Academy of Sciences, Azerbaijan
E. V. Morozov	Institute of Applied Mathematical Research of the Karelian Research Centre RAS, Russia
A. A. Nazarov	Tomsk State University, Russia
I. V. Nikiforov	Université de Technologie de Troyes, France
S. A. Nikitov	Kotelnikov Institute of Radio Engineering and Electronics of RAS, Russia
D. A. Novikov	ICS RAS, Russia
M. Pagano	University of Pisa, Italy
V. V. Rykov	Gubkin Russian State University of Oil and Gas, Russia
R. L. Smeliansky	Lomonosov Moscow State University, Russia
M. A. Sneps-Sneppe	Ventspils University College, Latvia
A. N. Sobolevski	Institute for Information Transmission Problems of RAS, Russia
S. N. Stepanov	Moscow Technical University of Communication and Informatics, Russia
S. P. Suschenko	Tomsk State University, Russia
J. Sztrik	University of Debrecen, Hungary
S. N. Vasiliev	ICS RAS, Russia
M. Xie	City University of Hong Kong, Hong Kong, China
A. Zaslavsky	Deakin University, Australia

Organizing Committee

V. M. Vishnevskiy (Chair)	ICS RAS, Russia
K. E. Samouylov (Vice Chair)	RUDN University, Russia
D. V. Kozyrev (Publication and Publicity Chair	ICS RAS and RUDN University, Russia
A. A. Larionov	ICS RAS, Russia

S. N. Kupriyakhina	ICS RAS, Russia
S. P. Moiseeva	Tomsk State University, Russia
T. Atanasova	IICT BAS, Bulgaria
I. A. Kochetkova	RUDN University, Russia

Organizers and Partners

Organizers

Russian Academy of Sciences (RAS), Russia
V.A. Trapeznikov Institute of Control Sciences of RAS, Russia
RUDN University, Russia
National Research Tomsk State University, Russia
Institute of Information and Communication Technologies of Bulgarian Academy of
   Sciences, Bulgaria
Research and Development Company "Information and Networking Technologies",
   Russia

Support

Information support was provided by the Russian Academy of Sciences. The conference
was organized with the support of the IEEE Russia Section, Communications Society
Chapter (COM19), and the RUDN University Strategic Academic Leadership Program.

Contents

Computer and Communication Networks

New Filtering Method to Reduce PAPR and OOBE of UFMC in 5G
Communication System 3
   *Yousif I. Hammadi, Riyadh Khlf Ahmed, Omar Abdulkareem Mahmood,
   and Ammar Muthanna*

A Novel Technique for Creating Optical Multi-carrier Generation Using
Nested Electro-Absorption Modulators 17
   *Mohammed Hasan Alwan, Yousif I. Hammadi, Mamoon A. Muhi,
   Omar Abdulkareem Mahmood, Alexey Tselykh,
   and Mohammed Saleh Ali Muthanna*

Discrete Time Markov Chain for Drone's Buffer Data Exchange
in an Autonomous Swarm 29
   P. Keyela, I. S. Yartseva, and Yu. V. Gaidamaka

Construction and Analysis of a Queueing Model with Service Prioritization
for 5G Systems with Customizable Network Slice Instances 41
   Y. Adou, E. Markova, and Yu. V. Gaidamaka

Spectrum and AI-based Analysis for a Flight Environment and Virtual
Obstacles Avoidance Using Potential Field Method for Path Control 54
   Ayham Shahoud, Dmitriy Shashev, and Stanislav Shidlovskiy

Clusters of Exceedances for Evolving Random Graphs 67
   Natalia M. Markovich and Maksim S. Ryzhov

Estimation of the Tail Index of PageRanks in Random Graphs 75
   Natalia M. Markovich and Maksim S. Ryzhov

Optimization of the Code Division Process Using Multilayer Orthogonal
Structures Based on M-Sequences 90
   D. Kukunin, A. Berezkin, and R. Kirichek

Research and Development of Data Compression Methods Based
on Neural Networks 103
   A. Berezkin, D. Kukunin, and R. Kirichek

Anomaly Electrocardiograms Automatic Detection with Unsupervised
Deep Learning Methods 117
  Eugene Yu. Shchetinin, Anastasia G. Glushkova,
  and Leonid A. Sevastianov

Using Neural Networks for Channel Quality Prediction in Wireless 5G
Networks 132
  Ekaterina Bobrikova, Anna Platonova, Ekaterina Medvedeva,
  Yu. V. Gaidamaka, and Sergey Shorgin

Analysis and Formalization of Requirements of URLLC, mMTC, eMBB
Scenarios for the Physical and Data Link Layers of a 5G Mobile Transport
Network 144
  Dmitry Aminev, Evgenia Bogdanova, and Dmitry Kozyrev

Traffic Arrival Model for Millimeter Wave 5G NR Systems 161
  E. M. Khayrov, V. A. Prosvirov, and Anna Platonova

Model for Analyzing Impact of Path Loss on eMBB Bit Rate Degradation
Under Priority URLLC Transmission in 5G Network 176
  Irina Kochetkova, Elena Makeeva, Anastasia Ageeva,
  and Andrey Gorshenin

Firewall Simulator Development for Performance Evaluation of Ranging
a Filtration Rules Set 190
  A. Yu. Botvinko and K. E. Samouylov

Analytical Modeling of Distributed Systems

Performance Analysis for Tethered HAP Systems: An Analytical Approach 205
  Dharmaraja Selvamuthu, Vidyottama Jain, and Raina Raj

Analysis of Power Management in a Tethered High Altitude Platform
Using $MAP/PH[3]/1$ Retrial Queueing Model 218
  Vidyottama Jain, V. M. Vishnevsky, Dharmaraja Selvamuthu,
  and Raina Raj

Investigation of a Finite-Source Retrial Queueing System with Two-Way
Communication, Catastrophic Breakdown and Impatient Customers Using
Simulation 231
  János Sztrik, Ádám Tóth, Ákos Pintér, and Zoltán Bács

Analysis of Retrial Queuing System with Limited Processor Sharing
Discipline and Changing Effective Bandwidth 243
  Alexander Dudin, Sergey Dudin, Olga Dudina, and Chesoong Kim

On the Distribution of the Number of Consecutively Lost Customers
in the *BMAP/PH/1/N* System 257
Valentina Klimenok and Alexander Dudin

Analysis of Tandem Retrial Queue with Common Orbit and MMPP
Incoming Flow 270
Anatoly Nazarov, Svetlana Paul, Tuan Phung-Duc, and Mariya Morozova

Average Cost Minimization in a Multi-server Retrial Queueing System
with a Controllable Reserve Group of Servers 284
Dmitry Efrosinin and Natalia Stepanova

Verification of Stability Condition in Unreliable Two-Class Retrial System
with Constant Retrial Rates 297
Ruslana Nekrasova, Evsey Morozov, and Dmitry Efrosinin

The Queueing System with Threshold-Based Direct and Inverse General
Renovation Mechanism 309
Viana C. C. Hilquias, I. S. Zaryadov, S. I. Matyushenko,
and T. A. Milovanova

Effective Algorithm of Estimation the Performance Measures of Group
of Servers with Dependence of Call Repetition on the Type of Call Blocking ... 324
Mikhail S. Stepanov, Sergey N. Stepanov, and Fedor S. Kroshin

Transient Behaviour of Finite-Source Single-Line Queueing Systems
with Jumps of Network Traffic 338
V. M. Vishnevsky, K. A. Vytovtov, E. A. Barabanova, G. K. Vytovtov,
and A. V. Dvorkovich

Probability Density of the Interval Duration Between Events
in the Generalized MAP with Its Incomplete Observability 349
Anastasia Keba and Ludmila Nezhel'skaya

Inventory Control with Returns and Controlled Markov Queueing Systems 361
S. Granin, V. Laptin, and A. Mandel

On Estimating the Average Response Time of High-Performance
Computing Environments 371
A. V. Gorbunova and V. M. Vishnevsky

Exponential Splitting Based Artificial Regeneration in Supercomputer
Queueing Model 385
Alexander Rumyantsev and Irina Peshkova

Modeling and Analysis of Multi-channel Queuing System Transient
Behavior for Piecewise-Constant Rates 397
   K. A. Vytovtov, E. A. Barabanova, and V. M. Vishnevsky

On the Reliability Estimation of the Gaussian Multi-phase Degradation
System 410
   Oleg Lukashenko

On Steady State Reliability and Sensitivity Analysis of a *k*-out-of-*n*
System Under Full Repair Scenario 422
   N. M. Ivanova

Author Index 435

Computer and Communication Networks

New Filtering Method to Reduce PAPR and OOBE of UFMC in 5G Communication System

Yousif I. Hammadi[1], Riyadh Khlf Ahmed[2],
Omar Abdulkareem Mahmood[2], and Ammar Muthanna[3,4(✉)]

[1] Bilad Alrafidain University College, Diyala, Iraq
[2] Department of Communications Engineering, College of Engineering,
University of Diyala, Diyala, Iraq
[3] Department of Telecommunication Networks and Data Transmission,
The Bonch-Bruevich Saint-Petersburg State University of Telecommunications,
Saint Petersburg 193232, Russia
ammarexpress@gmail.com
[4] Department of Applied Probability and Informatics, Peoples' Friendship University
of Russia (RUDN University), 117198 Moscow, Russia

Abstract. Several aspects of the new communication systems reflect the flexibility, coexistence, and diversity of 5G systems. Filtering communication window is a well-known method for generating desired Universal Filtered Multi Carrier (UFMC). Researchers are continuously developing new filtering methods to reduce PAPR and OOBE produced by the methods of traditional multicarrier in the communication systems, due to the current need for new technologies in 5G communication systems. When it comes to wireless communication systems, the UFMC is often viewed as the most significant technology, this study contributed in developing new filtering method to attain a better match with current necessities and requirements for new technologies. The proposed method entails deploying pulse shape windowing and evaluating windowing filters across multiple communication bands. Through assessment, this process reduces PAPR and OOBE while increasing spectral efficiency. The results showed the use of new filtering methods outperformed the previous proposed method -70 dBm for the OOBE reduction and 10 dB for PAPR reduction. It could be concluded that the new method produced significant results in UFMC systems, giving 5G wireless communication systems a competitive advantage.

Keywords: UFMC · PAPR · Filtering · OOBE

International Telecommunication Union (ITU) has classified the 5th generation (5G) wireless communication systems as Modern wireless telecommunications technology to allow for a wide range of services, including the main connecting technique that supports internet service with significantly enhanced throughput called enhanced Mobile Broadband (eMBB), further, the devices that provide

V. M. Vishnevskiy et al. (Eds.): DCCN 2022, LNCS 13766, pp. 3–16, 2022.
https://doi.org/10.1007/978-3-031-23207-7_1

communications between machines named massive Machine Type communications (mMTC), moreover, device types by which low round trip signal communication, low latency, which is so called Ultra-Reliable Low-Latency Communication (URLLC), which is also stands for very fast processing time or very low round trip time for the signal, Internet of Things (I-IoT), besides other situations [7]. That is, an adaptable waveform design is essential for a wide range of potential applications and scenarios. Due to the waveform pulse shape being the major air interface piece, the waveform shape appears to have been important for all these application situations as well. As a result, waveforms for 5G or even beyond must be built with adaptation and agility in consideration. Prior to 5G, many systems and standards, for instance the 4th generation, relied on OFDM technology and with diminutive peak power ratio, which is well-known as Peak to Average Power Ratio (PAPR), variant [22]. They do, however, have a number of downsides [5]. The use of UFMC (Universal Filtered Multi-Carrier) waveforms has a number of drawbacks, according to previous research, notably effective implementations (cheap cost, minimum complexity), simple adjustment on the recipient side, and seamless implementation with MIMO system architectures [4,12]. Because of the high PAPR, several academics have been working to develop better signals for 5G and beyond networks, the necessity for complete synchronization, and the substantial OOBEs that are generated by traditional UFDM systems [8,9,20]. In order to use 5G numerology [17], the current standard requires adaptive waveform characteristics, such as 15, 30, and 60 kHz subcarrier spacing and a specified symbol length [15]. It will also be necessary to use these waveforms in order to meet future requirements. When the subcarrier spacing varies from 15 to 30 kHz, for example, this leads in interaction with other spectral bands in the UFMC's orthogonality [16], which is undesirable. In order to minimize carrier interference, techniques including windowing, filtering, and guard band allocation are frequently employed [4]. But there are other waveforms available to meet the requirements of adaptation while enhancing OFDM's transmission range and elasticity [3]. For 5G and beyond, there is no known uniform dynamic waveform that can be utilized for all applications. Rectangle windows are used in the UFDM waveform to represent the sinc-shape of the frequency domain. Sidelobes (at 1/f, f is frequency) in the sinc-function are the fundamental drawback of UFMC for high OOBE, since they limit coexistence between adjacent resources and increase ACL (Adjacent Channel Leakage Ratio) (ACLR). Several windowing frequencies have been proposed to mitigate OOBE leakage because delay synchronization improves latency while increasing power consumption. Localization of amplitude and phase amplitude and the ability to enable short message transmissions are essential criteria for 5G [13]. In 5G subcarrier and subband waveforms, the use of filtering and windowing has been proposed as a means to decrease out-of-band releasing and permit asynchronous transporting [10,24]. The signal-to-noise ratio of conventional OFDM is significantly lower than that of subcarrier-based filtering methods, for instance the Filter Bank Multicarrier (FBMC) [11], where each subcarrier being filtered individually, and Generalized Frequency Division Multiplexing (GFDM)

[1]. Latency and system complexity are both negatively affected by the extended tail of filter impulse responses required by FBMC and GFDM filters in order to accomplish this performance increase. Subband filtering and windowing are acceptable solutions for waveforms such as 5G and beyond. While full band filtering may significantly minimize inter-carrier and inter-carrier interference (ICI), and inter-symbol interference (ISI) impacts on typical UFMC throughout almost entire communication band by employing only one individual filter [12]. The number of filter cassettes must be increased, hence, the OOBE restrictions will be exceeded, but we have to taking into account the complexity as well as latency [6]. It is common to see the UFMC waveform in 5G and future networks. An extensive body of research has demonstrated how UFMC improves traffic separation and resilience to time-frequency imbalance in 5G waveform implementations, fragmented spectrum support, sporadic network access, predominantly for IoT, the complexity ranges from low and medium, technical and information basis for OFDM while applying the filtering system to a collection of subchannels, and a short filter length. Digital signal processing uses window functions to reduce side lobe corruption at the subcarrier's border, which helps to improve overall signal quality. ICI and ISI may be distinguished from one another because Gaussian pulses behave identically to sinusoidal pulses in frequency as well as the time working domains, but fail spectacularly in orthogonality [21]. Likewise, the shape chosen must improve spectrum utilization, temporal and frequency division, as well as subcarrier orthogonality, among other things. It is just not feasible to have all of these characteristics simultaneously [2]. As a result, the waveform should be built in such a way that it provides a worthwhile trade-off between time and frequency segmentation, as well as increased spectrum performance over the present waveform. The aim of this paper was to assess the efficacy of a new filtering technique that utilizes variable windowing forms. The dependence on certain factors or elements in the frequency-time domain may have an effect on performance, OOBE of a multicarrier signal may now be managed with greater freedom because to the new method's assured trade-off for various pulse forms. For instance, the sidelobe of an OFDM subcarrier contributes to OOBE. Thus, edge subcarriers have been addressed using a variety of windowing techniques to reduce side-lobes, with smaller windows used for inner circumstances, such as Bohman windowing. Bohman windowing with varying roll-off values have been used to manage OOBE and provide further flexibility. Windowing shape variation is emphasized in this work in order to decrease OOBE, increase spectrum efficiency, and promote dynamic waveform concurrence in contrast to previous systems. Nonetheless, the total number of subcarriers will be separated as groups of subcarriers in UFMC methodology. These groups of subcarriers are called sub-bands. One sub-band range covers the whole frequency range of f-OFDM. There are the same number of subcarriers in each sub-band as there are subbands in FBMC, which means they are all the same size. The filter length and subband width of the UFMC system may both be readily modified to match the specific demands of individual users in the system. Because of its superior performance compared to other waveforms, the

UFMC is considered amongst the most exciting pulses for 5G technology. The effectiveness of a new filtering approach in the UFMC is the primary subject of this investigation. The performance of the system in terms of out-of-band (OOB) and bit error rate (BER) is compared in simulations utilizing a variety of UFMC system features. The UFMC system model is presented in further detail in the next section, which includes numerous side lobes.

1   System Model

The following equation is a representation of the UFMC signal at its most basic frequency bandpass level:

$$s[k] = \sum_{h=0}^{B-1} \sum_{l=0}^{L-1} \sum_{n=0}^{N-1} d_n^b g[l] e^{j2\pi k \frac{(n-1)}{N}} \tag{1}$$

Anywhere, across the nth subcarrier, the data will be sent, while Bth sub-band termed as d_n^b. On the other hand, the frequency-equivalent-windowing operation is given by g[l] of the corresponding time-domain finite impulse response (FIR) filter. Consequently, the size of the block can be determined according to the total size of the subband, N, and the filter length, L, as L+N-1. However, using the cyclic prefix, CP, the UF-OFDM will correspond to CP-OFDM, which provides improved resistance against ISI. Due to this, the transition zones, i.e., decay up/down, provide minor ISI mitigation in normal UFMC processes. Besides, Fig. 1 depicts a typical UFMC transmitter block schematic. UFMC transmitting side design with B subbands is shown in Fig. 1. In the method stated above, it has been designed. When the UFMC ith sub-module is executed, whereas, 1 to B are used, the baseband vector of the time-domain of (N+Nf-1) length, respectively, for sub-bands, is generated, where these contains the Quadrature Amplitude Modulation (QAM) capsules of dimension $ni \times 1$ is generated. Note that N-filter is the length of the filter, and N is the total number of extracts per symbol meant to justify all sub-bands without aliasing. Also, the support of single sub-band is constructed. Orthogonality the correlation is preserved because the symbols are transmitted back-to-back one after the other without any overlap. A more sophisticated receiver is required since there is no CP and the symbols are not circularly convoluted with the channel [14]. Traditional receivers use an FFT block twice as large as UFMC transmitters, however this is not always the case. When compared to CP-OFDM, UFMC offers superior frequency localization and resilience against time-frequency offsets [23]. Furthermore, as opposed to subcarrier-wise filtration, it has reduced filter lengths, making it somewhat more ideal towards low-latency workloads. However, due to their shorter length, these filters have less ability to suppress OOBE.

Accordingly, the pulse shaping feature of the UFMC waveform increased the asynchronous multi-user performance of the waveform. UFMC is backwards compatible with OFDM methods as well as other technologies. It has an extremely

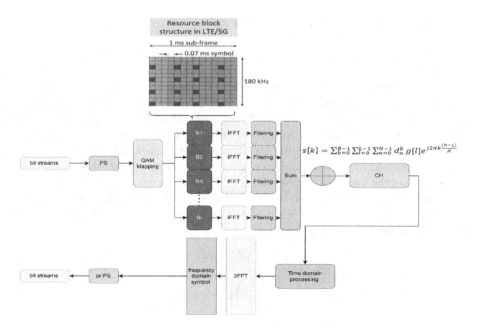

Fig. 1. The UFMC transmitter system

high spectral efficiency [21]. Improvements to the traffic flow will be made separately. Because of its resistance to asynchrony, resilience against spectrum fragmentation provides better assistance (time-frequency misalignments). Only for a very occasional basis, it is possible to gain access to the site (i.e. short bursts). It is a handy tool when used in conjunction with QAM. Low-to-medium degree of complexity is required. UFDM-related technology and know-how are utilized in a wide variety of applications. If you are a member of a CoMP, FBMC is the preferred method of communication. The filter length is reduced as a result of grouping subcarriers together for filtering. [2] The disadvantages of UFMC include its low complexity and noise enhancement. [23] CP is optional in order to reduce ISI; nevertheless, without CP, the impact of ICI and ISI is considerably greater. When there is a significant delay spread, multi-tap equalizers should be employed. A larger FFT size at the receiver -> corresponds to a higher level of complexity [19], as shown in the following Eqs. (2, 3).

$$\text{UFMC symbol} = \text{IFFT}\left\{\left[\boldsymbol{A}_{[\beta 1]}, \mathbf{s}_r^\tau, \boldsymbol{A}_{[\bar{B}2]}\right]^\tau\right\} \tag{2}$$

$$\mathbf{X}_{\text{UFMC symbolis.}} = \sum_{i=1}^{\bar{B}} \mathbf{x}_{\text{fuli}}^{(i)} \tag{3}$$

2   Proposed Filtering Method

One of the needs in next wireless technology systems have the potential to accommodate asynchronous modes. At next communication systems, the far more prominent waveforms employ either filtering or windowing somewhere at sender to minimize out-of-subband emissions (OOSBE). The Prototype Filter is the most important component of MCM systems. [15,17,22], and [13] have been cited as examples of various Prototype Filters for modulation schemes. In this paper, we discuss a new filter for MCM systems that is available. A Bohman window is formed by the convolution of two half-duration cosine lobes. Adding a term to make the very first differential zero at the perimeter gives you the product of a triangular window and one-time domain cycle of a cosine. As a result, the windows of the Bohmans begin to fall out $1/W^4$ [18]. Mathematically, the filter design method is sympathetic,

$$\eta(\kappa) = \eta(\kappa).\omega(\kappa) \tag{4}$$

$$w(k) = \left(1 - \left|\left(2 * \frac{k}{n} - 1\right)\right|\right) * \cos\left(\pi * \left| \left(2 * \frac{k}{n} - \right.\right.\right.$$
$$\left.\left.\left. 1\right) \right|\right) + \frac{\sin\left(\pi * \left|\left(2 * \frac{k}{n} - 1\right)\right|\right)}{p_i} \tag{5}$$

After modernization the Bohman windowing as explain Eq. 6 by using interleave rate coefficients with

$$W(k) =$$
$$\frac{-\left(\left(\frac{2 \times k - n}{n}\right) \mid -1\right) * \left(\pi + \sin\left(\pi \times abs\left(\frac{2kk-n}{n}\right)\right)\right) \times \cos\left(\pi \times abs\left(\frac{2 \times k - n}{n}\right)\right) + \sin\left(\pi \times abs\left(\frac{2 \times k - n}{n}\right)\right)}{\pi} \tag{6}$$

There was a significant difference between the two-time domain windowing as shown in the Fig. 2 and Fig. 4. Also, significant difference between the two-frequency domain windowing as shown in the Fig. 3 and Fig. 5. In addition, the signals are multiplied using a window function to smooth the discontinuity at the margins of the IFFT slices, which helps to prevent spectral leakage. As a consequence, the window function used has an impact on the quantity of signal and noise that enters each filter bank. As a result, the bandwidth of the selected window function influences the amount of noise that accumulates in each filter bank. In fact, the amount of cumulative noise produced by the window function of choice influences the amplitude estimation in each frequency bin. To put it another way, the noise floor exhibited at the FFT output changes depending on the window selected.

3   Results and Discussion

It has been determined that the OOBE, to free more spaces for short message signals and asynchronized communication, PAPR of UFMC systems are all significantly different from one another. To illustrate the optimal waveform for 5G

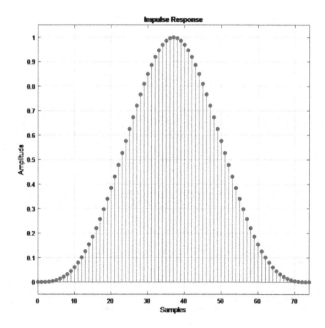

Fig. 2. Time domains performance of the Bohman windowing

wireless communication networks, these tests had a specific objective. In Fig. 6, the PSD is shown that is results from the UFMC structure, with 200 subcarriers. Instead, the entire bandwidth is grouped as resource blocks (RB), ten smaller bands, each of which has on its set of twenty-four subcarriers, resulting in fewer side lobes and a more consistent signal. The Fig. 6 shows that the UFMC system has a lower OOBE performance level than the CP-OFDM system compare with 3GPP slandered [5, 12].

Whereas, Table 1 displays the parameters and values adopted in the present research examination for further validation and understanding of the examination process.

The review of the previous literature indicated that CP-OFDM and UFMC produced equal performance at 20 dB SNR, but conventional UFMC required greater transmission power to reach the same results. For this research, it will be extremely beneficial to learn how to regulate the OOBE using an interleave windowing amplitude. Each subband's filter is critical in the UFMC system. It has the potential to significantly minimize OOBE while, enhancing system performance. In all UFMC systems of this work, data information is transferred on just 280 of the 512 frequency components that are available. 512 components are split into twenty subsets, each of which has fourteen subcarriers, resulting in a total of 512 subchannels. Using a filter reduces the UFMC system's OOB to a fraction of OOBE that resulted from the corresponding system, which is the CP-OFDM. The PSD of the out-of-band signal in the UFMC system decreases

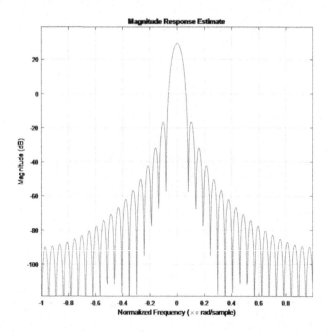

Fig. 3. Frequency-domains performance of the Bohman windowing

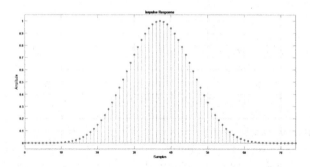

Fig. 4. Time domain performance of the modernization Bohman windowing

rapidly below −80 dB when it is received outside of the band. That is a perfect illustration of the UFMC system's benefits (Fig. 7).

The subband filter's cut off frequency attenuation is critical in all terms of the system performance. Attenuation between stopbands increases with signal strength, resulting in reduced out-of-band and consequently greater system performance as stopband attenuation increases. On the one hand, severe range quality is expected due to the significant cut off frequency absorption, therefore necessitates the employment of a good length filter, while on the other hand, a high degree of system complexity is required. As a result, the only method to achieve a compromise across system stability and sophistication is to choose the cut off frequency absorption and filter length that are adequate for the purpose.

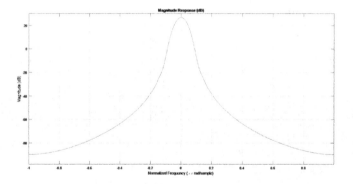

Fig. 5. Frequency domain performance of the modernization Bohman windowin

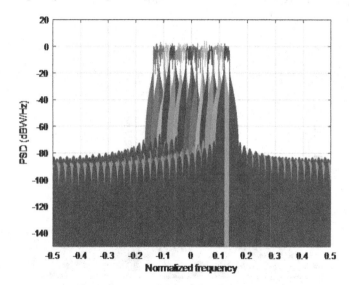

Fig. 6. Universal filtered multi-carrier PSD- Dolph-Chebyshev arrangement

The impact of cut off frequency absorption on the system is explored in the simulations for this study with the construction of the filter in order to get the desired results. Figure 8 depicts waveforms with a complementary cumulative distribution function (CCDF). These waveforms are that of the CP-OFDM signals, Conventional UFMC, and our novel UFMC structure. The proposed and traditional UFMC systems' PAPR performance was found to be superior to that of the 3GPP benchmark CP-OFDM system. While, the PAPR of the proposed system has achieved a significant enhancement to reduce the value of PAPR in comparison to the traditional system .

The process of out of band emission for the suggested system, as compared to the standard UFMC system, is depicted in the accompanying Fig. 9 and it clear that the proposed windowing scheme has supper reduction OOBE compare

Table 1. Suggested system's settings parameters

Parameter	Value
Window	Dolph-Chebyshev, proposed filter
B	6
A	0.6
I/FFT size	512
Sub-band size in terms of carriers	20
Time-domain size in terms of symbols	14
Size of Tx and Rx windows	$1/14*1/\Delta$
(Δ) sub-carrier spacing in KHz	15
Mapping order (QAM)	64-levels
Resource blocks (RBs)	14

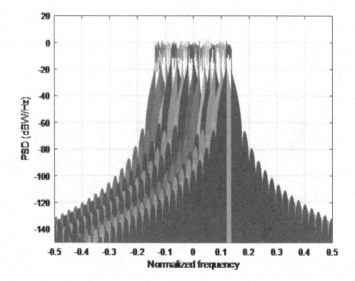

Fig. 7. Universal filtered multi-carrier PSD of proposed windowing method

with conventional UFMC and CP-OFDM. The Fig. 8 depicts two key realities. To begin, the suggested UFMC technique has a greater PAPR than the conventional CP-OFDM. Second, the suggested approach is capable to overcome the conventional obstacles and delivering results that are superior than traditional UFMC. Moreover, the windowing UFMC simulated results match the theoretical conclusions. One of the most fascinating discoveries was the use of out of band emission, which was able to achieve a dBm of -70 (Table 2).

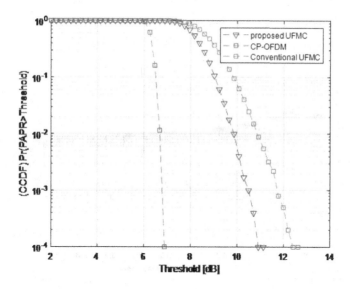

Fig. 8. Shown the PAPR performance for the proposed system and conventional UFMC system

Table 2. Table styles

	Technique	Key point indicator	Notes
[14]	Kaiser window (UFMC)	PSD	Decrease OOBE while maintaining PAPR
		CCDF	Abandonments of BER
		ACPR	Desertions of CCDF-CM
[25]	Windowing (CP-OFDM)	PSD	Decrease of OOBE without reducing error vector magnitude
		EVM	Decrease CCDF
			Desertions adjacent channel power leakage ratio
[26]	Windowing (warped waveforms)	PSD	Reduction of OOBE while maintaining the value of BER
		BER	Reduce of CCDF
			Reduce of adjacent channel power leakage ratio
Proposed UFMC	Windowing (UFMC)	PSD	
		CCDF	
		BER	
		OOBE	

When using the UFMC system, the bit error rate performance is shown in Fig. 10, respectively, for the 64 QAM constellations for proposed UFMC and UFMC chebwin. As the side lob attenuation grows, the system's performance improves. The system best performs can identify the side lob attenuation at 40 dB, but performance degrades when it is 5 dB. Side-lobe attenuation is recom-

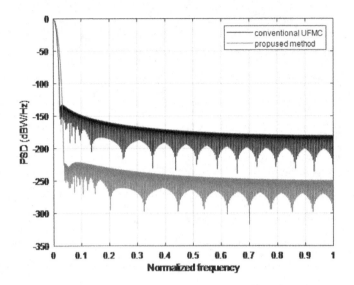

Fig. 9. PSD of proposed method with conventional UFMC

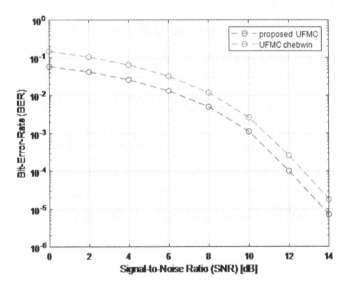

Fig. 10. BER of proposed method with conventional UFMC

mended because it can provide high performance with relatively minimal complexity in the context of the system's overall complexity. The more subbands and fewer subcarriers in a subband, the higher the performance. The following table compares the results of the current study with those of earlier studies.

4   Conclusion

5G waveforms are investigated in this research using the UFMC waveform as a possibility to generate new filtering method. The side lob attenuation of the filter in the UFMC system is researched since it is designed to reducing the OOB and PAPR of the system. The simulation findings suggest that the system performance improves with increasing side lob attenuation, and given the complexity of the system, attenuation of 10 dB or 20 dB is preferable. This research also investigates the UFMC system's performance with different subband widths and modulation constellations. System performance is boosted by the use of more subbands and filters in UFMC systems. As a result, subcarriers in the system are required to be broken down into more subbands. The moderate designs of the filter are suggested when complexity is taken into account to attain systems' advantages.

Acknowledgments. This work has been supported by the RUDN University Strategic Academic Leadership Program.

References

1. Abdallah, H.S.: Evaluation performance of multicarrier system using hybrid schemes to support 5G technology. In: 2018 5th International Conference on Electrical and Electronic Engineering (ICEEE), pp. 306–310. IEEE (2018)
2. Agarwal, P., Singh, A.P., Shukla, M.K.: Performance analysis of Fiedler-SLM mechanism in OFDM for PAPR reduction. Available at SSRN 3555729 (2020)
3. Ankaralı, Z.E., Peköz, B., Arslan, H.: Enhanced OFDM for 5G RAN. ZTE Commun. **15**(S1), 11–20 (2020)
4. BS, R., et al.: Performance analysis of OFDM, FBMC and UFMC modulation schemes for 5G mobile communication MIMO systems (2021)
5. Gopi, S., Kalyani, S.: An optimized SLM for PAPR reduction in non-coherent OFDM-IM. IEEE Wirel. Commun. Lett. **9**(7), 967–971 (2020). https://doi.org/10.1109/LWC.2020.2976935
6. Guo, Z., Liu, Q., Zhang, W., Wang, S.: Low complexity implementation of universal filtered multi-carrier transmitter. IEEE Access **8**, 24799–24807 (2020)
7. Hammoodi, A., Audah, L., Al-Jobouri, L., Mohammed, M.A., Aljumaily, M.S.: New 5G kaiser-based windowing to reduce out of band emission. Comput. Mater. Contin. **71**(2), 2721–2738 (2022). https://doi.org/10.32604/cmc.2022.020091. http://www.techscience.com/cmc/v71n2/45966
8. Hammoodi, A., et al.: Novel universal windowing multicarrier waveform for 5G systems. Comput. Mater. Contin. **67**(2), 1523–1536 (2021)
9. Hammoodi, A., Audah, L., Taher, M.A.: Green coexistence for 5G waveform candidates: a review. IEEE Access **7**, 10103–10126 (2019)
10. Hammoodi, A.T., Shawqi, F.S., Audaha, L., Qasim, A.A., Falih, A.A.: Under test Filtered-OFDM and UFMC 5G waveform using cellular network. J. Southwest Jiaotong Univ. **54**(5) (2019)
11. Hasan, A., Zeeshan, M., Mumtaz, M.A., Khan, M.W.: PAPR reduction of FBMC-OQAM using a-law and mu-law companding. In: 2018 ELEKTRO, pp. 1–4. IEEE (2018)

12. Hazareena, A., et al.: UFMC system performance analysis for 5G cellular networks. Turk. J. Comput. Math. Educ. (TURCOMAT) **12**(10), 162–167 (2021)
13. Hussain, G.A., Audah, L.: UFMC and f-OFDM: contender waveforms of 5G wireless communication system. Proc. Electr. Eng. Comput. Sci. Inform. **7**(2) (2020)
14. Hussain, K., López-Valcarce, R.: Optimal window design for W-OFDM. In: ICASSP 2020-2020 IEEE International Conference on Acoustics, Speech and Signal Processing (ICASSP), pp. 5275–5289. IEEE (2020)
15. Iwabuchi, M., et al.: 5G field experimental trial on frequency domain multiplexing of mixed numerology. In: 2017 IEEE 85th Vehicular Technology Conference (VTC Spring), pp. 1–5. IEEE (2017)
16. Kasmi, M., Mhatli, S., Bahloul, F., Dayoub, I., Oh, K.: Performance analysis of UFMC waveform in graded index fiber for 5G communications and beyond. Opt. Commun. **454**, 124360 (2020)
17. Marijanovic, L., Schwarz, S., Rupp, M.: A novel optimization method for resource allocation based on mixed numerology. In: ICC 2019-2019 IEEE International Conference on Communications (ICC), pp. 1–6. IEEE (2019)
18. Mukherjee, M., Shu, L., Kumar, V., Kumar, P., Matam, R.: Reduced out-of-band radiation-based filter optimization for UFMC systems in 5G. In: 2015 International Wireless Communications and Mobile Computing Conference (IWCMC), pp. 1150–1155. IEEE (2015)
19. Nadal, J., Nour, C.A., Baghdadi, A.: Novel UF-OFDM transmitter: significant complexity reduction without signal approximation. IEEE Trans. Veh. Technol. **67**(3), 2141–2154 (2017)
20. Ramli, K., Taher, M., Audah, L., Shah, N.S., Ahmed, M., Hammoodi, A., et al.: An enhanced partial transmit sequence based on combining hadamard matrix and partitioning schemes in OFDM systems. Int. J. Integr. Eng. **10**(3) (2018)
21. Rusek, F., Anderson, J.B.: Multistream faster than nyquist signaling. IEEE Trans. Commun. **57**(5), 1329–1340 (2009)
22. Shawqi, F.S., et al.: A new SLM-UFMC model for universal filtered multi-carrier to reduce cubic metric and peak to average power ratio in 5G technology. Symmetry **12**(6), 909 (2020). https://doi.org/10.3390/sym12060909
23. Thota, S., Kamatham, Y., Paidimarry, C.S.: Analysis of hybrid PAPR reduction methods of OFDM signal for HPA models in wireless communications. IEEE Access **8**, 22780–22791 (2020)
24. Yarrabothu, R.S., Nelakuditi, U.R.: Optimization of out-of-band emission using Kaiser-Bessel filter for UFMC in 5G cellular communications. China Commun. **16**(8), 15–23 (2019)

A Novel Technique for Creating Optical Multi-carrier Generation Using Nested Electro-Absorption Modulators

Mohammed Hasan Alwan[1], Yousif I. Hammadi[1], Mamoon A. Muhi[1],
Omar Abdulkareem Mahmood[2], Alexey Tselykh[3],
and Mohammed Saleh Ali Muthanna[3(✉)]

[1] Bilad Alrafidain University College, Al-Rafidain University College, Diyala, Iraq
[2] Department of Communications Engineering, College of Engineering,
University of Diyala, Diyala, Iraq
[3] Institute of Computer Technologies and Information Security,
Southern Federal University, Taganrog 347928, Russian Federation
muthanna@sfedu.ru

Abstract. In this paper, we present and quantitatively show a novel technique for the creation of an optically multi-carrier that is founded upon single and nested electro-absorption modulators (EAMs). These setups include single and four cascaded EAMs, and a radio-frequency (RF) signal. These stacked EAMs are triggered by a radio frequency signal that has various frequencies in each EAM. Where the EAM_1, EAM_2, EAM_3, and EAM_4 are driven by the RF signal with the frequency of ($f_e, f_e/2, f_e/4$, and $f_e/8$) GHz, respectively. The MATLAB program is used to develop and model the computational equations of the developed models, as well as to simulate the structures. We demonstrate that perhaps the amount of spectral lines is precisely dependent on the magnitude of the RF input signal to the EAM in our simulations. Additionally, the number of optical lines varies linearly with EAMs number. Furthermore, the number of optical lines that have a spectral flatness of 1dB also increases with increasing the number of EAMs.

Keywords: Nested electro-absorption modulators · Optical multi-carrier · Optical lines · Spectral flatness

1 Introduction

A number of co source for the structure of optical network systems including the dense wavelength division multiplexing (DWDM), optical time-division multiplexing (OTDM), and optical orthogonal frequency-division multiplexing (OOFDM) systems [1–6] is now widely used as an optical multi-carrier generator (OMG) in many applications. For these purposes, the produced comb must give a constant frequency with a fixed phase and width. Furthermore, spectral comb line flatness and cheap cost of installation are critical needs. Many ways

V. M. Vishnevskiy et al. (Eds.): DCCN 2022, LNCS 13766, pp. 17–28, 2022.
https://doi.org/10.1007/978-3-031-23207-7_2

for realizing OMGs have been suggested and implemented to far, including the use of a mode locked laser (MLL) [7,8]. The fundamental problem of the MLL approach [9] is the volatility of carrier frequency owing to the system's dependence on the laser cavity. A nonlinear effect in a complex nonlinear media is used in another approach [10]. However, to shape the spectrum, this implementation involves the use of a high-powered amplifier and an optical filter. Multiple optical carriers with exact channel spacing may be produced from a single seed light source using the optical modulation approach [11–14]. Some of the key benefits of this method are its cheap cost, easy fabrication, compact size, minimal noise [15], and great robustness. Several methods are employed to generate OMGs, including the use of optical modulation techniques. Normally, this modulation is based on each of two strategies: amplitude-phase hybrid modulation as well as intensity modulation. The number of created comb lines is proportional to the phase modulation in an amplitude-phase hybrid modulated signal, i.e., the amplitude of the electrical oscillator signal is controlled to tailor the quantity of developed comb lines [16–19]. Most of the drawbacks of this technology may be attributed to the restricted number of comb lines that can be formed as well as the usage of high-power auxiliary RF to activate the modulators. Through the use of domino intensity and phase modulations, Wu et al. actually launched an 8-GHz comb with 38 lines including an average spectral power change of 1 dB across a 10-GHz bandwidth [19]. When optical intensity modulators with relatively low potential are used to tackle the issue of high external RF power, the problem of high power may be overcome. Shang and colleagues [20,21] utilized two cascaded intensity modulators to create a total of 25 comb lines at a lower operating voltage. According to this study, a novel approach for tightly packed OMC is described that may be used for sophisticated transmission systems. The suggested expense OMC is composed of four EAMs that are stacked in a cascade fashion and are powered by a single microwave supply that operates at various frequencies (i.e., $f_i = f_e/2^{(i-1)}$). Accordingly, a tunable OMG can be achieved that gives the optical lines at the output within this empirical formula $N_i = (2^{(i-1)}) \times N_1 + 1$.

2   Systems Design and Mathematical Modeling

The configurations of the planned OMG settings are presented in this section. There's also a straightforward mathematical model that we believe best represents the system under consideration. However, whenever required, the definitions will be presented in order to reach the required aim of this research, which is the novel OMG design.

2.1   Single EAM Configuration

In order to begin the system description, designers will first provide an OMG configuration that would be based on a single EAM, as shown in Fig. 1. There are two parts to this device: a continuous-wave (CW) laser diode and an EAM that is powered by a sinusoidal RF signal well with frequency of f_e. Based on the quantum-confined stark effect in the multiple quantum well (MQW) semiconductor that is established the wave guide of the EAM, EAM modulates the intensity of a CW laser via the applied electric voltage and generates the optical lines at the output. To derive an analytical model that describes the generated optical lines at the output of the EAM, it can be started with the transfer function of the EAM which is expressed as:

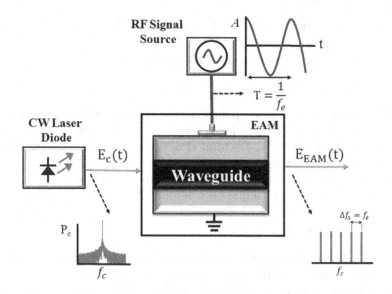

Fig. 1. Schematic diagram of the OMG source using a single EAM

$$h_{EAM}(t) = \sum_{k=1}^{\infty}\sum_{n=1}^{\infty}\left\{\frac{U_b^{(2n-1)}}{(2n-1)!}C_{(2k-1)(2n-1)}\frac{A^{2n-1}}{4^{n-1}}\cos\left[(2k-1)2\pi f_e t\right]\right.$$
$$\left.+\frac{U_b^{(2n)}}{(2n)!}C_{(2k)(2n)}\frac{A^{2n}}{2^{2n-1}}\cos\left[(2k)2\pi f_e t\right]\right\} \tag{1}$$

By using Euler's formula, Eq. 1 can be written as:

$$h_{EAM}(t) = \frac{1}{2}\sum_{k=1}^{\infty}\sum_{n=1}^{\infty}\left\{\frac{U_b^{(2n-1)}}{(2n-1)!}C_{(2k-1)(2n-1)}\frac{A^{2n-1}}{4^{n-1}}\right.$$
$$\times\left(\exp\left[j2\pi(2k-1)f_e t\right]+\exp\left[-j2\pi(2k-1)f_e t\right]\right)$$
$$\left.+\frac{U_b^{(2n)}}{(2n)!}C_{(2k)(2n)}\frac{A^{2n}}{2^{2n-1}}\times\left(\exp\left[j2\pi(2k)f_e t\right]+\exp\left[-j2\pi(2k)f_e t\right]\right)\right\} \tag{2}$$

where $U_b^{(\cdot)}$ is the series coefficients of the DC component, The amplitude of the radio frequency signal is denoted by the letter A. Furthermore, the series factor $C_k n$ is connected to the amplitude of radio frequency signal. The generated optical lines at the output of the EAM is expressed as:

$$E_{EAM}(t) = E_c(t)h_{EAM}(t)$$

$$= \frac{\sqrt{P_c}}{2} \sum_{k=1}^{\infty} \sum_{n=1}^{\infty} \left\{ \frac{U_b^{(2n-1)}}{(2n-1)!} C_{(2k-1)(2n-1)} \frac{A^{2n-1}}{4^{n-1}} \right.$$

$$\times \left(\exp\left[j2\pi\left[(2k-1)f_e + f_c\right]t\right] + \exp\left[-j2\pi\left[(2k-1)f_e - f_c\right]t\right] \right) \qquad (3)$$

$$+ \frac{U_b^{(2n)}}{(2n)!} C_{(2k)(2n)} \frac{A^{2n}}{2^{2n-1}} \times \left(\exp[j2\pi[(2k)f_e + f_c]t\right]$$

$$\left. + \exp[-j2\pi[(2k)f_e - f_c]t]\right)$$

where $E_c(t)$ is the electric field of the CW laser. P_c and f_c indicate the power and center frequency of the CW laser, respectively. f_e is the frequency of the RF signal. According to Eq. 3, the total of optical lines created by the EAM is dictated by the parameters Ub and A. Therefore, the number of carrier lines is controlled by A when U_b is set to be constant. Furthermore, the frequency separation here between optical fibers is determined by the RF signal frequencies.

2.2　Nested EAMs Configuration

To boost the amount of optical lines which must be created from a single OMG resource using a single EAM, the EAMs may be coupled in a cumulates way, which is referred to as nested EAMs, to maximize the number of optical lines generated from the source. In other words, the output optical field $E_o ut$ (t) is the consequence of multiplying the output fields of several modulators to produce the final optical field. Accordingly, the OMG source based on nested EAMs is proposed as demonstrated in Fig. 2. In this configuration, the laser diode is applied to generate a CW light, which is injected into serially cascading four EAMs. Each of these EAMs is powered by a radio frequency signal with a distinct frequency. Where the EAM_1, EAM_2, EAM_3, and EAM_4 are driven by the RF signal with the frequency of f_e GHz, $f_e/2$ GHz, $f_e/4$ GHz, and $f_e/8$ GHz, respectively. Firstly, the CW laser is modulated by the RF signal with the frequency of f_e and generated the optical lines (N_1) at the output of EAM_1. Then, these optical lines are sent to other cascaded EAMs. Where this setup can be mathematically modelled as following:

$$E_{out}(t) = E_c(t)h_1(t)h_2(t)\cdots h_M(t) = E_c(t)\prod_{i=1}^{M} h_i(t) \qquad (4)$$

where $h_i(t)$ is the transfer function of ith modulator. EAMs in the proposed OMG source are driven at different RF signal frequencies, where each EAM is driven by the frequency of $f_i = f_e/2^{(i-1)}$ Accordingly, Eq. 2 can be written as:

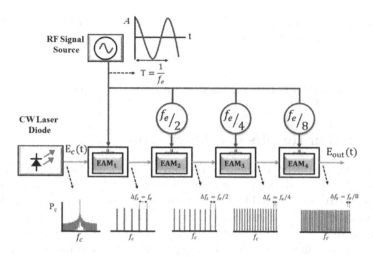

Fig. 2. Schematic diagram of the OMG source using nested EAMs.

$$
\begin{aligned}
h_i(t) = \frac{1}{2} \sum_{k=1}^{\infty} \sum_{n=1}^{\infty} \Bigg\{ & \frac{U_b^{(2n-1)}}{(2n-1)!} C_{(2k-1)(2n-1)} \frac{A^{2n-1}}{4^{n-1}} \\
& \times \left(\exp\left[j2\pi(2k-1)\frac{f_e}{2^{i-1}}t \right] + \exp\left[-j2\pi(2k-1)\frac{f_e}{2^{i-1}}t \right] \right) \\
& + \frac{U_b(2n)}{(2n)!} C_{(2k)(2n)} \frac{A^{2n}}{2^{2n-1}} \times \left(\exp\left[j2\pi(2k)\frac{f_e}{2^{i-1}}t \right] + \exp\left[-j2\pi(2k)\frac{f_e}{2^{i-1}}t \right] \right) \Bigg\}
\end{aligned}
$$
(5)

By substituting Eq. 5 into Eq. 4, the generated lines at the output of the proposed cascaded EAMs OMG source is expressed as:

$$
\begin{aligned}
E_{out}(t) = \frac{E_c(t)}{2^M} \prod_{i=1}^{M} \Bigg\{ \sum_{k=1}^{\infty} \sum_{n=1}^{\infty} \Bigg[& \frac{U_b^{(2n-1)}}{(2n-1)!} C_{(2k-1)(2n-1)} \frac{A^{2n-1}}{4^{n-1}} \\
& \times \left(\exp\left[j2\pi(2k-1)\frac{f_e}{2^{i-1}}t \right] + \exp\left[-j2\pi(2k-1)\frac{f_e}{2^{i-1}}t \right] \right) \\
& + \frac{U_b^{(2n)}}{(2n)!} C_{(2k)(2n)} \frac{A^{2n}}{2^{2n-1}} \\
& \times \left(\exp\left[j2\pi(2k)\frac{f_e}{2^{i-1}}t \right] + \exp\left[-j2\pi(2k)\frac{f_e}{2^{i-1}}t \right] \right) \Bigg] \Bigg\}
\end{aligned}
$$
(6)

When the number of cascaded EAMs is increased, according to Eq. 6, the number of optical lines grows, as seen in the inset of Fig. 2, which is based on Eq. 6. This is due to multiplication of many frequency components. As a result, the following empirical formula can be constructed to characterize the count of optical lines as a function of the amount of EAMs, so that each EAM is controlled

by an RF signal with a frequency of $f_e/2^{(i-1)}$, in which each EAM is directed
by an RF signal with a frequency of $f_e/2^{(i-1)}$.

$$N_i = (2^{(i-1)}) \times N_1 + 1 \qquad (7)$$

where N_i: is the number of optical lines for ith EAM. N_1: is the number of optical
lines for the first EAM.

3 Simulation Results and Discussion

In order to explore the performance of the OMG sources using optical modulators
that designed and analytically modelled in Sect. 2, VPI transmission Maker®
commercial software are used to simulate and investigate these sources. Optical
modulator-based OMG source is used the CW laser diode as an optical carrier
signal to modulate the RF signal and thus generates the optical lines at the
output which is called comb lines. In the presented simulation setups, the CW
laser diode was used with the simulation parameters are the center frequency
of 193.1 THz and the power of 20 dBm. An optical spectrum analyzer (OSA)
is used to visualize the generated optical lines and investigate the most important parameters of the OMG such as the number of optical lines, the frequency
spacing between these lines and the flatness of the generated lines.

3.1 Simulation Results of Single EAM Configuration

Simulation of the OMG origin with a singular EAM, as indicated in Fig. 1.
Throughout this hypothetical arrangement, the laser diode creates continuous-
wave (CW) light that is intensity modulated via the EAM by a sinusoidal RF
signal with a frequency of 25 GHz that is sent through the EAM. The simulated
setup of this OMG source was examined by cautiously adjusted the amplitude of
the RF signal (A) applied to the EAM, where A was varied from 0.5 V to 2.5 V.
Increased number of optical lines may be obtained by improving the amplitude
of the radio frequency signal supplied to the EAM. The magnitude of the Radio
wave given to the EAM had an effect on the generation of optical lines at the
EAM's output, which was investigated further. Figure 3 depicts the generated
optical lines from the OMG source designed with a single EAM and for different
amplitude of the RF signal (A), where A is varied from 0.5 V to 2.5 V. Figure 3(a)
depicts the created optical lines when A = 0.5 V, revealing the generation of 25
comb lines with something like a frequency separation of 25 GHz. This amount
of optical lines grows to 51 whenever the strength of the Radio signal doubles to
A=1 V, as shown in Fig. 3(b). The quantity of optical lines that come again from
EAM grows as the RF signal strength rises. This is shown in Fig. 3. Equation 3
says that by increasing how much power you put into the RF signal, you'll
also make the chirp parameter bigger, which is what this means. Figure 4 shows
the relationship between both the magnitude of the RF signal that is sent to
EAM and indeed the comb lines that are made at the production of EAM. This
indicates that the amount of optical lines can be seen to change linearly well
with amplitude of the RF signal that is used to power EAM.

0.35

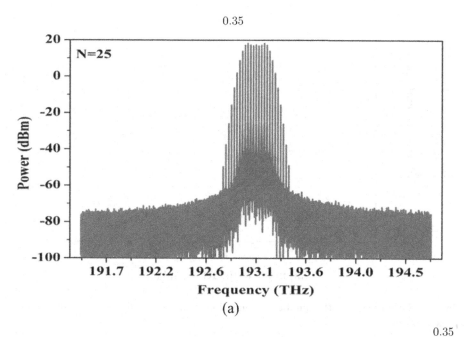

(a)

0.35

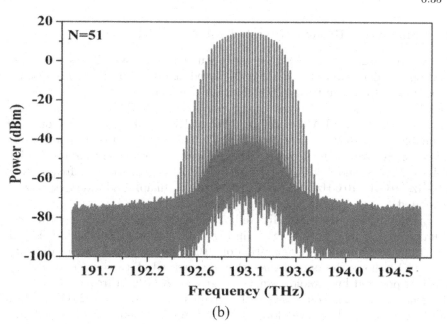

(b)

Fig. 3. Optical spectra of the generated optical lines by a single EAM at $\Delta f_s = 25$ GHz for amplitudes of the RF signal (a) A = 0.5 V, (b) A = 1 V

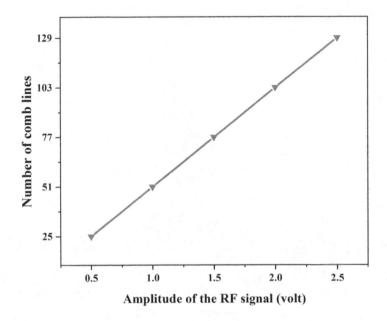

Fig. 4. Relation between the number of comb lines and the amplitude of the RF signal applied to EAM.

3.2 Simulation Results of Nested EAMs Configuration

To explore the gain of the designed OMG source using nested EAMs, the generated optical lines after each EAM is displayed, as shown in Fig. 5. Because it's easier to see, the quantity of optical lines and the spectral distance over them for each EAM are shown in an inset of the results in Fig. 5. EAMs are connected in cascaded, where each EAM driven by different RF signal frequency to construct a tunable OMG source as depicted in Fig. 2. This OMG source is simulated, In this simulated source, by kept the CW laser with its parameters and the EAM_1, EAM_2, EAM_3, and EAM_4 are driven by the RF signal with the frequency of 40 GHz, 20 GHz, 10 GHz, and 5 GHz, respectively, tunable and wide spectra were generated.

EAM_1 is powered by a 40 GHz radio frequency signal, and the optical spectrum created at its output is shown in Fig. 5(a). A total of 129 comb lines with something like a frequency separation of 40 GHz may be seen on this diagram. Figure 5(b) shows the optical lines that are created at the output of EAM_2, which is powered by a radio frequency signal of 20 GHz in frequency. Observe the generation of 259 comb lines with a frequency separation of 20 GHz, which can be observed to have been formed. Furthermore, the amount of optical lines with a spectroscopic flatness of 1 dB is presented in the panel of Fig. 5, where it would be noted that EAM_1, EAM_2, EAM_3, and EAM_4 provide 105, 211, 421, and 841 comb lines, accordingly. By these findings, it can be deduced that perhaps the count of optical lines increases significantly well with amount of

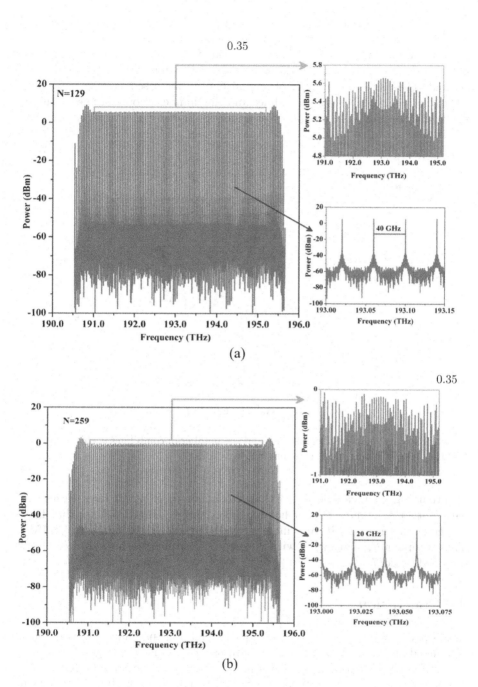

Fig. 5. Optical spectrum of the generated optical lines using cascaded EAMs OMG source at the output of (a) EAM_1, (b).

EAMs, so that each EAM is controlled by an RF signal with a frequency of $f_e/2^{(i-1)}$ and each EAM drives one optical line. This agrees with the mathematical analysis given in Eq. 7 and occurs due to due to the frequency spacing between the generated comb is decreased by $2^{(i-1)}$, while the optical bandwidth is fixed. Therefore, the number of optical lines is increased by $2^{(i-1)}$. Accordingly, the total number of optical lines and the number of optical lines within a spectral flatness of 1 dB are plotted as a function of the EAMs number as shown in Fig. 6.

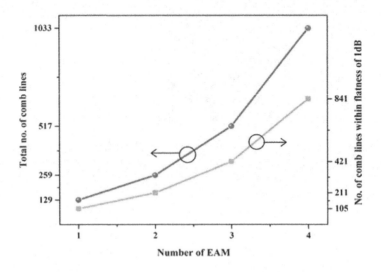

Fig. 6. The total number of comb lines and the number of comb lines within a spectral flatness of 1 dB versus the EAMs number.

From Fig. 6, it can be noted that the number of optical lines varies linearly with EAMs number. Furthermore, the number of optical lines that have a spectral flatness of 1dB also increases with increasing the number of EAMs. These results are in agreement with the empirical formula of the generated OMG obtained in Eq. 7.

4 Conclusion

In this paper, we suggest a novel technique for creating optical multi-carriers that is founded on the EAMs. It is necessary to use a sinusoidal wave generator with varied frequencies in order to make adjustable and wide-spectrum OMGs, which are powered by the EAMs. It has been shown that the suggested configuration can produce a wide band wavelength range with a high range of frequency lines and excellent flatness. EAMs in series have been shown to increase the flatness of optical spectra, and this has been shown in this paper. The frequency of the sinusoid, the numbers of EAM, and the strength of the RF signal all contribute to the band-width of the optical comb signal, which is defined as in the proposed configuration. At a 2 V RF voltage, we were able to get much more over 103 frequency lines, which is a significant improvement over previous findings. It is also possible to obtain frequency spacings of (40, 20, 10, and 5 GHz) and comb lines having power variations compared with fewer than one decibel with 129, 259, 517, and 1033 comb lines, respectively. The findings revealed a significant strategy for the development of OMG via the use of EAM.

Acknowledgement. The research is supported by postdoc fellowship granted by the Institute of Computer Technologies and Information Security, Southern Federal University, project No. P.D./22-01-KT.

References

1. Fresi, F., et al.: Advances in optical technologies and techniques for high capacity communications. J. Opt. Commun. Netw. **9**(4), C54–C64 (2017)
2. Adam, A.B.M., Wan, X., Wang, Z.: Energy efficiency maximization in downlink multi-cell multi-carrier NOMA networks with hardware impairments. IEEE Access **8**, 210054–210065 (2020)
3. Adam, A.B.M., Lei, L., Chatzinotas, S., Junejo, N.U.R.: Deep convolutional self-attention network for energy-efficient power control in NOMA networks. IEEE Trans. Veh. Technol. **71**(5), 5540–5545 (2022)
4. Kaymak, Y., Rojas-Cessa, R., Feng, J., Ansari, N., Zhou, M., Zhang, T.: A survey on acquisition, tracking, and pointing mechanisms for mobile free-space optical communications. IEEE Commun. Surv. Tutor. **20**(2), 1104–1123 (2018)
5. Adam, A.B.M., Muthanna, M.S.A., Muthanna, A., Nguyen, T.N., El-Latif, A.A.A.: Toward smart traffic management with 3D placement optimization in UAV-assisted NOMA IIoT networks. IEEE Trans. Intell. Transp. Syst. (2022)
6. Paul, B., Chiriyath, A.R., Bliss, D.W.: Survey of RF communications and sensing convergence research. IEEE Access **5**, 252–270 (2016)
7. Jaradat, A.M., Hamamreh, J.M., Arslan, H.: Modulation options for OFDM-based waveforms: classification, comparison, and future directions. IEEE Access **7**, 17263–17278 (2019)
8. Lowery, A.J., Zhuang, L., Corcoran, B., Zhu, C., Xie, Y.: Photonic circuit topologies for optical OFDM and Nyquist WDM. J. Lightwave Technol. **35**(4), 781–791 (2017)
9. Hammoodi, A., Audah, L., Taher, M.A.: Green coexistence for 5G waveform candidates: a review. IEEE Access **7**, 10103–10126 (2019)
10. Morosi, J., et al.: 25 Gbit/s per user coherent all-optical OFDM for Tbit/s-capable PONs. J. Opt. Commun. Netw. **8**(4), 190–195 (2016)
11. Guan, P., et al.: All-optical ultra-high-speed OFDM to Nyquist-WDM conversion based on complete optical Fourier transformation. J. Lightwave Technol. **34**(2), 626–632 (2016)
12. Muthanna, M.S.A., et al.: Deep reinforcement learning based trans-mission policy enforcement and multi-hop routing in QoS aware LoRa IoT networks. Comput. Commun. **183**, 33–50 (2021)

13. Muthanna, M.S.A., Wang, P., Wei, M., Abuarqoub, A., Alzu'bi, A., Gull, H.: Cognitive control models of multiple access IoT networks using LoRa technology cognitive systems research. Cogn. Syst. Res. **65**, 62–73 (2021). ISSN 1389-0417

14. Roberts, K., Zhuge, Q., Monga, I., Gareau, S., Laperle, C.: Beyond 100 Gb/s: capacity, flexibility, and network optimization. J. Opt. Commun. Netw. **9**(4), C12–C24 (2017)

15. Diamantopoulos, N.-P., Nishi, H., Kobayashi, W., Takeda, K., Kakitsuka, T., Matsuo, S.: On the complexity reduction of the second-order Volterra nonlinear equalizer for IM/DD systems. J. Lightwave Technol. **37**(4), 1214–1224 (2018)

16. Baryshev, V., Epikhin, V., Blinov, I., Donchenko, S.: Acousto-optic modulators in Raman-Nath diffraction regime as phase modulators in modulation transfer spectroscopy. In: 2016 IEEE International Frequency Control Symposium (IFCS), pp. 1–4. IEEE (2016)

17. Chaaf, A., et al.: Energy-efficient relay-based void hole prevention and repair in clustered multi-AUV underwater wireless sensor network. Secur. Commun. Netw. (2021)

18. Rafiq, A., Ping, W., Min, W., Muthanna, M.S.A.: Fog assisted 6TiSCH tri-layer network architecture for adaptive scheduling and energy-efficient offloading using rank-based Q-learning in smart industries. IEEE Sens. J. **21**(22), 25489–25507 (2021)

19. Sohn, S.I., Han, S.K.: Linear optical modulation in a serially cascaded electroabsorption modulator. Microw. Opt. Technol. Lett. **27**(6), 447–450 (2000)

20. Al-Falahy, N., AlMahamdy, M., Mahmood, A.M.: Performance analysis of millimeter wave 5G networks for outdoor environment: propagation perspectives. Indones. J. Electr. Eng. Comput. Sci. **20**(1), 214–221 (2020)

21. Eid, M.M., Rashed, A.N.Z.: Hybrid NRZ/RZ line coding scheme based hybrid FSO/FO dual channel communication systems. Indones. J. Electr. Eng. Comput. Sci. **22**(2), 866–873 (2021)

Discrete Time Markov Chain for Drone's Buffer Data Exchange in an Autonomous Swarm

P. Keyela[1], I. S. Yartseva[1]([⊠]), and Yu. V. Gaidamaka[1,2]

[1] Applied Probability and Informatics Department, Peoples' Friendship University of Russia (RUDN University), 6 Miklukho-Maklaya Street, 117198 Moscow, Russia
irina.s.yartseva@gmail.com, gaydamaka-yuv@rudn.ru
[2] Federal Research Center Computer Science and Control of the Russian Academy of Sciences, 44/2 Vavilova Street, 19333 Moscow, Russia

Abstract. The paper considers a network formed by a swarm of drones exchanging data in the absence of information about relative positioning, for example, when there are no connections to a satellite geolocation system and a cellular network. For the scenario where one of the drones is a receiver and the rest are transmitters, using the apparatus of the theory of random processes and the mathematical teletraffic theory, a method is proposed to estimate the main indicators of the efficiency of data exchange between drones, to analyze key performance metrics, namely, message loss probability, data transmission delay, spent energy per message transaction, and battery lifetime distribution.

Keywords: Drone swarm · Discrete time Markov chain · Loss probability · Transmission delay · Energy consumption · Battery lifetime

1 Introduction

Recently the design of wireless networks with terrestrial base stations (BS) using 5G technologies has been well studied, but the tasks of providing communication to remote areas lacking terrestrial infrastructure are still relevant. Researchers' attention is attracted to the analysis of 6G networks, taking into account the features of new technologies, such as mmWave or THzWave, mMIMO, etc. [1–4]. However, in 5G networks there are tasks that have not been solved even for 5G LTE network technology. These include the task of analyzing the efficiency of data exchange in a swarm of drones in the absence of information about mutual positioning. Such problems arise in cases where the drone swarm is being deployed in places where there are no connections to the GPS or GLONASS satellite geolocation system and to the cellular network, for example, inside natural and artificial formations that screen the signal passage (mountainous and wooded areas, caves, rooms, etc.) [5].

The reported study was funded by the Russian Science Foundation, project number 22-29-00694, https://rscf.ru/en/project/22-29-00694.

1.1   Drone Swarms Applications

Drone swarms, also widely known as Unmanned Aerial Vehicles (UAV) swarms very recently started gaining the attention of various industries because of the numerous benefits and applications they bring. Instead of using a single drone, more efforts are being invested to perfect the coordination of multiple drones to perform tasks in swarms, as the use of drone swarms is more efficient and offers more possibilities and functionalities [6]. In commercial domain drone swarms can be used in different ways to save time, to reduce labor and expenses. In agriculture, drone swarms are often used to watch over vast fields, helping to minimize time consumption and increase efficiency. Many delivery companies are already testing the benefits of using UAV swarms for their package delivery services [7,8]. More than in any other field, drone swarms are widely used for military reasons. Some authors note that swarms can be used "offensively and defensively, creating targeting dilemmas for sophisticated, expensive defensive systems" [9]. Other highly useful applications of drone swarms are emergency response and land surveying and mapping. In disaster management situations swarms can be used to deliver help, packages and supplies, track the current state of the situation, or even provide connectivity [10–12]. So drone swarms do find applications in many sectors and are still growing to capture other areas.

1.2   Basic Components of a Drone Swarm

The constraints and challenges to face and elements to take into account when it comes to drone swarms deployment are as wide and different as the scenarios in which swarms find applications and they are met at every level from the architecture conception and infrastructure building to software and network communication technologies used in [13]. A group of drone form a swarm when mechanisms are implemented to properly coordinate and control them. The primary elements to pay attention to when designing a drone swarm would be motion coordination and formation control, consensus, and flocking [14–17].

The problems of control, motion, and positioning of autonomous drone swarms are being tackled in different ways and by different researchers [18–24], and in this paper we focus on one of the components of these and many other tasks be exact the efficiency of data exchange between drones in a swarm. We consider a swarm of drones where one of them is a receiver and the rest are transmitters. Such a system model corresponds to the scenario of a swarm with a head drone or to a peer-to-peer network of drones at the stage of exchanging information about mutual positioning, when each network node periodically sends and receives broadcasts from all network nodes that are in its coverage zone. Furthermore, such a scenario is typical for the tasks of the Internet of Things, where the base station collects data from drone transmitters located in its coverage area [25]. The Quality of Service indicators we are interested in are data loss during transmission, data transmission delays, and energy efficiency, the latter metric being critical, since a finite battery limits lifetime. We evaluated these indicators using random processes and queuing theory, in particular, a queuing system with batch service and an embedded Markov chain.

2   System Model

Let us consider in detail the procedure for receiving and sending data by one of the drones of a swarm. As illustrated on Fig. 1, messages from other drones (say from N drones) of the swarm arrive to the buffer of the receiver in random time instants with an average intensity of λ messages per time unit and are accepted if there is empty space in the buffer. The receiver drone has a buffer of size L to store messages received from other drones. The content of buffer is decreasing each Δ time units after broadcasting max K messages from buffer according to first-come-first-served discipline. Therefore, the message arrival process is similar to batch arrivals with random batch size at the end of the time slot, where the batch is formed of all messages received during the current time slot. The metrics of interest are message loss probability, spent energy per message, message delay, and for the scenario of collecting data from the drone - drone's battery lifetime distribution.

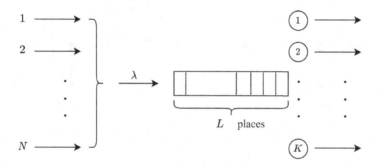

Fig. 1. Queueing system for Markov Chain model of the receiver's buffer

3   Mathematical Modeling of the System

We model the state of a drone's buffer as a Markov Chain (MC) with batch arrivals and batch service $M[x]|D[x]|K|L$ in discrete time with a constant time slot length Δ time units. Let t_n be the instant of departure of messages from the drone's buffer, $(t_{n-1}, t_n]-$ the nth slot, $t_0 = 0, t_n = t_{n-1} + \Delta$, $n \geq 1$. We denote $Q(n)$ the number of customers in queue (messages in buffer) at time instant $t_{n-1} + 0$, $n \geq 1$, $Q(1) = 0$. The incoming flow is assumed Poisson, and the number of customers arriving during a slot is a random variable distributed according to Poisson's law with the parameter $\lambda \Delta$.

We consider that the end of a slot is defined as an instant of departure of the customers from the queue [27,28]. An arrival of a batch at $t_n - 0$ is taken into account just before the end of a time slot. If the batch size exceeds the number of empty places in the queue, only customers for which there are space are accepted.

Since the number of the customers arrived in a time slot follows a Poisson law with parameter $\lambda\Delta$, the probability that k customers will arrive in the nth slot is given by (1):

$$A_k = \frac{(\lambda\Delta)^k}{k!}e^{-\lambda\Delta}, \quad k = 0, 1, 2, ... \tag{1}$$

Let us introduce the following notation illustrated in Fig. 2:

$A(n)-$ the batch size for the nth slot;
$\tilde{Q}(n)-$ the number of customers in the queue immediately after the arrival of a batch;
$S(n)-$ the number of customers served in the nth slot.

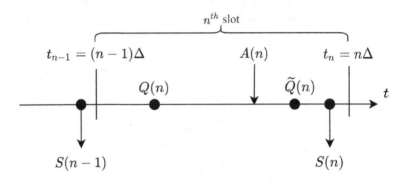

Fig. 2. Events in the nth slot

We have introduced the MC $Q(n)$ following [26]. Recall L being the buffer capacity and K is the number servers, then

$$\tilde{Q}(n) = min(Q(n) + A(n), L), \quad n \geq 1,$$

$$S(n) = min(\tilde{Q}(n), K), \quad n \geq 1,$$

and the value of $Q(n)$ can be found by

$$Q(n) = \tilde{Q}(n-1) - S(n-1), \quad n = 2, ... \tag{2}$$

The state space $\mathcal{Q}$ of the Markov Chain $Q(n)$ is

$$\mathcal{Q} = \{0, 1, 2, ..., L - K\} \tag{3}$$

and the matrix of transition probabilities $\mathbf{P} = ||p_{ij}||_{i,j\in\mathcal{Q}}$ has the following form:

1. $0 \leq i \leq K$

$$p_{ij} = \begin{cases} \sum_{k=0}^{K-i} A_k, & j = 0; \\ A_{K+j-i}, & 0 < j < L - K; \\ \sum_{k=L-i}^{\infty} A_k, & j = L - K; \end{cases} \tag{4}$$

2. $K < i \leq L - K$

$$p_{ij} = \begin{cases} 0, & 0 \leq j < i - K; \\ A_{K+j-i}, & i - K \leq j < L - K; \\ \sum_{k=L-i}^{\infty} A_k, & j = L - K. \end{cases} \tag{5}$$

Note that for the special case $L - K \leq K$ formula (4) for $0 \leq i \leq L - K$ is sufficient to determine p_{ij} and formula (5) is not needed.

Using the transition probabilities, we can numerically compute the stationary probabilities $\Pi = (\pi_0, \pi_1, ..., \pi_{L-K})$, where

$$\pi_i = \lim_{n \to \infty} P\{Q(n) = i\}, \quad i = 0, 1, ..., L - K.$$

4   Analytical of System Key Indicators

In this section, we investigate some important performance metrics of such a data transmission system.

4.1   Message Loss Probability

We examine the probability that at any time any targeted customer that arrives will be lost because it will not be accepted in the queue. Let the random variable $B(n)$ be the number of lost customers in the nth time slot after the arrival of a batch of $A(n)$ customers. We study the random variable $B(n)$ under the condition that there is at least one customer in the queue. So the probability mass function (pmf) of $B(n)$ looks as following:

$$f_B(i) = P\{B(n) = i | A(n) \geq 1\}, \quad i \in \{0, 1, ..., max(0, A(n) - K)\}.$$

Exactly i customers are lost in the nth time slot (i.e. $B(n) = i > 0$) if and only if

1. the state of the buffer is $Q(n) = k$, $k \in \mathcal{Q}$;
2. a batch of $A(n) = L - k + i$ customers arrives, $A(n) \geq 1$.

The illustration on Fig. 3 gives us a better understanding of the scenario.

Using the formula of conditional probability, we get the following:

$$f_B(i) = \left(\sum_{k=0}^{L-K} \pi_k A_{L-k+i} \right) \Big/ P\{A(n) \geq 1\} ;$$

$$P\{A(n) \geq 1\} = 1 - P\{A(n) = 0\} = 1 - e^{-\lambda \Delta}.$$

So, we have

$$f_B(i) = \left(\sum_{k=0}^{L-K} \pi_k A_{L-k+i} \right) \Big/ \left(1 - e^{-\lambda \Delta} \right). \tag{6}$$

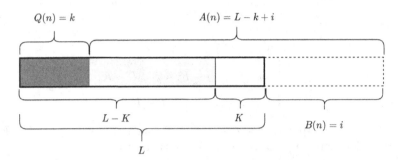

Fig. 3. Queue content and lost customers in the nth time slot

The probability that a random customer from the $L - k + i$ arrived customers will be among the i lost ones is equal to $\frac{i}{L-k+i}$. Putting it together with (6) and taking the summation over i we get the message loss probability as a probability of an incoming customer is lost:

$$P_{Loss} = \frac{1}{1 - e^{-\lambda\Delta}} \sum_{k=0}^{L-K} \pi_k \sum_{i=1}^{\infty} \frac{i}{L - k + i} A_{L-k+i}. \qquad (7)$$

We note that it may also be valuable to assess the probability that in an arbitrary time slot there is no loss of customers and this is the case when the number of arrived messages is less or equal to the number of empty places in the buffer of the drone:

$$P_{NoLoss} = f_B(0) = \sum_{k=0}^{L-K} \pi_k \sum_{j=1}^{L-k} A_j. \qquad (8)$$

4.2   Waiting Time Distribution

We consider an arbitrary slot in which we tag a specific arriving customer and check how long that customer will remain in the queue before being served. To help us tackle the task of obtaining the distribution of customer's waiting time in the queue, let us clarify some important issues: after the departure of $S(n-1)$ customers from the queue for service, there remains k customers in the buffer (i.e. $Q(n) = k$) occupying its first k places. This means that the arriving customers occupy the $L - k$ available places in the tail of the queue, and will not be served before those that were already in the queue.

Since there are K servers in the queueing system, we split the buffer into blocks of K places which gives us a total of $\lceil \frac{L}{K} \rceil$ blocks, where $\lceil . \rceil$ is the ceiling operator. For illustration, we number these blocks from 0 to $\lceil \frac{L}{K} \rceil - 1$ as shown in Fig. 4.

In the last block, there will be less than K places if L is not a multiple of K. If the tagged customer falls into the block B_i, it will wait exactly i slots before being served.

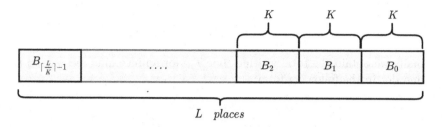

Fig. 4. Buffer splitting into blocks of size K

Let the random variable W, $W \in \{0, 1, 2, ..., \lceil \frac{L}{K} \rceil - 1\}$, be the number of time slots in the queue, i.e. the waiting time of the tagged customer. We will examine its pmf

$$f_W(i) = P\{W(n) = i\}.$$

The only way an incoming in the nth time slot customer will spend **zero slots** (i.e. $W(n) = 0$) in the queue is under the following conditions:

1. there were strictly less than K customers in the queue (i.e. $Q(n) = k < K$);
2. the tagged customer falls in B_0.

Taking into account these conditions $f_W(0)$ will have the following form:

$$f_W(0) = \sum_{k=0}^{K-1} \pi_k \sum_{j=1}^{K-k} A_j + \sum_{k=0}^{K-1} \pi_k \sum_{j=1}^{\infty} \frac{K-k}{K-k+j} A_{K-k+j}. \tag{9}$$

We similarly find pmf $f_W(i)$ for the other cases, where $W(n) = i$, $i > 0$:

$$
\begin{aligned}
f_W(i) = & \sum_{k=0}^{iK-1} \pi_k \sum_{m=1}^{K} \frac{m}{iK-k+m} A_{iK-k+m} \\
& + \sum_{k=0}^{iK-1} \pi_k \sum_{j=1}^{\infty} \frac{K}{(i+1)K-k+j} A_{(i+1)K-k+j} \\
& + \sum_{k=iK}^{min((i+1)K,L-K)-1} \pi_k \sum_{m=1}^{(i+1)K-k} A_m \\
& + \sum_{k=iK}^{min((i+1)K,L-K)-1} \pi_k \sum_{j=1}^{\infty} \frac{(i+1)K-k}{(i+1)K-k+j} A_{(i+1)K-k+j}.
\end{aligned}
\tag{10}
$$

For the total delay T_W for a message in the drone's buffer, we have to take into account the arrival and service offsets besides the waiting time $W(n)$ in the queue. We assume the arrival offset T_A and the service offset T_x to be random variables uniformly distributed in the slot of length Δ. Let it be denoted T_A, then

$$f_{T_A}(t) = f_{T_x}(t) = \frac{1}{\Delta}, \quad 0 < t \leq \Delta. \tag{11}$$

Finally we obtain the total waiting time T_W for an incoming message in the nth time slot in the following form:

$$f_{T_W}(t) = f_{T_A}(t) \sum_{i=0}^{\lceil \frac{L}{K} \rceil - 1} f_W(i) f_{T_x}(t). \tag{12}$$

4.3   Spent Energy per Message Transaction

To get the distribution of the required energy for a message's transaction, we first need to understand the different states of the energy expense of the drone transmitter. For every transaction, the sequence of states of the transmitter comprises the following steps:

1. searching the receiver within a constant time T_S with a constant power level P_{R_x};
2. accessing to the receiver in constant time T_{Ac} with a random power level P_{T_x} depending on the distance from the drone transmitter to the receiver;
3. waiting for start of transmission during the waiting time T_W, which is random, with a constant power level P_{R_x};
4. transmitting data for a constant time T_M with a random power level P_{T_x}.

Then the total energy needed to transmit the data is given by the following formula:

$$E_T = P_{R_x}.T_{R_x} + P_{T_x}.T_{T_x}. \tag{13}$$

Here

$$T_{R_x} = T_S + T_W, \quad T_{T_x} = T_{Ac} + T_M, \tag{14}$$

where P_{R_x} and T_{T_x} are constant values, T_{R_x} and P_{T_x} are random variables, and the probability density function (pdf) of T_{R_x} is given by

$$f_{T_{R_x}}(t) = f_{T_W}(t - T_S). \tag{15}$$

As for the random value P_{T_x}, it depends on the distance from the drone transmitter to the receiver and is given by

$$P_{T_x} = AD^\gamma, \tag{16}$$

where A is the transmitted power and γ the path loss exponent, D is the random distance from a transmitter drone uniformly distributed in the circle with radius R and receiver drone as its center.

From the pdf of the random variable D, which is $f_D(r) = \frac{2r}{R^2}, \quad 0 < r < R$, we obtain the pdf of P_{T_x} in the form of

$$f_{P_{T_x}}(y) = \frac{2(y/A)^{2/\gamma}}{\gamma R^2 y}, \quad y > 0. \tag{17}$$

Making the convolution of $f_{T_{R_x}}(t)$ and $f_{P_{T_x}}(y)$ taking into account (14) and (15) we get the distribution $f_{E_T}(t,y)$ of the energy spent per transaction according to (13).

4.4   Distribution of Drone Transmitter Lifetime

The operating cycle of a drone transmitter has two periods: a transaction period and a sleep period (when there is no message to transmit).

In the previous subsection, we explored the distribution of the energy E_T used during a transaction period. Let $E_{Idle} = P_{Idle}.T_{Idle}$ be the energy spent in a sleep period, where P_{Idle} is the constant power used and T_{Idle} the exponentially distributed time with parameter λ_I, which is determined by drone transmitter's data generation.

Now, let us consider a cycle duration T_C and the corresponding energy E_C for consecutive transaction and sleep periods $E_C = E_T + E_{Idle}$. Knowing $f_{E_T}(t,y)$ and $f_{E_{Idle}}(t)$, we get the pdf $f_{E_C}(t,y)$ of E_C as their convolution.

We denote as $E(t)$ the random variable of the spent energy by the drone transmitter at any given time t. We also make a reasonable assumption that the battery capacity of the drone transmitter U is much larger than the mathematical expectation of E_C ($U >> E[E_C]$), and this means that charging once the battery of the drone transmitter is sufficient to work for a large number of cycles.

Considering the central limit theorem, $E(t)$ can be approximated by a normally distributed random variable with parameters that can be found as in [29]:

$$\mu_{E(t)} = \frac{E[E_C]}{E[T_C]}t;$$

$$\sigma_{E(t)}^2 = \frac{\sigma^2[T_C]t}{(E[T_C])^3}. \tag{18}$$

Therefore, we can get the cumulative distribution function of the drone's lifetime T_L:

$$\begin{aligned} F_{T_L}(t) &= \mathrm{P}\{T_L \le t\} = 1 - \mathrm{P}\{T_L > t\} = \\ &= 1 - \mathrm{P}\{U - E(t) > 0\} = 1 - \mathrm{P}\{E(t) < U\} = \\ &= 1 - \Phi_{E(t)}(U), \end{aligned} \tag{19}$$

where $\Phi_{E(t)}$ is the normal distribution of the random variable of the spent energy $E(t)$ at the given time t, $t > 0$.

5   Conclusion

The proposed method can be used for estimating efficiency of data exchange between drones in a swarm with no information about drones mutual positions, for example at the stage of exchanging information about mutual positioning, when each network node periodically sends and receives broadcasts from all network nodes. The data exchange process was modeled in discrete time as a Markov Chain with batch arrivals and batch service. The quality indicators such as the message loss probability, the waiting delay, the needed energy for a message transaction, and the drone's battery lifetime have been evaluated analytically.

References

1. Giordani, M., Polese, M., Mezzavilla, M., Rangan, S., Zorzi, M.: Toward 6G networks: use cases and technologies. IEEE Commun. Mag. **58**(3), 55–61 (2020). https://doi.org/10.1109/MCOM.001.1900411
2. Zhang, Z., et al.: 6G wireless networks: vision, requirements, architecture, and key technologies. IEEE Veh. Technol. Mag. **14**(3), 28–41 (2019). https://doi.org/10.1109/MVT.2019.2921208
3. Bariah, L., et al.: A prospective look: key enabling technologies, applications and open research topics in 6G networks. IEEE Access **8**, 174792–174820 (2020). https://doi.org/10.1109/ACCESS.2020.3019590
4. Tang, S., Zhou, W., Chen, L., Lai, L., Xia, J., Fan, L.: Battery-constrained federated edge learning in UAV-enabled IoT for B5G/6G networks. Phys. Commun. **47**, 101381 (2021). https://doi.org/10.1016/j.phycom.2021.101381
5. Chen, S., Yin, D., Niu, Y.: A survey of robot swarms' relative localization method. Sensors **22**(12), 4424 (2022). https://doi.org/10.3390/s22124424
6. Campion, M., Ranganathan, P., Faruque, S.: UAV swarm communication and control architectures: a review. J. Unmanned Veh. Syst. **7**(2), 93–106 (2018). https://doi.org/10.1139/juvs-2018-0009
7. Amazon Prime Air prepares for drone deliveries. https://www.aboutamazon.com/news/transportation/amazon-prime-air-prepares-for-drone-deliveries. Accessed 13 July 2022
8. McFarland, M.: UPS drivers may tag team deliveries with drones. CNN Money (Washington), 21 (2017). http://money.cnn.com/2017/02/21/technology/ups-drone-delivery/index.html. Accessed 13 July 2022
9. The operational environment and the changing character of warfare, TRADOC Pamphlet 525-92, pp. 17–22, October 2019
10. Gkotsis, I., Kousouraki, A.C., Eftychidis, G., Kolios, P., Terzi, M.: Swarm of UAVs as an emergency response technology. In: Risk Analysis Based on Data and Crisis Response Beyond Knowledge, pp. 353–359. CRC Press (2019)
11. Suir, G.M., Reif, M.K., Hammond, S.L., Jackson, S.S., Brodie, K.L.: Unmanned aircraft systems to support environmental applications within USACE Civil Works (2018). https://erdclibrary.erdc.dren.mil/jspui/bitstream/11681/27428/1/ERDC%20SR-18-3.pdf

12. Roldán-Gómez, J.J., González-Gironda, E., Barrientos, A.: A survey on robotic technologies for forest firefighting: applying drone swarms to improve firefighters' efficiency and safety. Appl. Sci. **11**(1), 363 (2021). https://doi.org/10.3390/app11010363

13. Myjak, M.V.K., Ranganathan, P.: Unmanned aerial system (UAS) swarm design, flight patterns, communication type, applications, and recommendations. In: 2022 IEEE International Conference on Electro Information Technology (eIT), pp. 586–594. IEEE (2022). https://doi.org/10.1109/eIT53891.2022.9813866

14. Abdelkader, M., Güler, S., Jaleel, H., Shamma, J.S.: Aerial swarms: recent applications and challenges. Curr. Robot. Rep. **2**(3), 309–320 (2021). https://doi.org/10.1007/s43154-021-00063-4

15. Anderson, B.D., Yu, C., Fidan, B., Hendrickx, J.M.: Rigid graph control architectures for autonomous formations. IEEE Control Syst. Mag. **28**(6), 48–63 (2008)

16. Olfati-Saber, R., Fax, J.A., Murray, R.M.: Consensus and cooperation in networked multi-agent systems. Proc. IEEE **95**(1), 215–233 (2007)

17. Yu, W., Chen, G., Cao, M.: Distributed leader-follower flocking control for multi-agent dynamical systems with time-varying velocities. Syst. Control Lett. **59**(9), 543–552 (2010)

18. Watteyne, T., Augé-Blum, I., Dohler, M., Ubéda, S., Barthel, D.: Centroid virtual coordinates-a novel near-shortest path routing paradigm. Comput. Netw. **53**(10), 1697–1711 (2009)

19. Leong, B., Liskov, B., Morris, R.: Greedy virtual coordinates for geographic routing. In: 2007 IEEE International Conference on Network Protocols, pp. 71–80. IEEE (2007)

20. Filardi, N., Caruso, A., Chessa, S.: Virtual naming and geographic routing on wireless sensor networks. In: 2007 12th IEEE Symposium on Computers and Communications, pp. 609–614. IEEE (2007)

21. Karima, A., Mohammed, B., Azeddine, B.: New virtual coordinate system for improved routing efficiency in sensor network. Int. J. Comput. Sci. Issues (IJCSI) **9**(3), 59 (2012)

22. Dhanapala, D.C., Jayasumana, A.P.: Anchor selection and topology preserving maps in WSNs-a directional virtual coordinate based approach. In: 2011 IEEE 36th Conference on Local Computer Networks, pp. 571–579. IEEE (2011). https://doi.org/10.1109/LCN.2011.6115519

23. Anwit, R., Kumar, P., Singh, M.P.: Virtual coordinates routing using VCP-M in wireless sensor network. In: 2014 International Conference on Computational Intelligence and Communication Networks, pp. 402–407. IEEE (2014)

24. Samuylov, A., Moltchanov, D., Kovalchukov, R., Gaydamaka, A., Pyattaev, A., Koucheryavy, Y.: GAR: gradient assisted routing for topology self-organization in dynamic mesh networks. Comput. Commun. **190**, 10–23 (2022). https://doi.org/10.1016/j.comcom.2022.03.023

25. Petrov, V., et al.: Vehicle-based relay assistance for opportunistic crowdsensing over narrowband IoT (NB-IoT). IEEE Internet Things J. **5**(5), 3710–3723 (2017)

26. Moltchanov, D., Koucheryavy, Y.: D-BMAP/D/1/K queuing system with priorities. In: Proceedings of the International Congress on Ultra Modern Telecommunications and Control Systems, ICUMT 2010, Moscow, 18–20 October 2010, pp. 1–5 (2010)

27. Grover, R., Chaudhary, H., Sharma, G.: Geo/G/1 system: queues with late and early arrivals. In: Dash, S.S., Das, S., Panigrahi, B.K. (eds.) Intelligent Computing and Applications. AISC, vol. 1172, pp. 781–792. Springer, Singapore (2021). https://doi.org/10.1007/978-981-15-5566-4_70

28. Verdonck, F., Bruneel, H., Wittevrongel, S.: An all geometric discrete-time multi-server queueing system. In: Gribaudo, M., Sopin, E., Kochetkova, I. (eds.) ASMTA 2019. LNCS, vol. 12023, pp. 57–70. Springer, Cham (2020). https://doi.org/10.1007/978-3-030-62885-7_5
29. Alfa, A.S.: Queueing Theory for Telecommunications: Discrete Time Modelling of a Single Node System. Springer, New York (2010). https://doi.org/10.1007/978-1-4419-7314-6

Construction and Analysis of a Queueing Model with Service Prioritization for 5G Systems with Customizable Network Slice Instances

Y. Adou[1]([⊠]) [iD], E. Markova[1] [iD], and Yu. V. Gaidamaka[1,2] [iD]

[1] Peoples' Friendship University of Russia (RUDN University), 6 Miklukho-Maklaya St., Moscow 117198, Russian Federation
{1042205051,markova-ev,gaydamaka-yuv}@rudn.ru
[2] Federal Research Center Computer Science and Control of the Russian Academy of Sciences, 44-2 Vavilov St., Moscow 119333, Russian Federation

Abstract. The rapid deployment of the fifth generation (5G) of wireless systems since 2020 is profoundly transforming the telecommunication area, paving the way for new-generation networks that include more devices and services, are more reliable and enable faster communication. The network slicing (NS) technology, considered as one of the 5G systems' key features, allows the operation of several virtual networks called network slice instances (NSIs) on top of the same physical infrastructure, e.g. a base station. NSIs are specifically built to provide unique service types, i.e. the guaranteed bit rate (GBR), the best effort with minimum guaranteed (BG) or the best effort (BE). Furthermore, ensuring a satisfactory quality of service (QoS) level under NS technology is an important task, that may require to use the NSIs' isolation concept. In this paper, a queueing system model for analyzing the operation of three 5G NSIs under a preemption-based scheduler maintaining the required QoS levels is proposed. The system's key performance indicators (KPIs) are defined and computed. A numerical example of the system's functioning is provided.

Keywords: 5G · Wireless systems · NS · Preemption · Scheduling · NSI · Iterative method · KPI · Queueing theory · Network capacity · GBR · Service requirement

1 Introduction

Recently, the network slicing (NS) entered the phase of commercialization around the globe and, thus, is drawing the attention of researchers and scientists [10]. In all the 5G network's systems components, the NS technology is considered as one of the most promising [15]. In practice, NS offers flexible solutions to

This research was funded by the Russian Science Foundation grant number 22-79-10053 (https://rscf.ru/en/project/22-79-10053/).

manage/share efficiently the physical network resources through logical platforms known as network slice instances (NSIs), customizable and adaptable to the very specific needs of all sorts of tenants [1,10,15].

The services provided to NSIs' users have different quality of service (QoS) requirements, based on which they are divided into two categories—the real-time or guaranteed bit rate (GBR) services, e.g. video and voice calls, and the non-guaranteed bit rate (non-GBR) services, e.g. web browsing and file sharing [4,6, 12,15]. Depending on the presence or absence of a limitation on the minimum data rate, non-GBR services are divided into two categories—the best effort with minimum guaranteed (BG) and the best effort (BE) services [12,15]. To meet the QoS requirements, the physical infrastructure on which NSIs operate must be capable of meeting a service-level agreement (SLA) [15]. Moreover, adding to NS systems preemption schedulers based on the 3GPP priority's levels range may be of significance [10,11].

In the paper [16], the authors consider the operation of three NSIs under the resource reservation technique, with particular focus on the NSIs' flexible priority-based performance isolation, the fair QoS-aware resource allocation among users and the efficient use of resources. The NSIs were assumed to support each a unique service type from the GBR and non-GBR service categories, and their users—to have uniform data-rate requirements. In the paper [9], the authors aim to trade-off between isolation, efficiency, priority and customization in dynamic allocation of radio resources among the NSIs. They propose a simple and highly flexible scheme model under a two-layer scheduler with particular focus on the RAN slicing approach, where the radio resources' isolation is managed using the resource reservation technique. Note that the resource reservation technique is widely used by researchers [2,7,8,13], and apparently there is no paper on the subject of network slicing that does not consider it.

This paper's aim to construct and analyze a queueing system illustrating the functioning of three custom NSIs on top of one 5G base station (BS) under a preemption-based scheduler with consideration of the GBR SLA concept. Differently to the papers [2,7–9,13,16], the resource reservation technique is not considered here. One organizes the paper as follows. The Sect. 2 describes the model and presents the main considerations for its construction. The Sect. 3 formulates the mathematical model and proposes the formulas to compute the system's key performance indicators (KPIs). The Sect. 4 provides a numerical example of the system's functioning. Lastly, the Sect. 5 concludes the paper.

2   System Model

Let us consider an internet provider renting out its fifth generation (5G) base station (BS) to three tenants based on the network slicing (NS) technology, i.e. each tenant is assigned a unique fully customizable and logical wireless network—a network slice instance (NSI). Let C be the 5G BS's total network capacity and C_s—the s-th NSI's overall network capacity, $s \in \{1,2,3\}$, under the condition $C_1 + C_2 + C_3 \geq C$. We consider that the internet provider guarantees a minimum network capacity to each tenant [3,16], i.e. each tenant is assigned a guaranteed network capacity included in its overall network capacity. Let Q_s be the s-th NSI's guaranteed network capacity under the condition $Q_1 + Q_2 + Q_3 \leq C$.

Let us consider the Poisson arrival process of a unique request type at the s-th NSI with rate λ_s, $s \in \{1, 2, 3\}$. The arriving request requires b_s amount of resources for starting service under the condition $b_s \leq Q_s$. Let us consider the case when the service requirements b_s are uniform, i.e. $b_1 = b_2 = b_3 := b$, $b \in \mathbb{R}^+$. The average service time is exponentially distributed with the mean μ_s^{-1}.

A summary of the system's main notations is provided in the Table 1.

Table 1. The system's main notations

Notation	Description	Unit of measure
C	The BS's total network capacity	Capacity units (c.u.)
C_s	The s-th NSI's overall network capacity, $s \in \{1, 2, 3\}$, $C_1 + C_2 + C_3 \geq C$	c.u.
Q_s	The s-th NSI's guaranteed network capacity, $Q_s \leq C_s$, $Q_1 + Q_2 + Q_3 \leq C$	c.u.
λ_s	The arrival rate of requests at the s-th NSI, $\boldsymbol{\lambda} = (\lambda_1, \lambda_2, \lambda_3)$	Requests per time units (requests/t.u.)
μ_s^{-1}	The average service time of one request at the s-th NSI, $\boldsymbol{\mu} = (\mu_1, \mu_2, \mu_3)$	t.u.
b_s	The requirement for starting service of one request at the s-th NSI, $b_s \leq Q_s$, $b_1 = b_2 = b_3 := b$	c.u.
$\lfloor C_s/b \rfloor$	The maximum number of requests that may be admitted for service with the s-th NSI's overall network capacity, $\mathbf{N}^{\max} = (\lfloor C_1/b \rfloor, \lfloor C_2/b \rfloor, \lfloor C_3/b \rfloor)$	-
$\lfloor Q_s/b \rfloor$	The maximum number of requests that may be admitted for service with the s-th NSI's guaranteed network capacity, $\mathbf{N}^{\mathrm{g}} = (\lfloor Q_1/b \rfloor, \lfloor Q_2/b \rfloor, \lfloor Q_3/b \rfloor)$	-
n_s	The current number of requests at the s-th NSI, $\mathbf{n} = (n_1, n_2, n_3)$	-
$\mathbf{e}_s$	The s-th row of the identity matrix of size 3×3	-
$\mathbf{J}$	The three-dimensional all-ones vector	-

The radio admission control (RAC) scheme for accessing the s-th custom NSI is organized so that when the number n_s of servicing requests is less than $\lfloor Q_s/b \rfloor$, $s \in \{1, 2, 3\}$, and the amount of available resources at the BS is less than b, a preemption of one servicing request occurs at the $\hat{s}$-th NSI, $\hat{s} = \max \{\check{s} \in \{1, 2, 3\} : (\mathbf{n} - \mathbf{N}^{\mathrm{g}}) \cdot \mathbf{e}_{\check{s}} > 0\}$, once a new request arrives at that s-th NSI. Let us assign priority levels to the custom NSIs for clarifying the preemption method. Let the highest priority level "1" be assigned to the 1st NSI, the medium "2"—to the 2nd NSI and the lowest "3"—to the 3rd NSI.

One considers that upon arrival at its custom NSI a new request is bound to one path: direct admission, via preemption admission or blocking. Let us say the new request's admission is direct when the number n_s of servicing requests at its custom NSI is less than $\lfloor C_s/b \rfloor$ and the amount of available resources at the BS is greater than or equal to b, i.e. $(\mathbf{n} + \mathbf{e}_s - \mathbf{N}^{\max}) \cdot \mathbf{e}_s \leq 0 \wedge (\mathbf{n} + \mathbf{e}_s) \cdot b\mathbf{J} \leq C$, $s \in \{1, 2, 3\}$. Let us say the request's admission is via preemption when the number n_s of servicing requests at its custom NSI is less than $\lfloor Q_s/b \rfloor$ and the amount of available resources at the BS is less than b, i.e. $(\mathbf{n} + \mathbf{e}_s - \mathbf{N}^{\mathrm{g}}) \cdot \mathbf{e}_s \leq 0 \wedge (\mathbf{n} + \mathbf{e}_s - \mathbf{e}_{\hat{s}}) \cdot b\mathbf{J} \leq C$, $\hat{s} = \max \{ \check{s} \in \{1, 2, 3\} : (\mathbf{n} - \mathbf{N}^{\mathrm{g}}) \cdot \mathbf{e}_{\check{s}} > 0 \}$. Lastly, let us say the new request is blocked when the number n_s of servicing requests at custom its NSI is greater than or equal to $\lfloor Q_s/b \rfloor$ and the amount of available resources at the BS is less than b, i.e. $(\mathbf{n} + \mathbf{e}_s - \mathbf{N}^{\mathrm{g}}) \cdot \mathbf{e}_s > 0 \wedge (\mathbf{n} + \mathbf{e}_s) \cdot b\mathbf{J} > C$.

3   Mathematical Model

According to the defined RAC scheme and considering the Poisson distributed arrival processes, plus the exponentially distributed service times, one may describe the system's behavior using a three-dimensional Markov process $\mathbf{X}(t) = \{X_1(t), X_2(t), X_3(t), t > 0\}$, where $X_s(t)$—the number of servicing requests at the s-th NSI at time t, $s \in \{1, 2, 3\}$, over the system's state space:

$$\Omega = \left\{ \mathbf{n} \in \mathbb{N}^3 : (\mathbf{n} - \mathbf{N}^{\max}) \cdot \mathbf{J} \leq 0 \wedge \mathbf{n} \cdot b\mathbf{J} \leq C \right\}, \tag{1}$$

where $\mathbb{N}^3$—the set of all three-dimensional vectors with natural elements.

Let us introduce the main subsets [5] of the system's state space Ω for the model's further investigation:

- Ω_s^{dad}, $s \in \{1, 2, 3\}$—the subset with the system's states, where the admission of a new arriving request at the s-th NSI is direct:

$$\Omega_s^{\mathrm{dad}} = \{ \mathbf{n} \in \Omega : (\mathbf{n} - \mathbf{N}^{\max}) \cdot \mathbf{e}_s < 0 \wedge (\mathbf{n} + \mathbf{e}_s) \cdot b\mathbf{J} \leq C \}, \tag{2}$$

- Ω_s^{vpad}, $s \in \{1, 2, 3\}$—the subset with the system's states, where the admission of a new arriving request at the s-th NSI is via the preemption of one servicing request at the $\hat{s}$-th NSI, $\hat{s} = \max \{ \check{s} \in \{1, 2, 3\} : (\mathbf{n} - \mathbf{N}^{\mathrm{g}}) \cdot \mathbf{e}_{\check{s}} > 0 \}$:

$$\Omega_s^{\mathrm{vpad}} = \{ \mathbf{n} \in \Omega : (\mathbf{n} - \mathbf{N}^{\mathrm{g}}) \cdot \mathbf{e}_s < 0 \wedge (\mathbf{n} + \mathbf{e}_s) \cdot b\mathbf{J} > C \}. \tag{3}$$

For clarity, one may note that the main subsets' union $\Omega_s^{\mathrm{dad}} \cup \Omega_s^{\mathrm{vpad}}$, $s \in \{1, 2, 3\}$, regroups the system's states, where the admission of a new arriving request at the s-th NSI is possible. Therefrom, one may obtain the complement subset $\Omega_s^{\mathrm{block}}$, $s \in \{1, 2, 3\}$, with the system's states, where the admission of a new arriving request at the s-th NSI is impossible or blocked:

$$\Omega_s^{\mathrm{block}} = \{ \mathbf{n} \in \Omega : (\mathbf{n} - \mathbf{N}^{\mathrm{g}}) \cdot \mathbf{e}_s \geq 0 \wedge (\mathbf{n} + \mathbf{e}_s) \cdot b\mathbf{J} > C \}; \tag{4a}$$

or

$$\Omega_s^{\mathrm{block}} = \Omega \setminus \left(\Omega_s^{\mathrm{dad}} \cup \Omega_s^{\mathrm{vpad}} \right). \tag{4b}$$

Let us illustrate the logical relations between the system's main subsets (2), (3) with the Venn's diagram style. Figure 1 represents the Venn diagram of the system's main subsets, e.g. the subsets' intersection $\bigcap_{s=1}^{3} \Omega_s^{\text{vpad}} = \emptyset$ and the subsets' intersection $\bigcap_{s=1}^{3} \Omega_s^{\text{dad}}$ regroups the system's states, where the admission of any new arriving request at the system is direct.

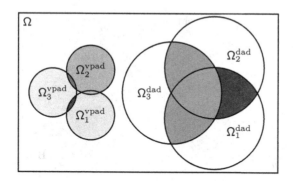

Fig. 1. The Venn diagram of the system's main subsets.

Based on the Venn diagram (Fig. 1) and considering the blocking subset (4), one may illustrate the system's state transition diagram as shown in Fig. 2 with the main subsets and the reference states.

Let us depict the transition diagram for a random state $\mathbf{n}, \mathbf{n} \in \Omega$, in the Fig. 3, based on which one may describe the discussed Markov process of the system's state space Ω with the system of equilibrium equations:

$$P\left(\mathbf{n}\right) \left(\boldsymbol{\lambda} \cdot \sum_{s=1}^{3} I_{\Omega_s^{\text{dad}} \cup \Omega_s^{\text{vpad}}} \left(\mathbf{n}\right) \mathbf{e}_s + \mathbf{n} \cdot \boldsymbol{\mu} \right) =$$

$$\boldsymbol{\lambda} \cdot \sum_{s=1}^{3} \left(P\left(\mathbf{n} - \mathbf{e}_s\right) H\left(\mathbf{n} \cdot \mathbf{e}_s\right) + P\left(\mathbf{n} - \mathbf{e}_s + \mathbf{e}_{\hat{s}}\right) I_{\Omega_s^{\text{vpad}}} \left(\mathbf{n} - \mathbf{e}_s + \mathbf{e}_{\hat{s}}\right) \right) \mathbf{e}_s +$$

$$\boldsymbol{\mu} \cdot \sum_{s=1}^{3} \left(P\left(\mathbf{n} + \mathbf{e}_s\right) I_{\Omega_s^{\text{dad}}} \left(\mathbf{n}\right) \left(\mathbf{n} + \mathbf{e}_s\right) \cdot \mathbf{e}_s \right) \mathbf{e}_s, \quad (5)$$

where $\hat{s} = \max \{ \breve{s} \in \{1,2,3\} : (\mathbf{n} - \mathbf{N}^{\text{g}}) \cdot \mathbf{e}_{\breve{s}} > 0 \}$, $P\left(\mathbf{n}\right)$, $\mathbf{n} \in \Omega$—the stationary probability that the system is in the state $\mathbf{n}$, $H\left(*\right)$—the Heaviside function, and $I_o\left(*\right)$—the Indicator function.

One notes that the Markov process describing the system's behavior is not reversible. Therefore, one may compute the system's stationary probability distribution $\mathbf{P} = \left(P\left(\mathbf{n}\right)\right)$, $\mathbf{n} \in \Omega$, i.e. a $|\Omega| \times 1$ matrix, using an iterative method [14] for solving the system of equilibrium equations:

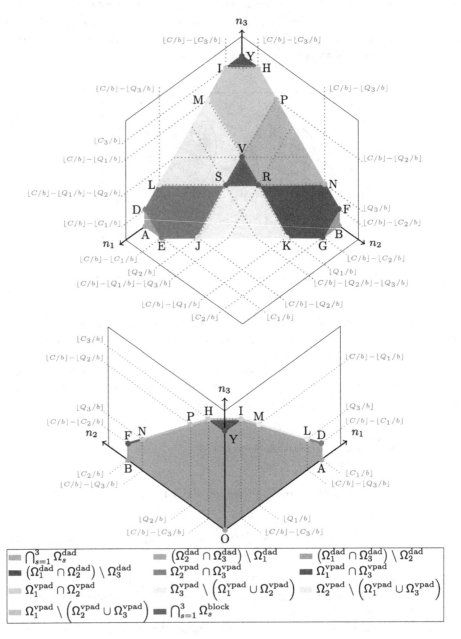

Fig. 2. The system's state transition diagram with the main subsets and the reference states.

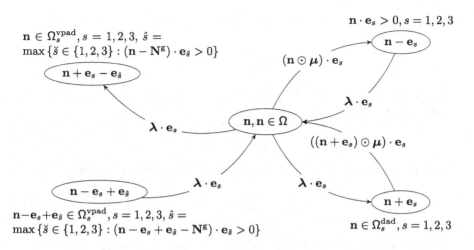

Fig. 3. The considered system's transition diagram for a random state $\mathbf{n}, \mathbf{n} \in \Omega$.

$$\mathbf{A}^\top \mathbf{P} = \mathbf{0}, \tag{6}$$

where $\mathbf{A}$—the infinitesimal generator of Markov process, i.e. a $|\Omega| \times |\Omega|$ matrix, whose entries $A(\mathbf{n} \in \Omega, \hat{\mathbf{n}} \in \Omega)$ are computed as follows: when $\mathbf{n} \neq \hat{\mathbf{n}}$

$$A(\mathbf{n}, \hat{\mathbf{n}}) = \begin{cases} \boldsymbol{\lambda} \cdot \mathbf{e}_s, \text{ if } \hat{\mathbf{n}} = \mathbf{n} + \mathbf{e}_s, & \text{s.t. } \mathbf{n} \in \Omega_s^{\text{dad}}, \\ \text{else if } \hat{\mathbf{n}} = \mathbf{n} + \mathbf{e}_s - \mathbf{e}_{\hat{s}}, & \text{s.t. } \mathbf{n} \in \Omega_s^{\text{vpad}}, \\ (\mathbf{n} \odot \boldsymbol{\mu}) \cdot \mathbf{e}_s, \text{ if } \hat{\mathbf{n}} = \mathbf{n} - \mathbf{e}_s, & \text{s.t. } \hat{\mathbf{n}} \in \Omega, \\ 0, \text{ otherwise}, & \text{i.e. } \hat{\mathbf{n}} \in \Omega \setminus \{\mathbf{n}\}, \end{cases}$$
$$s \in \{1, 2, 3\}, \ \hat{s} = \max\{\check{s} \in \{1, 2, 3\} : (\mathbf{n} - \mathbf{N}^{\text{g}}) \cdot \mathbf{e}_{\check{s}} > 0\}, \tag{7a}$$

when $\mathbf{n} = \hat{\mathbf{n}}$

$$A(\mathbf{n}, \mathbf{n}) = -\sum_{\hat{\mathbf{n}} \in \Omega \setminus \{\mathbf{n}\}} A(\mathbf{n}, \hat{\mathbf{n}}). \tag{7b}$$

After obtaining the system's stationary probability distribution $\mathbf{P}$, one may compute its key performance indicators (KPIs). Let us propose the following KPIs for the model's further investigation:

– The mean number of servicing requests at the s-th custom NSI, $s \in \{1, 2, 3\}$

$$\sum_{\mathbf{n} \in \Omega} P(\mathbf{n}) \, \mathbf{n} \cdot \mathbf{e}_s; \tag{8}$$

- The mean number of servicing requests at the system

$$\sum_{\mathbf{n}\in\Omega} P(\mathbf{n})\,\mathbf{n}\cdot\mathbf{J}; \tag{9}$$

- The direct admission probability for a new arriving request at the s-th custom NSI, $s\in\{1,2,3\}$

$$\sum_{\mathbf{n}\in\Omega_s^{\mathrm{dad}}} P(\mathbf{n}); \tag{10}$$

- The direct admission probability for a new arriving request at the system

$$\sum_{\mathbf{n}\in\bigcap_{s=1}^{3}\Omega_s^{\mathrm{dad}}} P(\mathbf{n}); \tag{11}$$

- The via preemption admission probability for a new arriving request at the s-th custom NSI, $s\in\{1,2,3\}$

$$\sum_{\mathbf{n}\in\Omega_s^{\mathrm{vpad}}} P(\mathbf{n}); \tag{12}$$

- The via preemption admission probability for a new arriving request at the system

$$\sum_{\mathbf{n}\in\bigcap_{s=1}^{3}\Omega_s^{\mathrm{vpad}}} P(\mathbf{n}); \tag{13}$$

- The blocking probability for a new arriving request at the s-th custom NSI, $s\in\{1,2,3\}$

$$\sum_{\mathbf{n}\in\Omega_s^{\mathrm{block}}} P(\mathbf{n}); \tag{14}$$

- The blocking probability for a new arriving request at the system

$$\sum_{\mathbf{n}\in\bigcap_{s=1}^{3}\Omega_s^{\mathrm{block}}} P(\mathbf{n}). \tag{15}$$

4 Numerical Example

One may conduct a numerical example of the system's functioning, focusing on the KPIs in relation with the 1st custom NSI—mean number (8), direct admission probability (10), via preemption admission probability (12) and blocking probability (14).

One may make the next preparations. Let C the BS's total network capacity be equal to 8.5 Gbps. Let C_1, C_2 and C_3 the 1st, 2nd and 3rd custom NSIs' overall network capacity and Q_1, Q_2 and Q_3 their guaranteed network capacity be respectively equal to 6.5 Gbps, 5 Gbps, 3 Gbps, 4.5 Gbps, 3 Gbps, and 1

Gbps. One may consider that the custom NSIs provide their users with video-based services, e.g. live-streaming, virtual reality (VR) [6], i.e. the requirement for starting service of one request equals 554.62 Mbps. Lastly, one may assume that the average service time of one request at each custom NSIs equals 1 s and therefore the offered load equation at each NSI may take the form $\lambda_s = \lfloor C_s/b \rfloor$, $s = \{1, 2, 3\}$. One may consider the scenario where the domain of the arrival rates λ_1, λ_2 and λ_3 of requests at the 1st, 2nd and 3rd custom NSIs are respectively the interval $[1, 2 * \lfloor C_1/b \rfloor]$ and the sets $\{\lfloor Q_2/b \rfloor, \lfloor C_2/b \rfloor\}$ and $\{\lfloor Q_3/b \rfloor, \lfloor C_3/b \rfloor\}$, i.e. see Figs. 4, 5, 6, 7.

The numerical example's parameters are summarized in the Table 2.

Table 2. The scenario's parameters

Notation	Value	Unit of measure
C	8.5	Gbps
C_1, C_2, C_3	6.5, 5, 3	Gbps
Q_1, Q_2, Q_3	4.5, 3, 1	Gbps
b	554.62	Mbps
λ_1	$1 \to 2 * \lfloor C_1/b \rfloor$	requests/sec
λ_2	$\lfloor Q_2/b \rfloor, \lfloor C_2/b \rfloor$	requests/sec
λ_3	$\lfloor Q_3/b \rfloor, \lfloor C_3/b \rfloor$	requests/sec
μ	$(1, 1, 1)$	sec

Let us examine the numerical results. In the Figs. 4, 5, 6 and 7 one may visualize how the increasing arrival rate of requests at the 1st NSI impacts its KPIs. First, one may examine the impact on the mean number of servicing requests at the 1st NSI, i.e. see Fig. 4. This KPI rapidly increases when the arrival rate of requests is closing to $\lfloor Q_1/b \rfloor$ from 1, and then gradually increases from that point, which may seem to be an inflection point, i.e. at $\lambda_1 = \lfloor Q_1/b \rfloor$. One may explain that global increase by the fact that the more requests arrive at the NSI, the more may obviously be servicing. Secondly, one may examine the impact on the direct admission probability, i.e. see Fig. 5. This KPI gradually decreases with an inflection point at $\lambda_1 = \lfloor Q_1/b \rfloor$, i.e. the more requests arrive at the NSI, the less may directly be admitted. Thirdly, one may examine the via preemption admission probability, i.e. see Fig. 6. This KPI increases at first and then suddenly and rapidly decreases, i.e. the turning point or global maximum is reached before the inflection point at $\lambda_1 = \lfloor Q_1/b \rfloor$. One may suggest that this KPI's behavior is due to the input parameters, i.e. the arrival rates of requests at the 2nd and 3rd NSIs are set very high, so obviously these NSIs are servicing more requests than what may be admitted with their guaranteed network capacities, and consequently, preemption might be mandatory for admitting new arriving requests at the 1st NSI. Lastly, one may examine the impact on the blocking probability, i.e. see Fig. 7. This KPI gradually increases with an inflection point at $\lambda_1 = \lfloor Q_1/b \rfloor$, i.e. the more requests arrive at the NSI, the more may be blocked.

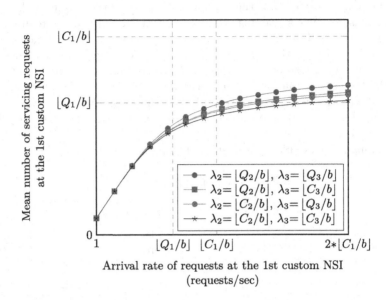

Fig. 4. Mean number of servicing requests at the 1st custom NSI vs. Arrival rate of requests at the 1st custom NSI.

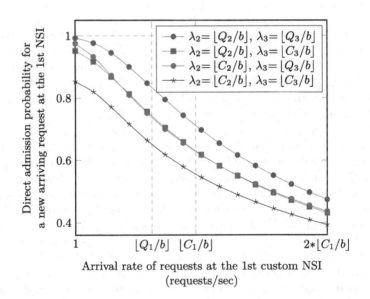

Fig. 5. Direct admission probability for a new arriving request at the 1st custom NSI vs. Arrival rate of requests at the 1st custom NSI.

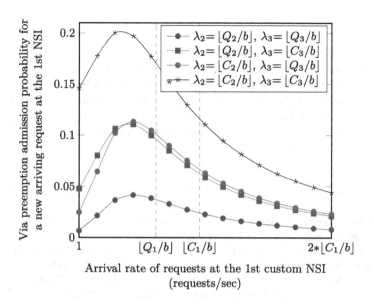

Fig. 6. Via preemption admission probability for a new arriving request at the 1st custom NSI vs. Arrival rate of requests at the 1st custom NSI.

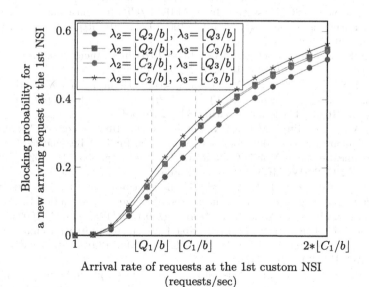

Fig. 7. Blocking probability for a new arriving request at the 1st custom NSI vs. Arrival rate of requests at the 1st custom NSI.

5   Conclusion

One proposed a queueing system model for analyzing the operation of three 5G network slice instances (NSIs) under a preemption-based scheduler, maintaining the required QoS levels. The model's general specifics were presented. The system's key performance indicators (KPIs) were defined and computed. A numerical example of the system's functioning was provided. The next stage of this paper may be to consider random requirements for starting users' service. One may later on investigate the operation of a random number of NSIs with random service requirements.

References

1. 5G industry campus network deployment guideline. Official Document NG.123, GSM Association (GSMA), October 2021. version 2.0
2. Aldemir, S., Şadi, Y., Erküçük, S., Okumuş, F.B.: NOMA-based radio resource allocation for machine type communications in 5G and beyond cellular networks. In: 2021 29th Signal Processing and Communications Applications Conference (SIU), pp. 1–4 (2021). https://doi.org/10.1109/SIU53274.2021.9477699
3. Basharin, G.P., Gaidamaka, Y.V., Samouylov, K.E.: Mathematical theory of teletraffic and its application to the analysis of multiservice communication of next generation networks. Autom. Control. Comput. Sci. **47**(2), 62–69 (2013). https://doi.org/10.3103/S0146411613020028
4. Chen, C., Tian, H., Nie, G.: Fairness resource allocation scheme for GBR services in downlink SCMA system. In: 2020 IEEE/CIC International Conference on Communications in China (ICCC), pp. 190–195 (2020). https://doi.org/10.1109/ICCC49849.2020.9238827
5. Devlin, K.: The Joy of Sets. Undergraduate Texts in Mathematics. Springer, New York (1993). https://doi.org/10.1007/978-1-4612-0903-4
6. Khatibi, S.: Radio Resource Management Strategies in Virtual Networks. Ph.D. thesis, University of Lisbon (2016). https://grow.tecnico.ulisboa.pt/wp-content/uploads/2016/08/Thesis_sina_khatibi_IST172360.pdf
7. Li, Y., Wang, Y., Jin, Y., Cheng, X., Xu, L., Liu, G.: Research on wireless resource management and scheduling for 5G network slice. In: 2021 International Wireless Communications and Mobile Computing (IWCMC), pp. 508–513 (2021). https://doi.org/10.1109/IWCMC51323.2021.9498806
8. Luu, Q.T., Kerboeuf, S., Kieffer, M.: Admission control and resource reservation for prioritized slice requests with guaranteed SLA under uncertainties. IEEE Trans. Netw. Serv. Manage. (2022). https://doi.org/10.1109/TNSM.2022.3160352
9. Marabissi, D., Fantacci, R.: Highly flexible RAN slicing approach to manage isolation, priority, efficiency. IEEE Access **7**, 97130–97142 (2019). https://doi.org/10.1109/ACCESS.2019.2929732
10. Meredith, J.M., Firmin, F., Pope, M.: Release 16 Description; Summary of Rel-16 Work Items. Technical report (TR) 21.916, 3rd Generation Partnership Project (3GPP), January 2022. version 16.1.0
11. Meredith, J.M., Soveri, M.C., Pope, M.: Management and orchestration; 5G end to end Key Performance Indicators (KPI). Technical specification (TS) 28.554, 3rd Generation Partnership Project (3GPP), March 2022. version 17.6.0

12. Mohgan, P.D., Das, D.: Efficient way of Non-GBR, high latency GTP-U packet transmission in 4G and 5G networks. In: 2021 IEEE International Conference on Electronics, Computing and Communication Technologies (CONECCT), pp. 1–4 (2021). https://doi.org/10.1109/CONECCT52877.2021.9622673
13. Naddeh, N., Jemaa, S.B., Eddine Elayoubi, S., Chahed, T.: Proactive RAN resource reservation for URLLC vehicular slice. In: 2021 IEEE 93rd Vehicular Technology Conference (VTC2021-Spring), pp. 1–5 (2021). https://doi.org/10.1109/VTC2021-Spring51267.2021.9448703
14. Stepanov, S.N.: Theory of Teletraffic: Concepts, Models, Applications [Teoriya teletraffika: kontseptsii, modeli, prilozheniya]. Goryachaya Liniya-Telekom, Moscow (2015). in Russian
15. Sultan, A., Pope, M.: Feasibility study on new services and markets technology enablers for network operation; Stage 1. Technical report (TR) 22.864, 3rd Generation Partnership Project (3GPP), September 2016. version 15.0.0
16. Yarkina, N., Gaidamaka, Y., Correia, L.M., Samouylov, K.: An analytical model for 5G network resource sharing with flexible SLA-oriented slice isolation. Mathematics. 8(7), 1177 (2020). https://doi.org/10.3390/math8071177

Spectrum and AI-based Analysis for a Flight Environment and Virtual Obstacles Avoidance Using Potential Field Method for Path Control

Ayham Shahoud$^{(\boxtimes)}$ ⓘ, Dmitriy Shashev ⓘ, and Stanislav Shidlovskiy ⓘ

Tomsk State University, 36 Lenin Ave, Tomsk 634050, Russia
ayhams86@gmail.com

Abstract. Computer vision-based navigation systems basically rely on the external environment texture for positioning. This research studies the selection of the computer vision navigation algorithm that suits better for a known flight environment. Two methods are used to analyze the flight environment images then a comparative study between them is done. While the first method depends on offline spectrum analysis using Fourier transform for the environment images, the second method uses artificial intelligence based on Convolutional Neural Network (CNN) to analyze such images. Usually, in traditional navigation tasks, failures in computer vision navigation systems are solved by depending on measurements from other systems that rises the cost, weight, and complexity. This work suggests avoiding the bad matching areas or these where vision failures are expected to take place, by treating them as virtual obstacles and avoiding them using the potential fields method. The previous suggestion could result in a stable standalone visual navigation and path tracking system, that is only if no conflict with the mission exists, which is assumed in this research. The two methods are implemented and tested on a path in a simulation environment consisting of a Robot Operating System (ROS), Gazebo simulator, and IRIS drone model. Results show that both methods give indicators to the environment structure, and help in selecting the efficient navigation algorithm. The CNN-based method offers a more detailed description of the environment in form of a meta-map (assistant map) which has been employed effectively to avoid areas over which positioning failures are expected.

Keywords: Obstacle avoidance · Scene matching · Flight environment · CNN · Frequency analysis · Potential fields · Path control

1 Introduction

Visual navigation systems are considered more independent than GPS which depends on radio signals from satellites. In spite of their independence from an

Supported by Tomsk State University Development Program (Priority-2030).

external provider, visual navigation systems rely on the external environment structure and texture which are in turn affected by the illumination, weather, seasons, and human activities. Generally, computer vision navigation systems are based either on matching and accumulating the changes caused by camera motion in the successively captured images like in Visual Odometry (VO), or on matching the captured image with a reference image that has been prepared offline like in scene matching. According to the matching technique, computer vision systems can be classified into a feature-based (that use keypoints) or correlation-based which use the whole image. Features might be corners, edges, lines, or points that are robust against variation in rotation and scale such as those detected by Scale Invariant Feature Transform (SIFT), Speed Up Robust Features (SURF), and (Oriented FAST and Rotated BRIEF) ORB [1].

Computer vision-based navigation systems are cheap and provide a lot of information about the surrounding environment, but they suffer from several problems such as performance instability with seasons, illumination, weather, and that image processing is a time-consuming process. The performance of the computer vision-based navigation system is subject to the surrounding environment (structure, illumination, shadows, seasons, forest environment, and urban), the equipment (camera, lenses, and processor), and the adopted methods and algorithms (Visual odometry, scene matching, features tracking, cross-correlation, optic flow, SIFT, corners, edges, and intensity). The image processing time and the equipment had been studied a lot and so many solutions now exist such as using Graphical Processing Units GPU, FPGA, using high-quality cameras and lenses, and suitable camera calibration, all these topics are out of the scope of this article [2].

As it is mentioned above, visual navigation depends on the environment's texture and structure i.e. as much as the environment contains stable and distinguishable texture (which is easy to detect and does not lead to ambiguity), it will result in a more robust navigation calculation. In areas where shadows and illumination variations dominate, the navigation algorithm parameters might need some continuous adaptation and tuning. Low structured areas such as wood, give flattened correlation results where no strong peak exists in the correlation output which leads to ambiguity or false matching. On another side, the existence of similar or identical features in the environment also might lead to ambiguity, especially with the existence of perturbation sources such as noise and weather. Human activity and seasons have a slow but strong effect on the environment texture, especially in the zones where snow lasts for long times. For example, a reference image captured in Spring might not be valid as a reference in Autumn or Winter and the opposite is right.

Generally, most navigation algorithms recommend that the environment satisfies "stillness prevails over movement in the scene". The previous condition is not always satisfied, but it exists some methods that help to exclude false matches (resulting because of the motion in the background in the environment) such as the Random Sample Consensuses (RANSAC) algorithm, or benefiting from other systems or knowledge about the environment to formulate constraints that help in that task. It is worth to be mentioned that in mutual aiding between

Inertial Navigation Systems (INS) and the vision systems, the INS measurements are used to limit the searching area for corresponding features, which in turn could help in solving the ambiguity problem. In navigation applications both the scale and angle (rotation) of the captured image change with the drone movements, so it is good to adopt algorithms that give good resistance against scale and rotation variation such as SIFT. The well-structured areas in urban are relatively less affected by volatile variations and even seasonal variations, examples of such areas are street intersections. The normalized cross-correlation is robust to illumination variation and works well in structured areas.

From the previous discussion, it is obvious that analyzing the flight environment can give good indicators of the suitable navigation algorithm and tools. Such analysis could increase the stability of the system from aside and the accuracy from another side. In case of failures in the calculation of the visual navigation position, a solution must be found to avoid loosing the drone, crushing it, causing damage to the surrounding, or stacking it in a bad area i.e., not suitable for visual navigation. Examples of such solutions: depending on another navigation system such as GPS which rise the cost, relying on an emergency plan "B" like maneuvering or loiter to leave the bad matching area, and sometimes it is good to land immediately.

In stand-alone navigation systems, it is good to avoid such areas according to the previous knowledge and analysis of the environment and that was adopted in this work. Some possible solutions in case of failure because of the unsuitable environment are shown in Fig. 1. For more explanation refer to Sect. 4.

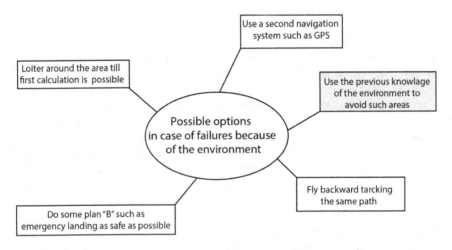

Fig. 1. The suggested solutions when a failure because of the flight environment occurs.

This research focuses on the selection of the suitable algorithms used in the calculation of computer vision-based navigation considering only the work environment texture. The goal is established by analyzing the environment to choose the algorithms that have higher detecting or matching efficiency, and to suggest some solutions in case of failures. The first method depends on frequency analysis of the flight area to estimate the general environment structure. The second method uses artificial intelligence based on CNN to classify such areas into classes that suit a certain type of navigation system and areas where bad matching is expected or bad navigation results. The potential field method is used to control the drone path to avoid crossing over areas where bad matching is expected to occur. The rest of this paper is organized as follows: Sect. 2 for the related studies, Sect. 3 for flight environment analysis, Sect. 4 for avoiding bad matching areas using potential field method, Sect. 5 for implementation and comparison, and Sect. 6 for the conclusion.

2   Related Study

Each computer vision navigation system works under certain circumstances. For example, VO needs a well-textured environment to sustain features tracking in the successive overlapped field of view. Navigation systems based on scene matching work well in well-structured areas when using cross-correlation techniques because the cross-correlation gives trusted results in such areas. In weak-textured environments maybe integration with another navigation system such as GPS should take place to overcome the failures. Structured areas usually contain lines (parallel or perpendicular) such as rivers, streets, and buildings. Some environments might be rich in features or keypoints but not well structured, for example, in woods, such features could be processed by SIFT for example [3].

Enhancing the matching process between two images is a matter of interest for many researchers to prevent mismatching. A preprocessing method for enhancing the correlation extremal between two images based on image entropy is presented in [4]. The matching between the two images was done after calculating the entropy for each image, which gave more accurate correlation results. Analyzing reference images to choose the best feature detector using frequency analysis is presented in [5]. Three types of detectors were presented, SURF, Line segment Detectors (LSD), and Maximally stable Extremely Region (MSER).

Using geometrical intersection as good matching areas is presented in [6,7], wherein [7], CNN was trained to classify good/bad matching areas and an assistant map was built to be used offline. Geometrical intersections were considered areas where good or true matchings are expected to occur, while weak structured areas such as homogeneous wide roofs and woods were considered bad or false matching areas. Also, a comparative study with the statistical methods that are based on variance calculations to detect false matching areas was done.

Using street geometry and dimension for image alignment and servoing is presented in [8]. The street intersections were used as landmarks for path following. Furthermore, street dimensions and orientations were used two align the

58   A. Shahoud et al.

captured image with the reference image. Landmark selection for scene matching with the knowledge of color histogram of the scene is presented in [9]. In [10], presented enhancing navigation performance of visual odometry during GPS outages, a GPS/INS/Vision integrated solution was proposed to provide continuous navigation. A GNSS/IMU/Vision ultra-tightly integrated navigation system for low-altitude aircraft is presented in [11]. In [12], while a multi-layer neural network was used to detect the edges in the captured images, the position was calculated using cross-correlation. The system was tested on Unmanned Aerial Vehicle (UAV), and matching edgy areas produced a reliable correlation result. Emergency selection of landing area for a drone is presented in [13]. Selecting the area was achieved using vision systems, then choosing the best site by analyzing high-precision 3D images took place. Another similar automated emergency landing system for drones based on computer vision with CNN and altimeter is presented in [14]. Obstacle avoidance using multi-point potential field for unmanned aerial vehicle in a dynamic environment is presented in [15].

In this work, a frequency analysis is applied to a reference image of the flight area. The results are employed to suggest suitable navigation and matching algorithms. Furthermore, CNN similar to that used in [7], is used to detect well-structured areas. After that, a comparison between the two methods took place. In addition to the comparison between the two analyzing methods, the main contribution of this paper is treating the bad matching areas represented on a meta-map (or the assistant map that is built depending on CNN output) as "obstacles" or "virtual obstacles" and avoiding crossing them using potential fields method. The proposed idea prevented mission failures or loosing the computer vision positioning without relying on another system. A general diagram of the designed system is shown in Fig. 2.

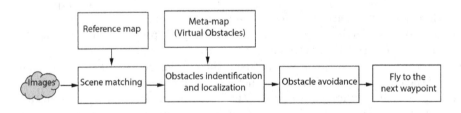

Fig. 2. The general system diagram.

3 Flight Environment Analysis

From the previous discussion, when the flight environment is known i.e. available aerial images are available from previous flights or from google maps, for example, it is good to benefit from them to choose suitable algorithms for navigation and matching. That could be established by many methods like frequency analysis, or depending on AI to analyze the environment images. Analyzing the environment can give good indicators about the structure of the environment and hence helps in selecting suitable algorithms to improve the overall system performance.

3.1 Environment Analysis Based on Frequency

Using 2D-FFT to analyze a digital image can give information about the nature of signals that exist in it, hence the structure of the environment. For example, an image with dominated parallel straight lines (maybe it includes streets or rivers) is expected to have a different spectrum than an image with perpendicular lines (buildings) or from a forest. In the first case, the spectrum is expected to be like "I-shape" i.e., 2D sinusoidal signal, while in the second one is expected to have "X-shape", which means 2 sinusoidal (maybe perpendicular) signals exist in the image, and the third one is expected to have "O-shape". For more details refer to [5].

It is obvious that images with "X-shape" 2D-FFT could be considered a well-structured area, and "I-shape" is not so well compared to "X-shape". "O-shape" areas might contain a lot of features or special points which could be discovered by SIFT, SURF, or ORB, but mostly a dominant structure does not exist, an example of this such area is the wood. In environments with an "O-shape", It will be useless to use scene matching with correlation techniques [6,7]. It is better to use image matching based on features,. In spite of that, ambiguity might occur in case of similar features in the area, especially in forests. In such environments, it is good to get assistance from other systems like GPS to avoid mission failures. Also, visual odometry using feature tracking works well, but also there will be a risk because of ambiguity, such risks could be partially treated using RANSAC.

In well-structured areas with "X-shape", scene matching based on cross-correlation techniques works well. In weak-structured areas with "I-shape", mismatching or false matching might take place when using the correlation technique. In such areas, it is better to use feature-based matching where, a few correspondent points with true matching are enough to match two images, after that, the position of the drone could be calculated in a georeferenced map using 2D-3D matching or 2D-2D matching techniques. Another solution for weak-structured areas is to combine both scene matching based on correlation technique and VO based on feature tracking, in this case, scene matching will compensate for the VO errors [8].

In this work, the flight environment is chosen to be urban. The spectrum for it is shown in Fig. 3. According to the spectrum shape, the area is well structured and suits scene matching using correlation techniques.

3.2 Environment Analysis Using CNN

Image classification using CNN is effective and used in many applications such as pattern recognition. In [7], it is used to classify images captured by a camera fixed on a drone into two classes: areas with a high probability to have a true match on a reference map based on their structure, and areas with a low probability to have a true match on a reference map (weak-structured areas). Also, presented

Flight Environment   **The Spectrum**

Fig. 3. The flight environment and its spectrum, it is obvious that the spectrum is of "x-shape"-like which is an indicator of a well-structured area.

the building of a meta-map that contains indicators or numbers that represent the probability that a drone located in the corresponding pixel of the reference map will have good or real matching on that map or not. Those numbers were binarized to zeroes and ones i.e., black and white.

This work adopts the same procedure to build the meta-map, but here to benefit from it to avoid loitering over bad matching areas (virtual obstacles) besides analyzing the flight area. According to the assistant map, if the area is well-structured so, no integration is needed with other systems, and scene matching based on the correlation technique works well, otherwise, the same aforementioned discussion in Subsect. 3.1 works. Building the meta-map using CNN offers a lot of options to deal with unsuitable areas for navigation (weak structured) such as: avoiding these areas and treating them as obstacles which are adopted in this paper.

In this work, CNN analyses are applied to the same flight area shown in Fig. 3. Visualizing the meta-map results shows that exist areas where if the drone is located in them, calculation of the drone position might not be possible, and these patches cannot be predicted using frequency analysis. A part of the meta-map that contains such bad matching areas is shown in Fig. 4.

4 Avoiding Bad Matching Areas Using Potential Field Method

In a standalone visual navigation system, failures in position calculation might lead to mission failure, drone crush, or damage to the surrounding objects. So, in addition, to choosing the suitable equipment (CPU, GPU, camera, lens..etc.), and choosing the best algorithms that suit the flight environment, it is good to have a solution (or decision) in case of position calculation failures. Such solutions were presented in Fig. 1, Using a second navigation system, emergency

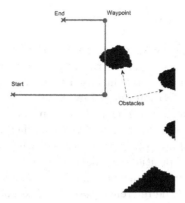

Fig. 4. The ideal path on the meta-map with the virtual obstacles. It is expected that the done will loose visual positioning over the planned path.

landing, loitering and trying to fix a solution, and flying backward following the same path. All the mentioned solutions might be costly and cause delays in the mission. Anyway, they might reduce the damages but they lead to mission failure. This research suggests avoiding such areas which might give false matching.

Avoiding areas where the camera fixed on the drone is expected to capture bad matching results is a good choice, but that is when enough information is available about the environment, and when the mission allows doing such warping in the path. One way to avoid these areas is to treat them as obstacles and depend on potential fields method to cross away from them. A repulsive field F^r is needed to repel the drone from areas with obstacles and an attractive field F^a is used to attract the drone to the next waypoint [15,16]. The equations that define the fields in case of one obstacle are presented in Eqs. 1, 2, 3, and 4.

$$F = F^a + F^r \tag{1}$$

$$F^a = k^a \frac{r^g - r}{|r^g - r|} \tag{2}$$

$$F^r = -k^r (\frac{1}{d} - \frac{1}{d^m})s : if\, d^m > d; else : F^r = 0 \tag{3}$$

$$s = \frac{r - o}{|r - o|} \tag{4}$$

where: F is the total field, F^a is the attractive field, F^r is the repulsive field, d^m represents the limit distance of the potential field's influence, d is the shortest distance to an obstacle, r^g is the position vector of the next waypoint, r is the position vector of the drone, o is the position vector of the obstacle, and k^a, k^r are constants.

5   Implementation and Comparison

The IRIS drone is equipped with a camera and GPS (as a reference) and flies over a path at an altitude of 100m constructed by several waypoints. The system is realized in a simulation environment based on ROS and Gazebo, and all programs were written using Python under Linux. The position is calculated using scene matching (with a reference image) based on normalized cross-correlation and the meta-map which was built using CNN classification. The path is intentionally chosen to contain areas (a patch of pixels) where if the drone is located in it, expected a failure in position calculation using the camera.

The calculated and the reference position in the horizontal plane is shown in Fig. 5. The calculated and the reference position on x-axis and y-axis are shown in Fig. 6 and Fig. 7, respectively. The drone avoided passing above the obstacle area because of the potential fields and continued toward the next waypoint. The RMS position error was 1.3m, and the continuity of position calculation was preserved. The important point in this contribution comes from the fact that for most applications warping the path during the mission is better than loosing control of the drone or mission failure. During the obstacle avoidance procedure and when the drone changes direction, it is noticeable that position errors have small jumps because of the drone (camera attached on it) vibration in these stages, but all that was in the accepted range.

Both frequency and CNN analysis give good indicators of the flight environment structure. They both help in choosing the suitable navigation methods and related algorithms concerning matching the images and calculating the drone position. The frequency analysis gives an overall estimation that does not count for individual local patches inside the environment as shown in this application, while the CNN meta-map gives a more detailed description. A comparison of results is shown in Table 1.

Table 1. Comparison between CNN analysis and frequency analysis for the flight environment.

Parameters	Frequency analysis	CNN analysis
Work mode	Only offline	Online (time-consuming) and offline using only the meta-map
Accuracy	Overall rough indicator of the structure, that does not account for all image patches	It has a better resolution and a detailed description of the whole map
Time-consuming and complexity	Easier (simpler) using 2D-FFT	It needs for a dataset, training the CNN, and building the meta-map, so it has more complexity
Applications	Visualizing the spectrum gives an estimate of the structure	It has a wide branch of applications. Besides the flight environment analysis, the meta-map could be used for many tasks, such as avoiding bad areas. It can be used also to choose the best loitering or landing areas but that need modifications

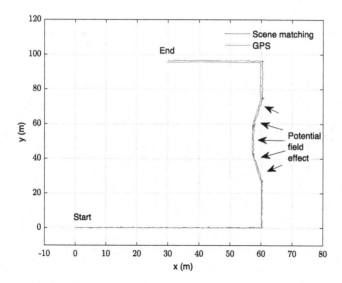

Fig. 5. The horizontal path calculated using scene matching with the reference path, the potential field effect on the path near the obstacle is obvious.

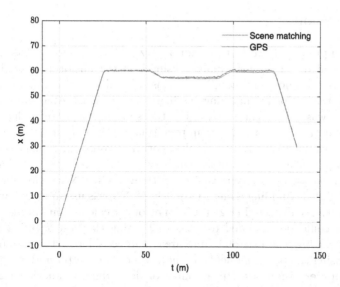

Fig. 6. The calculated position using scene matching with the reference position on x-axis.

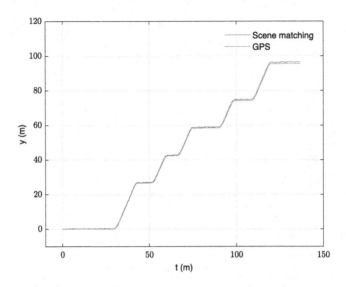

Fig. 7. The calculated position using scene matching with the reference position on y-axis.

6   Conclusion

The research presented two methods for analyzing flight environments. The first method depended on frequency analysis using Fourier transform of the flight area image, and the second one used CNN to classify well-structured areas and built a meta-map that helps in analyzing the flight environment. Analyzing the flight environment was beneficent in choosing the suitable navigation algorithm to enhance the overall navigation system performance. In both methods, analyzing the flight area helps in choosing suitable algorithms concerning image processing and position calculations. A comparison between both methods was done.

Although frequency analysis is simpler than CNN, CNN analysis is deeper and more detailed. Building a meta-map using CNN was useful in avoiding areas at which the drone expected large position errors or calculation failures, and the applications could be extended to involve choosing the best loitering or landing areas. Avoiding the bad matching areas and considering them as obstacles were done using the potential fields method, which was better and cheaper than loosing computer vision-based positioning or depending on another navigation system.

The system was realized in a simulation environment and a test was done on a predefined path in a 3D environment that contains bad matching areas. The overall system succeeded in avoiding obstacles where they existed, and the system performance was good in other areas with a position RMS error of 1.3m. In the future, we will focus on realizing and testing this system in the real world.

References

1. Ben-Afia, A., et al.: Review and classification of vision-based localisation techniques in unknown environments. IET Radar Sonar Navig. **8**(9), 1059–1072 (2014)
2. Szeliski, R.: Computer Vision: Algorithms and Applications. Text in Computer Science, Springer Science & Business Media, London (2010). https://doi.org/10.1007/978-1-84882-935-0
3. Guo, R., Zhou, D., Peng, K., Liu, Y.: Plane based visual odometry for structural and low-texture environments Using RGB-D sensors. IEEE Int. Conf. Big Data Smart Comput. (BigComp) **2019**, 1–4 (2019). https://doi.org/10.1109/BIGCOMP.2019.8679500
4. Tsvetkov, O., Tananykina, L.V.: A preprocessing method for correlation-extremal systems. Comput. Opt. **39**, 738–743 (2015). https://doi.org/10.18287/0134-2452-2015-39-5-738-743
5. Choi, S.H., Park, C.G.: Adaptive scene-matching algorithm based on frequency pattern analysis for aerial vehicle. In: 2019 12th Asian Control Conference (ASCC), pp. 1455–1459 (2019)
6. Zhao, Y., Wang, T.: A lightweight neural network framework for cross-domain road matching. Chinese Autom. Congr. (CAC) **2019**, 2973–2978 (2019). https://doi.org/10.1109/CAC48633.2019.8996270
7. Shahoud, A., Shashev, D., Shidlovskiy, S.: Detection of good matching areas using convolutional neural networks in scene matching-based navigation systems. In: Proceedings of the 31st International Conference on Computer Graphics and Vision, Nizhny Novgorod, Russia, 27–30 September 2021, pp. 443–452 (2021). https://doi.org/10.20948/graphicon-2021-3027-443-452
8. Shahoud, A., Shashev, D., Shidlovskiy, S.: Visual navigation and path tracking using street geometry information for image alignment and servoing. Drones **6**, 107 (2022). https://doi.org/10.3390/drones6050107
9. Jin, Z., Wang, X., Morelande, M., Moran, W., Pan, Q., Zhao, C.: Landmark selection for scene matching with knowledge of color histogram. In: Proceedings of the 17th International Conference on Information Fusion (FUSION), Salmanaca, Spain, 7–10 July 2014, pp. 1–8 (2014)
10. Liao, J., Li, X., Wang, X., Li, S., Wang, H.: Enhancing navigation performance through visual-inertial odometry in GNSS-degraded environment. GPS Solut. **25**(2), 1–18 (2021). https://doi.org/10.1007/s10291-020-01056-0
11. Zuo, Z., Yang, B., Li, Z., Zhang, T.: A GNSS/IMU/vision ultra-tightly integrated navigation system for low altitude aircraft. IEEE Sens. J. **22**(12), 11857–11864 (2022). https://doi.org/10.1109/JSEN.2022.3168605
12. Yol, A., Delabarre, B., Dame, A., Dartois, J., Marchand, E.: Vision-based absolute localization for unmanned aerial vehicles. In: 2014 IEEE/RSJ International Conference on Intelligent Robots and Systems, Chicago, IL, USA, pp. 3429–3434 (2014). https://doi.org/10.1109/IROS.2014.6943040
13. Bodunkov, N.E., Kim, N.V.: Autonomous landing-site selection for a small drone. Russ. Eng. Res. **41**(1), 72–75 (2021). https://doi.org/10.3103/S1068798X2101007X
14. Bektash, O., Pedersen, J.N., Ramirez Gomez, A.,Cour-Harbo, A.l.: Automated emergency landing system for drones: SafeEYE project. In: 2020 International Conference on Unmanned Aircraft Systems (ICUAS), pp. 1056–1064 (2020). https://doi.org/10.1109/ICUAS48674.2020.9214073

15. Subramanian, S., George, T., Thondiyath, A.: Obstacle avoidance using multi-point potential field approach for an underactuated flat-fish type AUV in dynamic environment. In: Ponnambalam, S.G., Parkkinen, J., Ramanathan, K.C. (eds.) IRAM 2012. CCIS, vol. 330, pp. 20–27. Springer, Heidelberg (2012). https://doi.org/10.1007/978-3-642-35197-6_3
16. Cho, J.-H., Pae, D.-S., Lim, M.-T., Kang, T.-K.: A real-time obstacle avoidance method for autonomous vehicles using an obstacle-dependent gaussian potential field. J. Adv. Transp. **2018**, 1–15 (2018). https://doi.org/10.1155/2018/5041401

Clusters of Exceedances for Evolving Random Graphs

Natalia M. Markovich$^{(\boxtimes)}$ and Maksim S. Ryzhov

V.A. Trapeznikov Institute of Control Sciences, Russian Academy of Sciences,
Profsoyuznaya Str. 65, 117997 Moscow, Russia
markovic@ipu.rssi.ru, nat.markovich@gmail.com, maksim.ryzhov@frtk.ru

Abstract. Evolution of random undirected graphs by the clustering attachment (CA) without node and edge deletion and with uniform node deletion is investigated. The CA causes clusters of consecutive exceedances of the modularity over a sufficiently high threshold. The modularity is a measure that allows us to divide graphs into communities. It shows the connectivity of nodes in the community. An extremal index (a local dependence measure) approximates the mean cluster size of exceedances over a sufficiently high threshold. Considering the change of the modularity at each evolution step, the extremal index of the latter random sequence indicates the consecutive large connectivity of nodes and thus, it reflects the community appearance during the network evolution. This allows to consider the community structure of the network from perspectives of the extreme value analysis. By simulation study we show that estimates of the extremal index of the modularity and the tail index of node degrees depend on the CA parameters. The latter estimates are compared both for evolution without node and edge deletion and with uniform node deletion.

Keywords: Random graph · Evolution · Modularity · Clustering attachment · Extremal index · Tail index

1 Introduction

Network evolution attracts interest of researchers due to numerous applications [1]. The popular mechanism to model growing network is preferential attachment (PA) applied both to directed and undirected graphs, see [2,3] among others. The attachment of new nodes starts from a seed network. A newly appended node may connect to m_0 existing nodes. By the PA a newly appended node chooses an existing node i randomly from the network with a probability proportional to its degree k_i (i.e. the number of its edges): $P_{PA}(i) = k_i / \sum_j k_j$ [4]. The PA

The reported study was funded by the Russian Science Foundation RSF, project number 22-21-00177 (recipient N.M. Markovich, conceptualization, methodology development, formal analysis, writing–original draft preparation; recipient M.S.Ryzhov, software, data validation).

models provide a "rich-get-richer" mechanism since the earliest appended nodes get more edges. This leads to a power law node degree distribution with index α_{TI}. The power law distribution is determined as

$$P\{k_i = j\} = Cj^{-\alpha_{TI}}, \quad C > 0, \quad \alpha_{TI} > 0.$$

Another idea applied to undirected graphs is the clustering attachment (CA) [4,5]. The CA to an existing node i is provided proportional to its clustering coefficient c_i

$$P_{CA}(i) \propto c_i^{\alpha} + \epsilon. \tag{1}$$

Here $\alpha \geq 0$ and $\epsilon \geq 0$ are attachment parameters. The clustering coefficient c_i is the probability that two random neighbors of node i are connected [5]. It is determined by

$$c_i = \frac{2\Delta_i}{k_i(k_i - 1)}. \tag{2}$$

Here, Δ_i is the number of links between neighbors of node i or, equivalently, the random number of triangles involving node i. Since $k_i(k_i - 1)/2$ is the maximum number of triangles that may exist for node i, $c_i < 1$ holds. ϵ is mostly a dominating term in (1). It is assumed that each new node attaches to $m_0 \geq 2$ existing nodes.

It is expected in [4] that the CA does not lead to a power-law degree node distribution, but to an exponential tail of the node degree distribution in contrast to the PA. We aim to check this hypothesis by estimation of the node tail index (TI). This can be simply explained by (1) and (2). Really, attaching m_0 new nodes to a node i leads to increasing its node degree k_i, but Δ_i may remain the same as in the previous step of the evolution. Then c_i and hence, the $P_{CA}(i)$ decrease at further attachments. The distribution of the number of triangles and the clustering coefficient is a subject of a vast research, see [5].

In many real-world networks, objects have the tendency to cluster together in groups. Since a triangle in a graph is the most clustered subgraph consisting of three nodes, the number of triangles shows a quality of the cluster, [6].

Another important observation is that the CA gives rise to a specific cluster structure of the community appearance during the network growing. A community consists of nodes that are strongly connected with each other and weakly connected with nodes from other communities [7]. To be precise, the communities arise sequentially during the evolution that forms the cluster structure of the modularity Q [4]. Let us recall that the modularity is a measure of the connectivity degree of nodes [8,9].

We analyse fluctuations of the modularity those are changing in time t when the CA proposed in [4] is used for the growth and evolution of the network. The latter fluctuations may happen in clusters of consecutive exceedances of the normalized modularity

$$\psi(t) = Q(t)/\langle Q \rangle - 1$$

over a sufficiently high threshold u, where $\langle Q \rangle$ denotes the average of the modularity values over some evolution period. The clustering of $\psi(t)$ is enhanced when α in (1) grows, see Fig. 1(b) in [4].

The evolved modularity series allows us to study clustering properties of random modularity sequences instead of clustering of graphs. Note that nodes in the graph cannot be definitely enumerated. The clusters of exceedances correspond to consecutive large values of the modularity and hence, to strongly connected communities in the network at some evolution steps.

To study clusters of exceedances of the modularity sequences we attract the results of the extreme value theory. The extremal index (EI) measures a local dependence in the sense that it approximates the reciprocal of the mean cluster size [10]. Considering the change of the modularity at each evolution step, the extremal index of the latter random sequence indicates the consecutive large connectivity of nodes and thus, it reflects the community appearance during the network evolution. This allows to consider the community structure of the network from perspectives of the extreme value analysis.

The TI shows the heaviness of the distribution tail. It will be estimated by the Hill's estimator [11] with a bootstrap selected number of largest order statistics k [12].

Our objective is to establish an impact of parameters α and ϵ in (1) on the estimated EI of the modularity series and the TI of the node degrees. We check these relations for two cases: (1) an uniformly chosen node is deleted every time when a new node is added; (2) nodes and edges are not deleted when the network is growing during the evolution. In the first case a total number of nodes in the network remains thus fixed. In the second case, the numbers of nodes and edges grow up.

Our results are based on the simulation study. Undirected graphs are considered.

The paper is organized as follows. Related definitions and results are given in Sect. 2. Main results are provided in Sect. 3. We finalize with conclusions in Sect. 4.

2   Related Works

2.1   Clustering Attachment for Graph Evolution

The CA simulates an evolutionary growth of graphs starting from an initial seed graph $G_0 = (V_0, E_0)$. V_0 and $\|V_0\|$ are a set of vertices and their number, respectively. Similar notations E_0 and $\|E_0\|$ are applied for edges. A new node i is appended to m_0 existing nodes. $m_0 \geq 2$ may be taken as in [4]. $m_0 = 2$ means that the new node i may get Δ_i that is equal to 0 or 1 in (2). ϵ denotes a nonzero attachment probability to node i that is proportional to $P_{CA}(i)$ when $c_i = 0$ holds.

2.2 Modularity

For an undirected graph $G = (V, E)$, the modularity Q is a measure to partition the network into communities [8]. Q shows how many edges exist within communities and between them:

$$Q = \frac{1}{2e} \sum_{ij} \left[A_{ij} - \frac{k_i k_j}{2e} \right] \text{II}(i, j), \tag{3}$$

$e = \frac{1}{2} \sum_{ij} A_{ij}$ is a number of edges in G, A is an adjacency matrix, $\text{II}(i, j)$ is equal to 1 when nodes i and j belong to the same community, k_i is a degree of node i. We use the Greedy Modularity Maximization Algorithm (GMMA). It is a hierarchical agglomeration algorithm for detecting communities which is faster than many competing algorithms: its running time on a network with n vertices and m edges is $O(md \log(n))$, where d is a depth of the dendrogram describing the community structure [8].

2.3 Extremal Index

A stationary sequence $\{Y_n\}_{n \geq 1}$ of random variables (rvs) with distribution function (df) $F(x)$ and $M_n = \max_{1 \leq j \leq n} Y_j$ is said to have EI $\theta \in [0, 1]$ if for each $0 < \tau < \infty$ there is a sequence of real numbers $u_n = u_n(\tau)$ such that

$$\lim_{n \to \infty} n(1 - F(u_n)) = \tau \quad \text{and} \quad \lim_{n \to \infty} P\{M_n \leq u_n\} = e^{-\tau\theta}$$

hold [10]. The EI plays a key role in the extreme value analysis since it allows to get a limit distribution of sample maximum of rvs when the latter are dependent, i.e. $P\{M_n \leq u_n\} = F^{n\theta}\{u_n\} + o(1)$ as $n \to \infty$. The EI is equal to one for independent identically distributed rvs. The converse is incorrect. As closer θ to zero, as stronger the local dependence or clustering. The EI measures the local clustering tendency of high threshold exceedances. Its reciprocal $1/\theta$ approximates the mean number of exceedances per cluster (the mean cluster size).

To estimate θ we use the intervals estimator proposed in [13] that is one of the most effective and simple known estimators. Taking the exceedance times $1 \leq S_1 < ... < S_{N_u} \leq n$ of $\{Y_n\}_{n \geq 1}$, the observed interexceedance times are $T_i = S_{i+1} - S_i$ for $i = 1, ..., N_u - 1$. $N_u = \sum_{i=1}^{n} \text{II}\{Y_i > u\}$ is the number of observations exceeding a predetermined high threshold u. We denote $L \equiv L(u) = N_u - 1$. The intervals estimator is defined as

$$\hat{\theta}_n(u) = \begin{cases} \min(1, \hat{\theta}_n^1(u)), & \text{if } \max\{T_i : 1 \leq i \leq L\} \leq 2, \\ \min(1, \hat{\theta}_n^2(u)), & \text{if } \max\{T_i : 1 \leq i \leq L\} > 2, \end{cases} \tag{4}$$

where

$$\hat{\theta}_n^1(u) = \frac{2(\sum_{i=1}^{L} T_i)^2}{L \sum_{i=1}^{L} T_i^2}, \quad \hat{\theta}_n^2(u) = \frac{2(\sum_{i=1}^{L}(T_i - 1))^2}{L \sum_{i=1}^{L}(T_i - 1)(T_i - 2)}. \tag{5}$$

The intervals estimator requires a choice of u as parameter. We find u by discrepancy method proposed in [14].

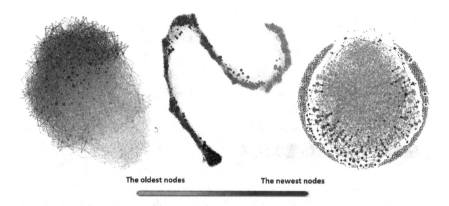

Fig. 1. The seed graph with number of nodes 5000 obtained by the CA (2) and (6) with $\alpha = 1$, $\epsilon = 0$ and $m_0 = 2$ (left) and the graph obtained from the seed graph by the same CA without node and edge deletion (middle) and applying an uniform node deletion (right) after $5 \cdot 10^4$ evolution steps. The point size is proportional to the node degree.

3   Main Results

Let $G(t) = (V(t), E(t))$ denote the graph at evolution step t. We consider the CA with $m_0 = 2$ throughout. The normalized measure

$$P_{CA}(i, t) = \frac{c_i^{\alpha}(t) + \epsilon}{\sum_{j \in V(t)} c_j^{\alpha}(t) + \|V(t)\| \epsilon} \tag{6}$$

is used instead of (1). Examples of a seed graph and its evolution by the CA are shown in Fig. 1. The evolution without node and edge deletion causes a kind of a barbell graph, see Fig. 1 (middle). This graph consists of a path of well conducted cliques those are weak connected with each other due to bottlenecks [15]. An uniform node deletion at each step of the evolution causes a large number of isolated nodes since the number of triangles decreases, see Fig. 1 (right).

In Fig. 2 we observe a cluster structure of time series $\psi(t)$ against evolution steps t for four pairs of CA parameters (α, ϵ). The CA evolution without node and edge deletion and with the uniform node deletion are compared. $\langle Q \rangle$ denotes the average over evolution steps over the interval $t \in [10^4, 5 \cdot 10^4]$. Spike trains in Fig. 2 indicate evolution steps that are accompanied by a creation of a new triangle. It leads to increasing of the modularity and thus, to the appearance its clusters of exceedances.

The case $\epsilon = 0$ is specific. If additionally node i does not belong to any triangle of nodes and $c_i = 0$ holds not for all $i \in V(t)$, then $P_{CA}(i, t) = 0$ follows by (6). The latter implies that new nodes cannot be attached to node i.

The clustering dynamic against the parameter α in (1) both without node and edge deletion and with uniform node deletion is shown in Fig. 3. There is a difference between two strategies of the deletion. Figure 3 shows that the EIs

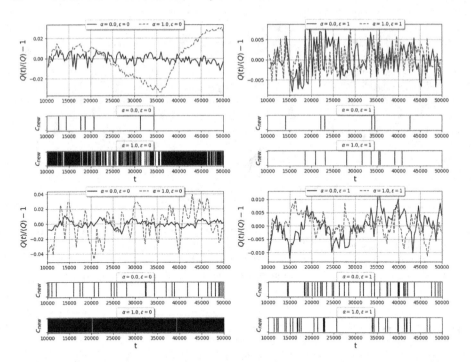

Fig. 2. The evolution of the normalized graph modularity against the CA evolution steps (top); and spike trains denoting injections of a new triangle when new nodes are appended and their clustering coefficient c_{new} is equal to one (two bottom lines): without node and edge deletion (upper three lines) and with uniform node deletion (bottom three lines).

for $\epsilon = 0$ tend to decrease without the deletion and to increase with the node deletion as α increases and they are closer to zero than the EIs for $\epsilon = 1$ which tend to be stable. The smaller EIs, the stronger clustering (or local dependence) of modularity sequences. For any positive value of $\epsilon \geq 1$, the EI is stable with regard to $\alpha > 0$. $(\alpha, \epsilon) = (0, 0)$ leads to a large EI. Figure 3 allows to compare the EIs for the uniform node deletion and the case without node and edge deletion. The latter case leads to the EI close to one that means an almost disappearance of clusters of exceedances for $\epsilon = 1$ (the same may be observed for any positive $\epsilon > 1$). The latter also relates to the weak conductance of the graph, see Fig. 1 (middle).

The uniform node deletion causes a bit stronger clustering for $\epsilon = 1$ since the EI is about 0.8. For $\epsilon = 0$ the situation is opposite: the EI is close to zero and the clustering of the modularity is strong. The local dependence is stronger when the CA is provided without the node and edge deletion.

The TIs for $\epsilon = 0$ tend to increase as α increases as far as the TIs are stable and have larger values for $\epsilon = 1$. This means that the distribution tails become lighter as α increases for $\epsilon = 0$. The uniform node deletion impacts on the TIs.

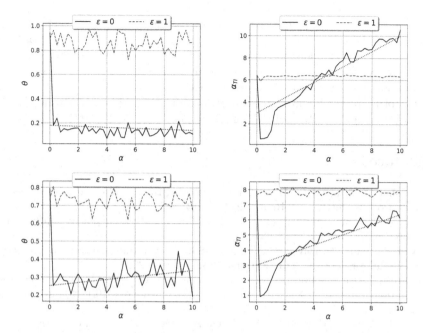

Fig. 3. Intervals estimates (4), (5) of the EIs for normalized modularity $Q/\langle Q \rangle - 1$ (left) and tail indices of node degrees α_{TI} (right) averaged over 30 simulated graphs evolved by the CA without the node and edge deletion (top) and with the uniform node deletion (bottom).

The TIs for positive ϵ and $\epsilon = 0$ tend to the same values as α increases. For the CA without node and edge deletion the TIs grow faster that leads to lighter tails than for positive ϵ. This may be explained by the smaller clustering coefficient than for the CA with node deletion.

$\alpha = 0$ corresponds to a constant $P_{CA}(i, t)$ for any i and any ϵ. This implies that a node i can be attached to existing nodes irrespective of the number of its triangles Δ_i.

4 Conclusion

The evolution of undirected graphs by the CA with uniform node deletion and without node and edge deletion is studied. Our novelty is to measure the dynamics of clusters in the graphs by finding the EI of the graph modularity sequence formed during the evolution.

The EI of the graph modularity and the TI of node degrees are estimated. The estimates are compared by the CA parameters. By simulation, it is found that $\epsilon = 1$ in (1) leads to stable large values of the EI and the TI. This happens since the clustering coefficient does not impact on the CA attachment. It means a weak clustering of the modularity and a weak heaviness of node degree distribution tails.

$\epsilon = 0$ causes a strong clustering since the EIs that are close to zero, and the heaviness of the tail of the node degree weakens since the TI α_{TI} increases as $\alpha > 0$ increases. The similar study is provided for the CA without node and edge deletion. The $\epsilon = 0$ leads to a stronger clustering and lighter node degree tail than for the CA with uniform node deletion.

Our future research may concern to a random parameter m_0, the using chains instead of triangles in (2) and the consideration of $\alpha \in (0, 1]$ in (1).

References

1. Ghoshal, G., Chi, L., Barabási, A.L.: Uncovering the role of elementary processes in network evolution. Sci. Rep. **3**, 2920 (2013)
2. Norros, I., Reittu, H.: On a conditionally Poissonian graph process. Adv. Appl. Prob. (SGSA) **38**, 59–75 (2006)
3. Wan, P., Wang, T., Davis, R.A., Resnick, S.I.: Are extreme value estimation methods useful for network data? Extremes **23**, 171–195 (2020)
4. Bagrow, J., Brockmann, D.: Natural emergence of clusters and bursts in network evolution. Phys. Rev. X. **3**, 021016 (2012). https://doi.org/10.1103/PhysRevX.3. 021016
5. Gao, P., Van Der Hofstad, R., Southwell, A., Stegehuis, C.: Counting triangles in power-law uniform random graphs. Electron. J. Comb. **27**(3), 1–21 (2020)
6. Weedage, L., Litvak, N., Stegehuis, C.: Locating highly connected clusters in large networks with HyperLogLog counters. J. Compl. Netw. **9**(2), cnab023 (2021)
7. Fortunato, S.: Community detection in graphs. Phys. Rep. **486**(3), 75–174 (2010)
8. Clauset, A., Newman, M.E., Moore, C.: Finding community structure in very large networks. Phys. Rev. E **70**(6), 066111 (2004). https://doi.org/10.1103/PhysRevE. 70.066111
9. Dugué, N., Prerz, A.: Directed Louvain : maximizing modularity in directed networks. Research Report] Université d'Orléans. hal-01231784 (2015)
10. Leadbetter, M.R.: Extremes and local dependence in stationary sequences. Zeitschrift für Wahrscheinlichkeitstheorie und Verwandte Gebiete **65**(2), 291–306 (1983). https://doi.org/10.1007/BF00532484
11. Hill, B.M.: A simple general approach to inference about the tail of a distribution. Ann. Statist. **3**, 1163–1174 (1975)
12. Markovich, N.M.: Nonparametric Analysis of Univariate Heavy-Tailed data: Research and Practice. Wiley, Chichester, West Sussex (2007)
13. Ferro, C.A.T., Segers, J.: Inference for clusters of extreme values. J. R. Statist. Soc. B. **65**, 545–556 (2003)
14. Markovich N.M., Rodionov I.V.: Threshold selection for extremal index estimation. ArXiv e-prints, arXiv:2009.02318 (2020)
15. Censor-Hillel, K., Shachnai, H.: Partial information spreading with application to distributed maximum coverage. In: Proceedings of the 29th ACM SIGACT-SIGOPS symposium on Principles of distributed computing (PODC 2010), pp. 161–170. ACM, New York, USA (2010). https://doi.org/10.1145/1835698.1835739

Estimation of the Tail Index
of PageRanks in Random Graphs

Natalia M. Markovich$^{(\boxtimes)}$ⓘ and Maksim S. Ryzhovⓘ

V.A. Trapeznikov Institute of Control Sciences, Russian Academy of Sciences,
Profsoyuznaya Str. 65, 117997 Moscow, Russia
markovic@ipu.rssi.ru, nat.markovich@gmail.com, maksim.ryzhov@frtk.ru

Abstract. Superstar nodes to which a large proportion of nodes attach
in the evolving graphs are considered. We attract results of the extreme
value theory regarding sums and maxima of non-stationary random
length sequences to predict the tail index of the PageRanks and Max-
linear models as influence measures of superstar nodes. To this end, the
graphs are divided into mutually weakly dependent communities. Maxima and sums of the PageRanks over communities are used as weakly
independent block-data. Tail indices of the block-maxima and block-sums
and hence, of the PageRanks and the Max-linear models are found to be
close to the minimum tail index of series of representative nodes taken
from the communities. The graph evolution is provided by a linear preferential attachment. The tail indices are estimated by data of simulated
and real temporal graphs.

Keywords: Random graph · Tail index · PageRank · Max-linear
model · Superstar node · Community · Evolution · Preferential
attachment

1   Introduction

Random graphs attract the attention of researchers due to numerous applications
including complex networks and communication systems. The randomness of
such graphs consists in the random numbers of incoming and outgoing links of
the nodes that are called in- and out-degrees, respectively. Random graphs are
subject to heterogeneity of distributions of node indices and their dependence
structure.

A well-known feature of random graphs such as Web graphs is that the in- and
out-degrees are power law distributed. The latter distribution is determined as
$P\{X = j\} = Cj^{-\alpha}$, $C > 0$, $\alpha > 0$. Despite the in- and out-degrees are discrete
random variables (r.v.s), their distribution can be approximated by regularly

The reported study was funded by the Russian Science Foundation RSF, project number 22-21-00177 (recipient N.M. Markovich, conceptualization, methodology development, formal analysis, writing–original draft preparation; recipient M.S. Ryzhov, software, data validation).

V. M. Vishnevskiy et al. (Eds.): DCCN 2022, LNCS 13766, pp. 75–89, 2022.
https://doi.org/10.1007/978-3-031-23207-7_7

varying heavy-tailed distributions [1]. The distribution tail of a non-negative
r.v. X is called regularly varying with the tail index (TI) α if it holds

$$\overline{F}(x) = P\{X > x\} = x^{-\alpha}\ell(x), \tag{1}$$

where $\ell(x)$ is a slowly varying function, i.e. $\lim_{x\to\infty} \ell(tx)/\ell(x) = 1$ holds for
any $t > 0$. The smaller TI the heavier the distribution tail. A non-homogeneous
dependence between nodes and non-stationary heavy tails complicate a plausible
statistical analysis.

The PageRank (PR) is a more general measure of the node influence than
the in-degree. The PRs of random Web pages are derived to be regularly varying
distributed [1,2]. Empirical studies stated that the TIs of the in-degrees and PRs
of Web pages have values of α smaller than 2. By properties of regularly varying
distributions, this means that the variance of the in-degrees and PRs is infinite.
Another measure of the node influence is the Max-Linear Model (MLM) [3].

Our aim is to evaluate TIs of the PRs and MLMs of *superstar nodes* connected by a large number of incoming links to other nodes in evolving random
graphs. The edge and node structure of such graphs is changed in time. We
use results of the extreme value theory regarding sums and maxima of non-
stationary random length sequences to find the latter TIs that is a novelty. To
this end, we divide the evolving random graphs into communities by a Directed
Louvain's algorithm [4] which constitute weakly dependent subgraphs and take
representative nodes from each community. A community consists of nodes that
are strongly connected with each other and weakly connected with nodes from
other communities [5,6]. The node indices in the communities may also be dependent and non-stationary distributed in terms of scaling the number of nodes. We
estimate the TIs of the block-maxima and block-sums of the node PRs over
the communities and compare it with the minimum TI of the PRs among the
representative series. The graph evolution is provided by a linear preferential
attachment (PA) [7]. We study simulated evolving graphs with stationary distributed in- and out-degrees and homogeneous dependence structure and real
temporal graphs. The Hill's and QQ plot estimators (see, [8]) to estimate the TI
are used.

In Sect. 2.1 we recall theoretical results obtained in [9,10] and discuss their
application to find the TIs of the PRs and MLMs of the superstars. Required
methods are recalled in Sect. 3. In Sect. 4.1, 4.2 TIs are estimated by data of
simulated stationary graphs evolved by the PA and of real temporal graphs. We
finalize with conclusions in Sect. 5.

2 Theory and Its Interpretation for Random Graphs

2.1 Tail Index of Random Sums and Maxima

In [1,2], the PR R of a randomly chosen Web page (i.e. a vertex of a Web graph)
is presented as a solution of a fixed-point problem

$$R \overset{D}{=} \sum_{j=1}^{N} A_j R_j + Q, \tag{2}$$

assuming that $\{R_j\}$ are independent identically distributed (i.i.d.) copies of R and $E(Q) < 1$ holds. Q is an arbitrary positive r.v. that means a personalization or the user preference in terms of the PR. N models a number of incoming links to a page (the in-degree). The A_js are assumed to be independent and distributed as some r.v. A with $E(A) < 1$. $=^D$ denotes the equality in distribution. The PRs of nearest neighbors of an underlying node which have in-coming links to this node form the random sum in (2) by the original definition of the PR given in [11]. It is stated in [1,2] that the stationary distribution of R is regularly varying and its TI is determined by the most heavy-tailed distributed term in the regularly varying distributed pair (N, Q). By replacing the sum by maximum one can obtain similar results with regard to the MLM that is the solution of the following equation

$$R =^D \left(\bigvee_{j=1}^{N} A_j R_j \right) \vee Q. \tag{3}$$

In [9,10] the TI of sums and maxima of random length weighted non-stationary sequences was found. The latter sequences can be considered as the rows of a doubly-indexed array of r.v.s $(Y_{n,i} : n, i \geq 1)$ in which the "row index" n corresponds to time, and the "column index" i corresponds to the level. The "column" series are assumed to be stationary distributed with regularly varying tails and the TIs $\{k_i\}_{i \geq 1}$. One of the "column" series is assumed to have a minimum TI k_1. An arbitrary dependence between "column" series is allowed. It was found that the TI of both sums and maxima over rows is equal to k_1 [9]. The same may be true if there are a random number of such the most heavy-tailed "column" series with k_1 [10]. The results remain true if the TIs of elements in the "columns" are different, apart from those "columns" with k_1 [10].

2.2 Random Sums and Maxima in Random Graphs

The results in [9,10] can be applied to the sums and maxima in the right-hand sides of (2) and (3). This scheme allows to find TIs of the PRs and the MLMs of superstar nodes within communities, to which a large proportion of nodes of the community attach. In our empirical study we exclude isolated nodes from the consideration. The PR or the MLM of a superstar node is calculated by sum or maximum of PRs of its nearest neighbors having incoming links to the latter node. The superstar may have a few followers in other communities but we disregard them. By theory and our simulation study, the TI of their PRs (or the MLMs) may be approximated by the TI of the most heavy-tailed representatives of communities, i.e. by the minimum TI of the representative series.

In terms of graphs the random length "row" sequences can be considered as communities of nodes. The "column" series consist of nodes that are taken from each community as their representatives, see Fig. 1.

As the PRs in the communities may be non-stationary distributed and arbitrary dependent that is natural for practice, the representative series with the minimum TI must be stationary distributed. The latter property may not be fulfilled in heterogeneous random graphs. One has to test the stationarity of the representative series with the minimum TI. The representative series may be formed by branches of all possible trees rooted at each node of the communities. To reduce a number of the representative series we take the ith maxima of each community as a member of the ith representative series. The first maxima are excluded, i.e. $i \geq 2$.

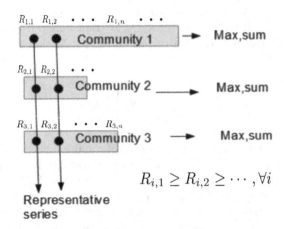

Fig. 1. The communities as the "row" sequences and the representative node series as the "column" series: the maxima and sums are taken over PRs in the communities.

3   Required Methods

3.1   Estimators of the Tail Index

Let us recall the Hill's estimator and the QQ plot used to estimate the tail index. Let $X_{(1)} \leq X_{(2)} \leq ... \leq X_{(n)}$ be the order statistics of the sample $X^n = \{X_1, X_2, ..., X_n\}$. The Hill's estimator of the extreme value index γ that is a reciprocal to the tail index α, $\gamma = 1/\alpha$, is defined as

$$\hat{\gamma}_0 = \frac{1}{k} \sum_{i=1}^{k} \ln X_{(n-i+1)} - \ln X_{(n-k)}. \tag{4}$$

The Hill's estimator is not calculated by the whole sample but only by the k largest order statistics that is very natural for modelling of the distribution tail. The key parameter k has to be determined by the sample beforehand. There are numerous methods to find k. We use the bootstrap method described in [12]. The Hill's estimator is derived under the assumption that the observations are independent and identically regularly varying distributed. The independence and stationarity are crucial requirements that are not precisely fulfilled for random graphs in practice.

The classical QQ plot is used to fit a distribution model to the underlying data. It can be used to estimate the tail index by a sample X^n of r.v.s which are distributed by Eq. (1) [8]. Let $X_{(1)} \geq X_{(2)} \geq ... \geq X_{(n)}$ be the order statistics in the decreasing order. We apply a logarithmic transformation to the data and define the set of points

$$S_n = \{\left(-\log \frac{j}{n+1}, \log X_{(j)}\right), j = 1, 2, ..., k\} \tag{5}$$

such that $k \to \infty$, $k/n \to 0$ as $n \to \infty$. By Proposition 4.1 in [8] S_n converges to

$$T_n = \{\left(x, \frac{x}{\alpha}\right) ; x \geq 0\} + \left(-\log \frac{k}{n+1}, \log X_{(k)}\right)$$

as $n \to \infty$. The method is to make a QQ plot after a logarithmic transformation of the data by comparing the k largest order statistics with the theoretical quantiles of the exponential distribution. This provides an asymptotically linear plot of slope $1/\alpha$ starting from the point $\left(-\log \frac{k}{n+1}, \log X_{(k)}\right)$. Here, k can be taken equal to $n^{2/3}$.

3.2 Directed Louvain's Algorithm

In order to divide a graph into non-overlapping and weakly connected communities, one can use a Directed Louvain's Algorithm [4]. The latter is an improvement of the well-known Louvain's Algorithm based on the maximization of the modularity

$$Q = \frac{1}{m} \sum_{ij} \left(A_{ij} - \frac{k_{out}(i)k_{in}(j)}{m}\right) \delta(\sigma_i, \sigma_j).$$

Here, $m = \|E\|$ is a number of edges in the directed graph, $k_{out}(i)$ and $k_{in}(j)$ refer to the out-degree and in-degree of node i, respectively. (A) denotes an adjacency matrix, where A_{ij} refers to an edge from node i to node j. $\delta(\sigma_i, \sigma_j)$ is equal to 1 when nodes i and j belong to the same community.

The algorithm is convenient for large directed networks. In the paper we use this algorithm to divide graphs into communities. The community size is random.

3.3 Mann-Whitney U Test

The null hypothesis is that r.v.s from two parts of the sample have equal distributions. Let us recall that the Mann-Whitney U statistic is defined as

$$U = \sum_{i=1}^{n} \sum_{j=1}^{m} S(X_i, X_j),$$

with

$$S(X,Y) = \begin{cases} 1, & X > Y, \\ 1/2, & X = Y, \\ 0, & X < Y. \end{cases}$$

Here, $X_1, ..., X_n$ and $Y_1, ..., Y_m$ are assumed to be i.i.d. r.v.s of the first and second samples, respectively. Since for large samples U statistic is approximately normally distributed, the critical area for the null hypothesis is defined as

$$U \leq m_U - \sigma_U \cdot z, \qquad U \geq m_U + \sigma_U \cdot z,$$

where the mean and standard deviation of U are

$$m_U = \frac{mn}{2}, \quad \sigma_U = \sqrt{\frac{mn(m+n+1)}{12}},$$

respectively. For the 5% critical level, the normal quantile $z = 1.96$ is taken. If the U-value falls into the interval,

$$(m_U - \sigma_U \cdot z, m_U + \sigma_U \cdot z), \tag{6}$$

then the null hypothesis is not rejected.

4 Practice

4.1 A Study of Stationary Simulated Graphs

We consider stationary evolved graphs that are created by the PA $\alpha-$, $\beta-$ and $\gamma-$ schemes proposed in [7] starting from a seed network containing at least one node. The TIs of the in- and out-degrees of these graphs are calculated by formula (2.9) in [7]

$$\alpha_{in} = \frac{1 + \delta_{in}(\alpha + \gamma)}{\alpha + \beta}, \quad \alpha_{out} = \frac{1 + \delta_{out}(\alpha + \gamma)}{\beta + \gamma}, \tag{7}$$

$(\alpha, \beta, \gamma, \delta_{in}, \delta_{out})$ is a set of parameters of the PA schemes. α is the probability to create a new edge from a newly appearing node to an existing node. The reverse edge is generated with the probability γ. β is the probability to create a new edge between two existing nodes. The TI of the PR is not yet theoretically obtained. To estimate the TI we use the Hill's and the QQ-plot estimators.

In our experiment, the PA schemes with different sets of parameters (α, β, γ) and $\delta_{in} = \delta_{out} = 1$ shown in Table 1 were applied. The number of nodes in the evolved graphs were taken equal to $n = 10^4$. For each set (α, β, γ), we repeatedly simulated 100 graphs. The mean of the TI estimates of the in- and out-degrees and PRs calculated over 100 samples and $(\alpha_{in}, \alpha_{out})$ calculated by (7) are presented. The parameter k in (4) is calculated by the bootstrap method [12] where the number of bootstrap re-samples were taken equal to 100.

Since TIs of the in- and out-degrees are known, we calculate their mean squared errors (MSE) over the simulated graphs. Since TIs of the PRs are unknown (but they are close to those of the in-degrees) the bootstrap estimates of their MSEs are calculated, see Table 1. By Table 1 one can conclude that the QQ plot estimator (5) has better accuracy than the Hill's estimator. The QQ plot method is faster than the Hill's estimator coupled with the bootstrap to calculate k since the bootstrap requires the calculation over a sufficiently large number of re-samples. The estimates of the PR are close to that for the in-degree.

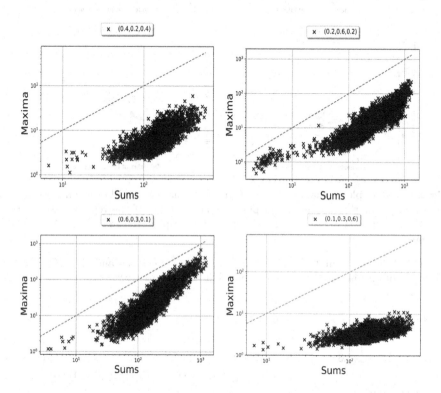

Fig. 2. The scatter plots 'maxima-versus-sums' of the PRs of nodes for the networks evolved by the PA with parameters described in Table 1.

The obtained graphs are divided into communities by the Directed Louvain's Algorithm [4]. The communities are, in fact, weakly dependent data blocks of random lengths. Scatter plots in Fig. 2 demonstrate two classes of the communities. The first class consists of the communities with a few nodes and, hence, relatively small maxima and sums of their node PRs. The second class contains the most of communities of similar sizes with larger sums and maxima. In Fig. 3, the QQ-plot estimates of the block-sums against the minimum TIs of the PRs of representative series tend to a diagonal trend. Although the TIs of the block-maxima also demonstrate a trend, the latter may deviate from that for the block-sums.

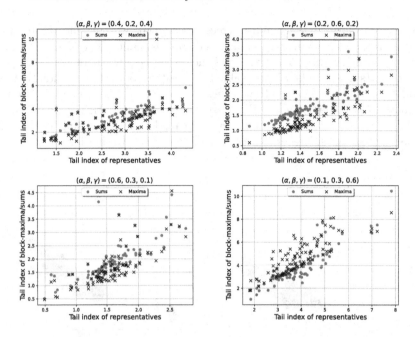

Fig. 3. The QQ plot estimates of TIs of the block-maxima and block-sums over communities against the minimum QQ plot estimates of TIs of the representative series for PRs: each point corresponds to one of the 100 graphs evolved by the PA (α, β, γ)-schemes.

Table 1. The mean and the MSE (in brackets) of the Hill's and the QQ plot (5) estimates of the in- and out-degrees and PR TIs for networks evolved by the PA schemes.

PA, $\delta_{in} = \delta_{out} = 1$		Hill			QQ plot		
(α, β, γ)	$(\alpha_{in}, \alpha_{out})$	$\hat{\alpha}_{in}$	$\hat{\alpha}_{out}$	$\hat{\alpha}_{PR}$	$\hat{\alpha}_{in}$	$\hat{\alpha}_{out}$	$\hat{\alpha}_{PR}$
(0.4,0.2,0.4)	(3,3)	2.083	2.092	2.665	2.613	2.593	2.766
		(0.969)	(0.961)	(0.033)	(0.175)	(0.184)	(0.021)
(0.2,0.6,0.2)	(1.75, 1.75)	1.596	1.592	1.753	1.626	1.62	1.745
		(0.032)	(0.033)	(0.005)	(0.019)	(0.021)	(0.003)
(0.6,0.3,0.1)	(1.889, 4.25)	1.657	3.181	1.245	1.72	3.347	1.357
		(0.068)	(1.231)	(0.004)	(0.033)	(0.853)	(0.003)
(0.1,0.3,0.6)	(4.25, 1.889)	3.186	1.67	4.53	3.334	1.72	4.792
		(1.224)	(0.061)	(0.128)	(0.876)	(0.033)	(0.061)

4.2 A Study of Real Graphs

We analyze four temporal graphs, i.e. graphs that change with time [13]. An application of such graphs is given by gossiping and in general of information dissemination. Graphs MTH, ASK, SPR and WIKI taken from [14] were studied.

Table 2. Temporal networks from [14] and their communities.

Name	Nodes, Temporal edges	Description	Number of Communities	Mean size of Communities (std)
sx-Mathover Flow (MTH)	24818, 506550	Comments, questions, and answers on Math Overflow	150	165.453 (634.666)
sx-askubuntu (ASK)	159316, 964437	Comments, Questions, And answers On Ask Ubuntu	4521	35.239 (485.063)
sx-Superuser (SPR)	194085, 1443339	Comments, Questions, And answers On Super User	3467	55.981 (723.008)
Wiki-talk-Temporal (WIKI)	1140149, 7833140	Users editing Talk pages on Wikipedia	47885	23.81 (1144.492)

Table 3. Descriptive statistics of the in- and out-degrees and PR in the temporal networks.

Graph	PageRank Min, max	Mean, std	In-degree Min, max	Mean, std	Out-degree Min, max	Mean, std
MTH	0.373, 148.421	6.426,21.473	1,5378	199.12, 789.85	1,5931	270.247,987.152
ASK	0.326,509.944	2.622,16.6	1, 4926	19.335,196.061	1, 8729	37.2,384.168
SPR	0.307,764.489	3.087,23.36	1,9576	27.071,290.929	1,26996	57.781,711.496
WIKI	0.89, 503.311	1.828, 5.428	1, 31949	8.229, 244.1324	1, 264905	20.033,1309.492

In Table 2 the description of the graphs and their communities is shown. By Tables 2 and 3 one can see that the WIKI graph contains the largest number of nodes and edges and as a consequence, the largest maxima of the in- and out-degrees. However, the maximum of its PR is not the largest one and it is comparable with that of the ASK graph. The scatter plots of the block-sums and maxima calculated over the communities are shown in Fig. 4.

The communities can be attributed to two classes by block-data.[1] The first class contains communities whose maxima and sums nearly coincide and its points hug the linear line $y = x$. The maxima and sums may then have similar

---

[1] The classification into the classes of communities is caused by the Loivan's algorithm. The classification is not so explicit for evolving simulated graphs in Fig. 2 that might be due to stationary distributions of their in- and out-degrees.

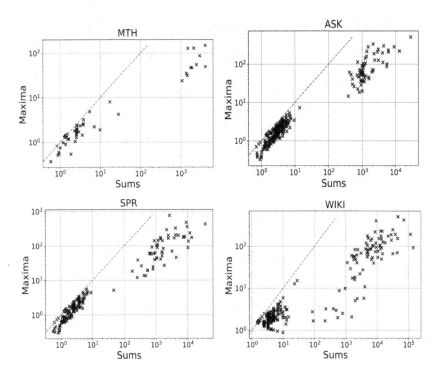

Fig. 4. The scatter plots 'maxima-versus-sums' of the PRs of nodes for the temporal networks described in Table 2.

distributions. The second class consists of the rest of communities. Their sums are larger than the maxima. The scatter plot of the WIKI graph is specific. It contains nodes belonging visually to a "buffer area" between two classes. These nodes were assigned to the first class due to their links to a few significant nodes of this class. The maxima and sums of the second class have mostly the heavier tails than those of the first class because their TIs are smaller, see Table 4. This implies that the second class dominates in the networks which is confirmed by Table 5. The sums have somewhat heavier distribution tails than the maxima.

We compare the TIs of the sums and maxima of the PRs over communities in Table 4, 5 with the minimum TIs among the representative series in Table 6. One can conclude that the block-maxima and sums over all communities in Table 5 have similar values of the TIs as the minimum TIs among the representative series taking into account the confidence intervals. This result is in agreement with results in [9,10] that we aimed to check.

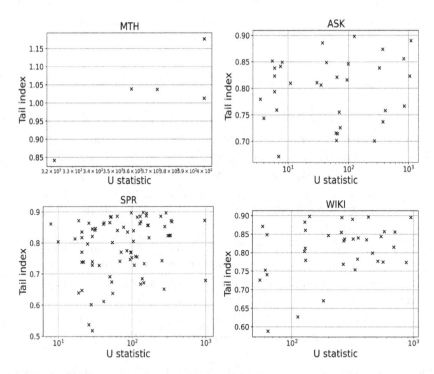

Fig. 5. The smallest QQ plot estimates of TIs of the PRs against the U-statistics of the Mann-Whitney U test for representatives series.

It remains to check a stationarity of the representative series with minimum TIs that is required by theory in [9,10], Sect. 2.1. To this end, we apply the Mann-Whitney U test. The null hypothesis is that r.v.s from two parts of the representative series have equal distributions. Note that the r.v.s in the representative series can be considered as weak dependent because so are the communities. The representative series are divided into two samples according to their belonging to two classes in Fig. 4. Considering the representative samples, we show the smallest QQ plot estimates of the TIs of the PRs versus their U-statistics in Fig. 5. The estimates are taken such that they fall in the confidence intervals of the minimum TI estimates in the right column of Table 6. Figure 6 shows that the values of the U statistics fall into the interval (6). This means the U statistics are not in the critical area. This confirms that the null hypothesis about the distribution similarity of two parts of the representative series with the minimum TIs corresponding to two classes is not rejected for all underlying temporal graphs.

Since the real graphs may be huge we use a set of their subgraphs. Namely, each graph is divided into communities by the Directed Louvain's algorithm [4] and the largest communities with the number of nodes more than 1000 are used further. The latter communities are again divided into communities and

Table 4. TIs of the block-maxima and sums of the PRs over communities of two classes in the temporal networks with the bootstrap confidence intervals in brackets.

	Class 1				Class 2			
	Maximum		Sum		Maximum		Sum	
Graph	Hill	QQ Plot	Hill	QQ Plot	Hill	QQ Plot	Hill	QQ Plot
MTH	1.899 (0.993, 3.642)	2.004 (1.953, 2.055)	1.172 (0.103, 2.836)	1.205 (1.061, 1.348)	1.271 (0.164, 2.808)	1.354 (1.17, 1.537)	1.309 (0.322, 2.346)	2.573 (2.464, 2.682)
ASK	4.692 (3.595, 5.792)	5.055 (5.053, 5.058)	3.751 (2.171, 5.484)	2.942 (2.935, 2.95)	2.409 (0.316, 3.291)	2.573 (2.533, 2.613)	1.081 (0.348, 2.383)	1.083 (0.86, 1.306)
SPR	5.009 (3.202, 5.049)	5.28 (5.275, 5.281)	4.123 (2.697, 6.376)	3.268 (3.261, 3.274)	1.735 (0.172, 3.284)	1.698 (1.601, 1.795)	1.003 (0.368, 2.578)	1.446 (1.306, 1.586)
WIKI	7.002 (6.134, 10.569)	3.198 (3.196, 3.202)	3.017 (2.413, 4.402)	1.792 (1.782, 1.803)	1.987 (0.587, 3.847)	2.192 (0.587, 3.847)	1.233 (0.349, 2.503)	1.158 (0.978, 1.337)

Table 5. TIs of sequences of maximum and sums of the PRs over all communities in the temporal networks with the bootstrap confidence intervals in brackets.

	Maximum over all communities		Sum over all communities	
Graph	Hill	QQ plot	Hill	QQ plot
MTH	1.276 (0.907,2.513)	0.563 (0.093,1.218)	1.409 (0.15,2.615)	0.310 (0.08,0.602)
ASK	1.125 (0.701,2.977)	0.701 (0.568,0.834)	1.123 (0.263,2.891)	0.396 (0.016, 0.808)
SPR	1.204 (0.770,3.855)	0.657 (0.491,0.822)	0.925 (0.358, 3.135)	0.371 (0.011,0.776)
WIKI	1.505 (1.117, 5.819)	1.315 (1.296, 1.334)	0.652 (0.063, 4.209)	0.645 (0.564, 0.726)

their block-maxima and sums were calculated. The number of the communities is quite different which may worsen the accuracy of the TI estimation for the block-maxima and sums if the number is not large enough as for the MTH graph. We compare the TIs of the sums and maxima of the PRs over communities with the minimum TIs among the representative series. One may conclude that the

Table 6. Minimum TI estimates of the PRs found among representative series with the bootstrap confidence intervals in brackets, where the minimum was selected among Hill's estimates with the corresponding QQ plot estimates and vice versa.

Graph	Number of Series	Minimum tail index by Hill's estimate		Minimum tail index by QQ plot estimate	
		Hill	QQ plot	Hill	QQ plot
MTH	4258	0.613	0.659	0.848	0.843
		(0.375,1.745)	(0.042,1.361)	(0.015,1.4)	(0.093,1.218)
ASK	23079	0.445	1.235	1.156	0.671
		(0.438,1.747)	(1.145,1.324)	(0.465,1.994)	(0.567,0.833)
SPR	26838	0.459	1.149	1.152	0.517
		(0.119,1.594)	(1.027,1.27)	(0.43,1.919)	(0.354,0.848)
WIKI	159042	0.602	1.917	1.046	0.587
		(0.178,1.393)	(1.87,1.963)	(1.008,3.293)	(0.358,0.894)

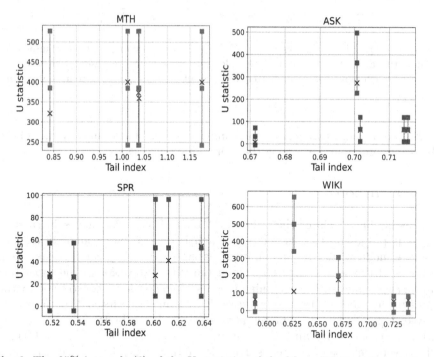

Fig. 6. The 95% intervals (6) of the U-statistics of the Mann-Whitney U test shown by vertical intervals against the five smallest values of the QQ plot estimates of the TIs of the PRs for representatives series marked by crosses.

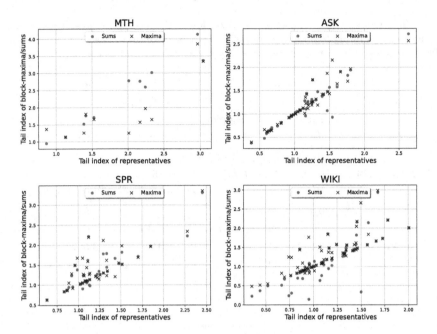

Fig. 7. The QQ plot estimates of the TIs of the block-maxima and block-sums over communities against the minimum QQ plot estimates of the representative series for PRs where each point corresponds to one of the selected subgraphs of temporal graphs.

block-maxima and sums over all communities in Fig. 7 have similar values of the TIs as the minimum TIs among the representative series.

5   Conclusion

The prediction of TIs of the PRs and MLMs of superstar nodes in random graphs is proposed. To this end, the graph is divided into communities. The superstar node within a community is assumed to have incoming links from all nodes of the community. The TIs of the PRs and MLMs of the superstars may be approximated by the minimum PR's TI among series constructed by the representative nodes of the communities. The obtained results confirm the theory derived in [9,10]. The ith representative series is chosen by the ith PR maxima within each community. Since communities are of a random size, some nodes may fall in several representative series that leads to their dependence. The latter does not contradict the assumptions in [9,10]. A number of the representative series those have a minimum TI may be random. The advantage of such approach is that the communities constitute weak connected subgraphs and hence, the r.v.s of the representative series are weakly dependent that is mostly required for the TI estimation. By the theory [9,10] the representative series with a minimum TIs have to be stationary distributed with regularly varying tails. The r.v.s of the rest of representative series may be distributed with different TIs larger than the minimum one.

Among estimators of the TI, the QQ plot estimator seems to be the most trustable metric by our study of the simulated evolving networks.

References

1. Volkovich, Y.V., Litvak, N.: Asymptotic analysis for personalized web search. Adv. Appl. Prob. **42**(2), 577–604 (2010)
2. Jelenkovic, P.R., Olvera-Cravioto, M.: Information ranking and power laws on trees. Adv. Appl. Prob. **42**(4), 1057–1093 (2010)
3. Markovich, N.M., Ryzhov, M., Krieger, U.R.: Nonparametric analysis of extremes on web graphs: PageRank versus max-linear model. Commun. Comput. Inf. Sci. **700**, 13–26 (2017)
4. Dugué, N., Perez, A.: Directed Louvain : maximizing modularity in directed networks // [Research Report] Université d'Orléans, hal-01231784 (2015)
5. Fortunato, S.: Community detection in graphs. Phys. Rep. **486**(3), 75–174 (2010)
6. Mester, A., Pop, A., Mursa, B.-E.-M., Grebla, H., Diosan, L., Chira, C.: Network analysis based on important node selection and community detection. Mathematics **9**, 2294 (2021)
7. Wan, P., Wang, T., Davis, R.A., Resnick, S.I.: Are extreme value estimation methods useful for network data? Extremes. **23**, 171–195 (2020)
8. Das, B., Resnick, S.I.: QQ plots, random sets and data from a heavy tailed distribution. Stoch. Model. **24**(1), 103–132 (2008)
9. Markovich, N., Rodionov, I.: Maxima and sums of non-stationary random length sequences. Extremes **23**(9), 451–464 (2020)
10. Markovich, N.: Extremes of Sums and Maxima with Application to Random Networks (2021). arXiv:math.PR/2110.04120
11. Brin, S., Page, L.: The anatomy of a large-scale hypertextual web search engine. Comput. Netw. ISDN Syst. **30**, 107–117 (1998)
12. Markovich, N.: Nonparametric Analysis of Univariate Heavy-Tailed Data: Research and Practice. John Wiley & Sons, New York (2007)
13. Michail, O.: An introduction to temporal graphs: an algorithmic perspective. In: Zaroliagis, C., Pantziou, G., Kontogiannis, S. (eds.) Algorithms, Probability, Networks, and Games. LNCS, vol. 9295, pp. 308–343. Springer, Cham (2015). https://doi.org/10.1007/978-3-319-24024-4_18
14. Leskovec, J., Krevl, A.: SNAP Datasets: Stanford Large Network Dataset Collection (2014).http://snap.stanford.edu/data/

Optimization of the Code Division Process Using Multilayer Orthogonal Structures Based on M-Sequences

D. Kukunin$^{(\boxtimes)}$(ID), A. Berezkin(ID), and R. Kirichek(ID)

Bonch-Bruevich Saint-Petersburg State University of Telecommunications,
Bolshevikov Ave. 22 Build. 1, 193232 St. Petersburg, Russia
`coux@yandex.ru`

Abstract. The work contains a description of methods for creating orthogonal code constructions based on recurrent maximum length sequences. These constructions are multilayer structures that are formed when the signal spectrum is expanded by direct sequences. The method of direct spread spectrum is the most effective among the known methods, it allows not only to increase the signal base, but also to organize multiple access with code separation of channels in the future. Therefore, the main purpose of this work is to optimize code structures containing a multiplexed signal. Optimization in this case concerns increasing the energy efficiency of the signal transmitted in the general spectrum. Optimization methods are proposed that do not affect the result of demultiplexing the compacted signal at reception. At the same time, the use of special mathematical algorithms for processing multilayer orthogonal structures using the dual basis of the Galois field allows you to control the integrity of the data, providing a high degree of noise immunity of the signal.

Keywords: CDMA · M-sequence · Galois field

1 Introduction

Data transmission using tethered high-altitude unmanned telecommunication platforms ensures the construction of a new generation of networks, for which energy efficiency is of particular importance. Energy efficiency in wireless networks is achieved through various means, among which a special place is occupied by the technology of spread spectrum of signal, which in this work is used to organize the code division of channels.

The advantages of multiple access technology based on code division or code channel division over time and frequency division are obvious [1]. Suffice it to recall that TDMA (Time Division Multiple Access) and FDMA (Frequency Division Multiple Access) rely on physical quantities that, one way or another, are their common resource. At the same time, time is the most valuable resource, and the frequency range, as a rule, is regulated by legislative and legal restrictions.

V. M. Vishnevskiy et al. (Eds.): DCCN 2022, LNCS 13766, pp. 90–102, 2022.
https://doi.org/10.1007/978-3-031-23207-7_8

Multiple access based on the CDMA (Code Division Multiple Access) is not a new technology, its foundations were laid in the last century. The CDMA communication system, known as IS-95, used Walsh [2] codes as spreading the spectrum of sequences.

Walsh codes are orthogonal. They are analogs of digital sinusoids built on the basis of Rademacher functions. The main disadvantage of Walsh sequences can be considered, first of all, an insufficiently "good" autocorrelation function, which does not allow them to fully withstand phase failures. This work suggests using structures based on recurrent M-sequences to expand the signal spectrum and further code compaction.

2   Forward and Inverse M-Sequences

The method of spread spectrum of the signal by a direct sequence, depending on the tasks being solved, uses various variants of the code combinations [3–7]. A good autocorrelation function reduces the probability of phase failures, at the same time a large code distance between combinations provides high noise immunity [8,9].

The most important factor in the joint use of sequences suitable for spread spectrum is their orthogonality. It is this property that makes it possible to produce code compaction and transmit independent streams of information in a common frequency band.

By their nature, maximum length sequences [10–12] are quasi-orthogonal, but code constructions based on a linear sum of direct and inverse M-sequences are orthogonal with respect to them. This allows you to combine them into a common frequency spectrum and transmit in one time interval.

The direct M-sequence $\{U\}$ can be constructed in various ways, for example, using a scheme based on a recurrent shift register with outputted adders (see Fig. 1).

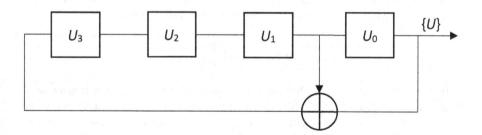

Fig. 1. Direct M-sequence generator based on $P(x) = x^4 + x + 1$.

The scheme (see Fig. 1) is based on the characteristic polynomial $P(x)$. To construct an inverse M-sequence R, this polynomial is multiplied by $(x + 1)$. Thus, the inverse sequence construction scheme for $P(x) = (x^4 + x + 1)(x + 1) = x^5 + x^4 + x^2 + 1$ will have the form (see Fig. 2).

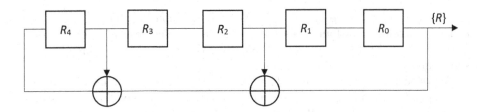

Fig. 2. Generator of the inverse M-sequence based on $P(x) = x^5 + x^4 + x^2 + 1$.

3 Orthogonal Structures Based on M-Sequences

For example, consider a bipolar code construction, which is a linear sum of the direct (see Fig. 1) and reverse M-sequence (see Fig. 2), based on $P(x) = x^4 + x + 1$: $S = (+2\ -2\ 0\ +2\ 0\ +2\ -2\ -2\ -2\ 0\ 0\ 0\ +2\ 0\ 0)$.

It is not difficult to make sure that the signal S is orthogonal to all M-sequences that are not included in this linear sum. Also, the signal S is orthogonal to the signals that are formed by these M-sequences.

The orthogonality of two signals is determined by their scalar multiplication, which is zero:

$$\int^T S_i(t)S_j(t)dt = 0, i \neq j, \tag{1}$$

where $S_i(t)$ and $S_j(t)$ are orthogonal code constructs that do not affect each other over a period of time T.

It should also be noted that to generate the S signal satisfying the condition (1), M-sequences with initial phases formed by elements of the $GF(2^4)$ were used: ε^5 (direct) and ε^9 (inverse).

Let's decompose the first k-element section of the signal S by the dual basis of the field $GF(2^k)$. The elements of the dual basis $\{\lambda\}$ are connected to the elements of the left power basis $\{\alpha\}$ via the trace function [9]:

$$T(\lambda_i \alpha_j) = \begin{cases} 0, & \text{if } i \neq j \\ 1, & \text{if } i = j. \end{cases} \tag{2}$$

In this example, $k = 4$, that is, the subject of processing is a fragment of the signal $(+2\ -2\ 0\ +2)$. Graphically, the process of decomposition by the basic functions ε^{14}, ε^2, ε, 1 satisfying the condition (2) can be represented in the polar coordinate system (see Fig. 3), where the signals are given in classical form, $x(t) = A\cos(\omega t + \varphi_0)$.

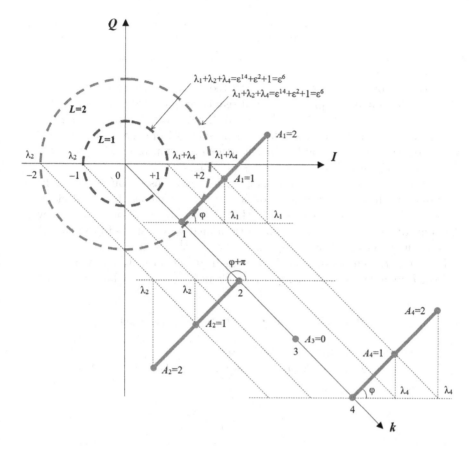

Fig. 3. Graphical representation of the decomposition process of the k-element section of the signal S on a dual basis.

As can be seen from Fig. 3, the results of processing the first section of the S signal are placed at two energy levels (L_1 and L_2) and represent equal sums of the field vectors. This result can be explained by the uniform distribution of energy between the direct and inverse M-sequences that make up the total signal S. At the same time, it should be noted that the value of the vector obtained at both energy levels corresponds to the sum of the initial phases of the sequences of maximum length included in the compacted signal. That is, the amount $\lambda_1 + \lambda_2 + \lambda_4 = \varepsilon^{14} + \varepsilon^2 + 1 = \varepsilon^6$ corresponds to the equality of $\varepsilon^5 + \varepsilon^9 = \varepsilon^6$. This, in turn, determines the elements of the series obtained in the process of sequential processing of k-element sections of the signal S. Thus, each subsequent section, when decomposed by a dual basis, forms a Galois field, similar to the processing of a section of a classical unipolar M-sequence. That is, when processing the second section ($-2\ 0\ +2\ 0$) at both levels, the element $\varepsilon^{14} + \varepsilon = \varepsilon^7$ will be received, the third - ε^8, the fourth - ε^9 and so on.

This property is more clearly manifested in the case of processing a multilayer signal, for example, for $S = (0\ -2\ +2\ 0\ +2\ 0\ -6\ +6\ +2\ +2\ -4\ -4\ 0\ +2\ 0)$, which includes four direct M-sequences with initial phases: ε, ε^3, ε^8, ε^{12} and four inverses with phases: ε^4, ε^5, ε^9, ε^{14}.

Processing of the first section, as expected, forms a field element that corresponds to the sum of the phases of all M-sequences included in the signal S: $\varepsilon + \varepsilon^3 + \varepsilon^8 + \varepsilon^{12} + \varepsilon^4 + \varepsilon^5 + \varepsilon^9 + \varepsilon^{14}$. One of the densest sections $(+6\ +2\ +2\ -4\ )$ in this case, it should form a field element shifted from the initial phase of ε^5 by 7 steps, that is, ε^{12}. Graphically, the decomposition process of this site can also be represented in polar coordinates (see Fig. 4).

Figure 4 shows that after processing with a dual basis, double energy levels are formed with the same content of field vectors, that is, the set of elements of even ("blue") levels coincides with the set of elements of odd ("red") levels. The sums of these sets are equal and form a field vector shifted from the initial phase of the total signal by 7 steps: $\varepsilon^{14} + \varepsilon^3 + \varepsilon^{11} = \varepsilon^{12}$.

Indeed, if the entire signal S is processed, it becomes clear that the energy levels of L include the same pairs of field elements (see Table 1).

Table 1. The result of complete processing signal.

L	Number of the k-element section														
	1	2	3	4	5	6	7	8	9	10	11	12	13	14	15
6				1	ε^4	ε^5	ε^{13}	ε^{14}							
5				1	ε^4	ε^5	ε^{13}	ε^{14}							
4				1	ε^4	ε^5	ε^{13}	ε^3	ε^4	ε^5	ε^{13}	ε^{14}			
3				1	ε^4	ε^5	ε^{13}	ε^3	ε^4	ε^5	ε^{13}	ε^{14}			
2	ε^5	ε^6	ε^7	ε^8	ε^9	ε^{10}	ε^{11}	ε^{11}	ε^{11}	ε^{12}	ε^6	ε^7	ε^2	ε^3	ε^4
1	ε^5	ε^6	ε^7	ε^8	ε^9	ε^{10}	ε^{11}	ε^{11}	ε^{11}	ε^{12}	ε^6	ε^7	ε^2	ε^3	ε^4
$\sum_{1,3,5}$	ε^5	ε^6	ε^7	ε^8	ε^9	ε^{10}	ε^{11}	ε^{12}	ε^{13}	ε^{14}	1	ε	ε^2	ε^3	ε^4
$\sum_{2,4,6}$	ε^5	ε^6	ε^7	ε^8	ε^9	ε^{10}	ε^{11}	ε^{12}	ε^{13}	ε^{14}	1	ε	ε^2	ε^3	ε^4

Thus, there is a variant of optimizing the compacted signal S in order to increase its energy efficiency. Removing even or odd layers is equivalent to halving the signal level. In fact, this is the averaging of energy between the direct and inverse M-sequences that are part of it. After this operation, the signal takes the form: $S_{opt} = (0\ -1\ +1\ 0\ +1\ 0\ -3\ +3\ +1\ +1\ -2\ -2\ 0\ +1\ 0)$.

Its processing at the reception by a dual basis will also make it possible to accurately distinguish the initial phase and all subsequent elements of the field. As will be shown later, determining the M-sequences included in the signal will not require bringing the S_{opt} to its original form, since it inherits all orthogonal properties from the original signal.

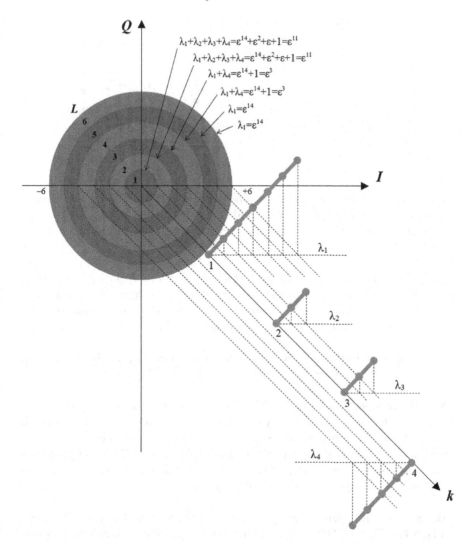

Fig. 4. Graphical representation of the decomposition process of a dense section of the signal S on a dual basis. (Color figure online)

Let's consider an example with a compacted signal that combines all even direct and odd inverse M-sequences (the same polynomial $P(x) = x^4 + x + 1$ is used). The exception, in this case, is the M-sequence formed from a single element of the field. Thus, the total signal will look like: $S = (+4\ -4\ +4\ -2\ 0\ 0\ +2\ -2\ 0\ +2\ -4\ +6\ -6\ +6\ -6)$. Halving the signal level transforms it into the form: $S_{opt} = (+2\ -2\ +2\ -1\ 0\ 0\ +1\ -1\ 0\ +1\ -2\ +3\ -3\ +3\ -3)$. Performing sequential processing of k-element sections of S_{opt} at the reception, we obtain the results presented in Table 2.

Table 2. The result of full processing S_{opt}.

L	Number of the k-element section														
	1	2	3	4	5	6	7	8	9	10	11	12	13	14	15
3									1	ε^4	ε^{10}	ε^{11}	ε^{12}	ε^{13}	ε^{14}
2	ε^{12}	ε^{13}	ε^{14}						1	ε^4	ε^{10}	ε^{11}	ε^{11}	ε^{11}	ε^{11}
1	ε^{11}	ε^{12}	ε^{13}	ε^3	ε^4	ε^5	ε^6	ε^9	ε^{10}	ε^{11}	ε^{11}	ε^{11}	ε^{11}	ε^{11}	ε^{11}
$\sum$	1	ε	ε^2	ε^3	ε^4	ε^5	ε^6	ε^7	ε^8	ε^9	ε^{10}	ε^{11}	ε^{12}	ε^{13}	ε^{14}

The phases obtained as a result of processing (see Table 2), represent a series of consecutive elements of the Galois field GF(2^4). The first phase, as expected, corresponds to the sum of all elements, except for a single one (not included in the signal): $\varepsilon + \varepsilon^2 + \varepsilon^3 + \varepsilon^4 + \varepsilon^5 + \varepsilon^6 + \varepsilon^7 + \varepsilon^8 + \varepsilon^9 + \varepsilon^{10} + \varepsilon^{11} + \varepsilon^{12} + \varepsilon^{13} + \varepsilon^{14} = 1$.

The selection of direct and inverse M-sequences at reception, as mentioned above, will not require the return of the optimized signal to its original form. It will be enough to define its scalar multiplication with all address combinations.

Consider an example with a longer polynomial $P(x) = x^8 + x^4 + x^3 + x^2 + 1$.

The dual basis in this case will be a number of elements of the Galois field GF(2^8): $\lambda_1 = \varepsilon^{252}$, $\lambda_2 = \varepsilon^{251}$, $\lambda_3 = \varepsilon^{45}$, $\lambda_4 = \varepsilon^{98}$, $\lambda_5 = \varepsilon$, $\lambda_6 = 1$, $\lambda_7 = \varepsilon^{254}$, $\lambda_8 = \varepsilon^{253}$.

Now the number of address combinations is 255 M-sequences. Let the direct combinations be used to transmit the symbol "0" and the inverse ones encode the symbol "1". At the same time, for convenience, we uniquely compare the channel numbers to the degrees of the Galois field elements GF(2^8), which form the initial phases of the corresponding address combinations. That is, channel 10 will use the M-sequence with the initial phase ε^{10}, channel 100 - ε^{100} and so on.

We organize the transmission "0" in randomly selected channels: 12, 46, 83, 95, 144, 168, 174, 196, 203, 241. Similarly, we use random channels to transmit "1": 3, 9, 38, 75, 102, 134, 156, 215, 226, 253.

The result of adding 10 direct and 10 inverse M-sequences will be the signal:
$S = (-6\ -4\ +4\ -2\ +2\ +2\ 0\ 0\ -2\ -2\ -4\ -8\ 0\ +10\ -4\ +2\ +2\ +4\ +2\ +4\ +8$
$-2\ 0\ 0\ 0\ -2\ 0\ -6\ 0\ -4\ +2\ -4\ -6\ 0\ +2\ -4\ 0\ -10\ -2\ +2\ +4\ -2\ +2\ -10\ 0\ +6$
$-2\ -4\ 0\ +2\ +2\ +6\ +2\ 0\ -2\ +6\ +2\ +2\ +6\ -2\ +4\ -10\ +8\ 0\ -6\ -2\ -4\ -4\ -2$
$+6\ -4\ -2\ 0\ +6\ -4\ -4\ 0\ -2\ +6\ +4\ +4\ -4\ +4\ +4\ +10\ -2\ +2\ +4\ +6\ -4\ -2$
$-4\ -2\ +4\ +6\ -2\ -2\ +6\ -2\ -4\ +8\ +2\ -12\ -2\ +4\ -8\ +8\ 0\ -2\ +4\ 0\ +10\ +2$
$+6\ -6\ +2\ +2\ +2\ +6\ +4\ -8\ -4\ 0\ -2\ 0\ -6\ +2\ +6\ +2\ 0\ -4\ -4\ +2\ +2\ 0\ +2\ -4$
$0\ +4\ -4\ -4\ +4\ -4\ +6\ -4\ -4\ 0\ -10\ -6\ -2\ 0\ +4\ -4\ -2\ +4\ 0\ +2\ +8\ -2\ -2$
$+6\ -4\ +4\ +4\ +8\ +4\ -8\ -2\ -2\ -8\ 0\ +6\ -4\ +4\ -6\ +4\ 0\ +2\ +2\ +4\ -2\ -6\ -6$
$0\ 0\ +2\ +4\ +4\ -4\ -4\ -4\ +2\ +8\ +2\ -4\ +2\ +2\ -4\ +6\ -6\ +8\ -6\ 0\ -6\ +2\ +8\ 0$
$+2\ -4\ +10\ +6\ +8\ 0\ 0\ +4\ +2\ +6\ -6\ -6\ -6\ -4\ -6\ +10\ -4\ +6\ -6\ -2\ -2\ -4$

−6 0 +2 −2 −2 +4 +6 +4 0 −4 −2 0 +4 +4 0 −6 −2 +4 +2 −6 −8 0 +4 −2
−10 −2).

The optimized signal, as mentioned above, assumes a reduction in the amplitude level of the signal S by half: S_{opt} = (−3 −2 +2 −1 +1 +1 0 0 −1 −1 −2
−4 0 +5 −2 +1 +1 +2 +1 +2 +4 −1 0 0 0 −1 0 −3 0 −2 +1 −2 −3 0 +1 −2 0
−5 −1 +1 +2 −1 +1 −5 0 +3 −1 −2 0 +1 +1 +3 +1 0 −1 +3 +1 +1 +3 −1
+2 −5 +4 0 −3 −1 −2 −2 −1 +3 −2 −1 0 +3 −2 −2 0 −1 +3 +2 +2 −2 +2
+2 +5 −1 +1 +2 +3 −2 −1 −2 −1 +2 +3 −1 −1 +3 −1 −2 +4 +1 −6 −1 +2
−4 +4 0 −1 +2 0 +5 +1 +3 −3 +1 +1 +1 +3 +2 −4 −2 0 −1 0 −3 +1 +3 +1
0 −2 −2 +1 +1 0 +1 −2 0 +2 −2 −2 +2 −2 +3 −2 −2 0 −5 −3 −1 0 +2 −2
−1 +2 0 +1 +4 −1 −1 +3 −2 +2 +2 +4 +2 −4 −1 −1 −4 0 +3 −2 +2 −3 +2
0 +1 +1 +2 −1 −3 −3 0 0 +1 +2 +2 −2 −2 −2 +1 +4 +1 −2 +1 +1 −2 +3
−3 +4 −3 0 −3 +1 +4 0 +1 −2 +5 +3 +4 0 0 +2 +1 +3 −3 −3 −3 −2 −3 +5
−2 +3 −3 −1 −1 −2 −3 0 +1 −1 −1 +2 +3 +2 0 −2 −2 −1 0 +2 +2 0 −3 −1 +2
+1 −3 −4 0 +2 −1 −5 −1).

It can be represented graphically in comparison with the main signal (see
Fig. 5).

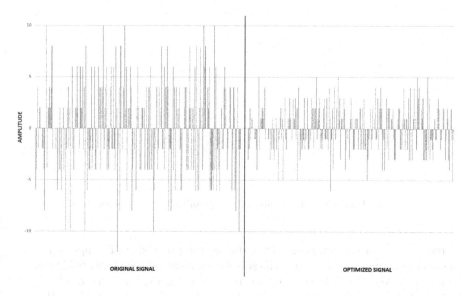

Fig. 5. A signal based on twenty M-sequences and its optimization.

By sequentially processing k-element sections of S_{opt} with the dual basis
of the Galois field $GF(2^8)$, you can make sure that a number of field elements
appear, starting with the vector ε^{117}, which is the sum of the phases of twenty
previously selected M-sequences. Thus, the equality is fulfilled: $\varepsilon^{117} = \varepsilon^{12} + \varepsilon^{46}$
$+ \varepsilon^{83} + \varepsilon^{95} + \varepsilon^{144} + \varepsilon^{168} + \varepsilon^{174} + \varepsilon^{196} + \varepsilon^{203} + \varepsilon^{241} + \varepsilon^{3} + \varepsilon^{9} + \varepsilon^{38} + \varepsilon^{75}$
$+ \varepsilon^{102} + \varepsilon^{134} + \varepsilon^{156} + \varepsilon^{215} + \varepsilon^{226} + \varepsilon^{253}$.

Figure 6 shows the processing of the first five ($t = 1..5$) sections of S_{opt} that form a fragment of the signal (-3 -2 $+2$ -1 $+1$ $+1$ 0 0 -1 -1 -2 -4). The sums of the basic functions λ at all energy levels l are displayed in detail.

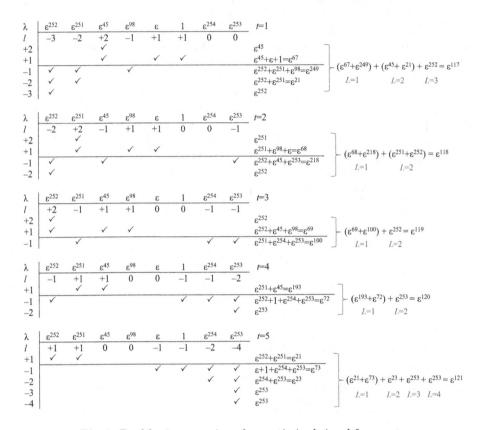

Fig. 6. Dual basis processing of an optimized signal fragment.

Indeed, as can be seen from Fig. 6, during the processing of a bipolar signal fragment by a dual basis, a sequential series of elements of the Galois field $GF(2^8)$ is allocated. And the first element of this series is the expected vector ε^{117}.

It is obvious that interference affecting the signal should lead to distortion of the results of its processing at reception. But, considering that S_{opt} is processed by the dual basis as equidistant code, it can be argued about its high noise immunity.

At reception, the optimized signal S_{opt} is multiplied by all address sequences in the same way as the original signal S. Previously, we associated channel numbers with the degree of field elements that generate address sequences. We will allocate information, for example, for 95, 156 and 62 channels. Thus, when multiplying S_{opt} by M-sequences U_{95}, U_{156}, U_{62} with phases ε^{95}, ε^{156} and ε^{62}, we should get the normalized values $+1$ (the direct address sequence was transmitted), -1 (the inverse address sequence was transmitted) and 0 (nothing was transmitted), accordingly. Taking into account the features of the orthogonal structure we have constructed and the fact that the signal amplitude has been halved, we will normalize the result of scalar multiplication by the value $(n + 1)/2 = 128$, where n is the period of M-sequences.

Information detection on channel number 95:

$S_{opt} \times U_{95} = [\ (-3)(-1) + (-2) + 2(-1) + (-1) + 1 + 1 + (-1) + (-1) +$
$(-2) + (-4)(-1) + 5 + (-2)(-1) + (-1) + 1 + 2 + 1 + 2(-1) + 4(-1) +$
$(-1)(-1) + (-1)(-1) + (-3)(-1) + (-2) + (-1) + (-2)(-1) + (-3) + 1 +$
$(-2)(-1) + (-5)(-1) + (-1)(-1) + 1 + 2(-1) + (-1) + 1 + (-5)(-1) + 3$
$+ (-1) + (-2)(-1) + (-1) + 1 + 3 + (-1) + (-1)(-1) + 3 + (-1) + 1 +$
$3 + (-1)(-1) + 2(-1) + (-5)(-1) + 4 + (-3)(-1) + (-1)(-1) + (-2) +$
$(-2)(-1) + (-1)(-1) + 3 + (-2) + (-1)(-1) + 3(-1) + (-2)(-1) + (-2)$
$+ (-1)(-1) + 3(-1) + 2(-1) + 2(-1) + (-2)(-1) + 2 + 2(-1) + 5 + (-1)$
$+ (-1) + 2(-1) + 3 + (-2) + (-1)(-1) + (-2)(-1) + (-1) + 2(-1) + 3 +$
$(-1)(-1) + (-1) + 3 + (-1) + (-2)(-1) + 4(-1) + 1 + (-6) + (-1) + 2$
$+ (-4) + 4(-1) + (-1)(-1) + 2 + 5 + (-1) + 3 + (-3)(-1) + 1 + (-1) +$
$(-1) + 3(-1) + 2(-1) + (-4)(-1) + (-2) + (-1)(-1) + (-3)(-1) + 1 + 3$
$+ 1 + (-2)(-1) + (-2) + 1 + (-1) + (-1) + (-2)(-1) + 2(-1) + (-2)(-1)$
$+ (-2)(-1) + 2 + (-2) + 3 + (-2) + (-2)(-1) + (-5)(-1) + (-3)(-1) +$
$(-1)(-1) + 2 + (-2) + (-1) + 2(-1) + 1 + 4(-1) + (-1) + (-1) + 3 + (-2)$
$+ 2 + 2 + 4 + 2(-1) + (-4) + (-1) + (-1) + (-4)(-1) + 3(-1) + (-2) +$
$2 + (-3) + 2(-1) + 1 + (-1) + 2 + (-1)(-1) + (-3)(-1) + (-3)(-1) + 1$
$+ 2 + 2 + (-2) + (-2)(-1) + (-2)(-1) + 1 + 4 + (-1) + (-2) + 1 + (-1)$
$+ (-2) + 3 + (-3)(-1) + 4(-1) + (-3) + (-3)(-1) + (-1) + 4 + (-1) +$
$(-2) + 5 + 3 + 4 + 2 + (-1) + 3 + (-3)(-1) + (-3)(-1) + (-3) + (-2)(-1)$
$+ (-3)(-1) + 5 + (-2)(-1) + 3 + (-3)(-1) + (-1)(-1) + (-1) + (-2) +$
$(-3)(-1) + (-1) + (-1)(-1) + (-1) + 2 + 3 + 2 + (-2)(-1) + (-1)(-1)$
$+ 2(-1) + 2 + (-3)(-1) + (-1) + 2(-1) + (-1) + (-3)(-1) + (-4)(-1) +$
$2(-1) + (-1) + (-5)(-1) + (-1)(-1)\ ]/128 = 1$

Information detection on channel number 156:

$S_{opt} \times U_{156} = [\ (-3)(-1) + (-2) + 2(-1) + (-1)(-1) + (-1) + 1 + (-1) +$
$(-1) + (-2)(-1) + (-4)(-1) + 5(-1) + (-2) + (-1) + (-1) + 2(-1) + (-1)$
$+ 2(-1) + 4(-1) + (-1) + (-1)(-1) + (-3) + (-2)(-1) + (-1) + (-2) +$
$(-3)(-1) + (-1) + (-2) + (-5) + (-1)(-1) + (-1) + 2 + (-1) + 1 + (-5)$
$+ 3(-1) + (-1)(-1) + (-2)(-1) + (-1) + 1 + 3(-1) + 1 + (-1) + 3(-1) +$
$(-1) + (-1) + 3(-1) + (-1)(-1) + 2 + (-5) + 4(-1) + (-3)(-1) + (-1) +$
$(-2) + (-2) + (-1) + 3(-1) + (-2) + (-1) + 3(-1) + (-2)(-1) + (-2)(-1)$
$+ (-1)(-1) + 3(-1) + 2(-1) + 2 + (-2) + 2 + 2 + 5(-1) + (-1) + (-1) +$
$2(-1) + 3(-1) + (-2)(-1) + (-1) + (-2) + (-1) + 2(-1) + 3(-1) + (-1)$
$+ (-1)(-1) + 3 + (-1) + (-2) + 4 + 1 + (-6) + (-1) + 2(-1) + (-4) + 4$
$+ (-1)(-1) + 2(-1) + 5 + 1 + 3 + (-3)(-1) + 1 + 1 + (-1) + 3 + 2(-1) +$
$(-4)(-1) + (-2)(-1) + (-1) + (-3) + 1 + 3 + (-1) + (-2) + (-2) + (-1) +$
$1 + (-1) + (-2) + 2(-1) + (-2)(-1) + (-2) + 2(-1) + (-2)(-1) + 3(-1) +$

$(-2) + (-2) + (-5) + (-3) + (-1) + 2 + (-2)(-1) + (-1) + 2(-1) + (-1)$
$+ 4(-1) + (-1) + (-1)(-1) + 3(-1) + (-2) + 2(-1) + 2 + 4(-1) + 2(-1)$
$+ (-4) + (-1) + (-1)(-1) + (-4) + 3(-1) + (-2) + 2 + (-3) + 2 + (-1)$
$+ (-1) + 2(-1) + (-1)(-1) + (-3) + (-3)(-1) + (-1) + 2(-1) + 2(-1) +$
$(-2)(-1) + (-2) + (-2)(-1) + 1 + 4(-1) + (-1) + (-2)(-1) + 1 + (-1) +$
$(-2) + 3 + (-3) + 4(-1) + (-3) + (-3) + 1 + 4(-1) + 1 + (-2)(-1) + 5(-1)$
$+ 3 + 4 + 2(-1) + (-1) + 3(-1) + (-3) + (-3) + (-3)(-1) + (-2)(-1) +$
$(-3)(-1) + 5 + (-2) + 3(-1) + (-3)(-1) + (-1) + (-1) + (-2) + (-3)(-1)$
$+ (-1) + (-1)(-1) + (-1) + 2(-1) + 3 + 2 + (-2) + (-1) + 2(-1) + 2$
$+ (-3) + (-1) + 2(-1) + 1 + (-3)(-1) + (-4) + 2 + (-1) + (-5)(-1) +$
$(-1)(-1)]/128 = -1$

Information detection on channel number 62:

$S_{opt} \times U_{62} = [\ (-3) + (-2)(-1) + 2 + (-1)(-1) + (-1) + 1 + (-1) +$
$(-1)(-1) + (-2)(-1) + (-4) + 5 + (-2) + (-1) + (-1) + 2(-1) + (-1) +$
$2(-1) + 4 + (-1)(-1) + (-1)(-1) + (-3)(-1) + (-2)(-1) + 1 + (-2)(-1)$
$+ (-3)(-1) + 1 + (-2)(-1) + (-5) + (-1) + (-1) + 2 + (-1) + 1 + (-5)$
$+ 3(-1) + (-1) + (-2)(-1) + 1 + 1 + 3 + (-1) + (-1)(-1) + 3(-1) + 1$
$+ 1 + 3(-1) + (-1)(-1) + 2(-1) + (-5) + 4 + (-3)(-1) + (-1) + (-2) +$
$(-2) + (-1)(-1) + 3 + (-2)(-1) + (-1)(-1) + 3(-1) + (-2) + (-2) + (-1)$
$+ 3 + 2 + 2(-1) + (-2) + 2(-1) + 2 + 5 + (-1)(-1) + 1 + 2(-1) + 3 +$
$(-2)(-1) + (-1) + (-2) + (-1)(-1) + 2(-1) + 3(-1) + (-1) + (-1)(-1) +$
$3(-1) + (-1)(-1) + (-2) + 4(-1) + (-1) + (-6) + (-1) + 2(-1) + (-4)(-1)$
$+ 4(-1) + (-1) + 2(-1) + 5(-1) + (-1) + 3(-1) + (-3)(-1) + 1 + (-1) +$
$1 + 3 + 2(-1) + (-4)(-1) + (-2) + (-1)(-1) + (-3) + (-1) + 3 + (-1) +$
$(-2) + (-2) + (-1) + (-1) + 1 + (-2) + 2 + (-2)(-1) + (-2)(-1) + 2(-1)$
$+ (-2) + 3(-1) + (-2) + (-2)(-1) + (-5)(-1) + (-3) + (-1)(-1) + 2(-1)$
$+ (-2)(-1) + (-1)(-1) + 2 + (-1) + 4 + (-1)(-1) + (-1) + 3 + (-2) +$
$2 + 2(-1) + 4 + 2 + (-4)(-1) + (-1)(-1) + (-1)(-1) + (-4)(-1) + 3(-1)$
$+ (-2)(-1) + 2(-1) + (-3) + 2 + 1 + (-1) + 2 + (-1)(-1) + (-3)(-1) +$
$(-3)(-1) + 1 + 2 + 2(-1) + (-2)(-1) + (-2) + (-2)(-1) + 1 + 4 + 1 +$
$(-2) + 1 + 1 + (-2) + 3(-1) + (-3) + 4 + (-3) + (-3)(-1) + (-1) + 4 +$
$1 + (-2)(-1) + 5 + 3 + 4(-1) + 2(-1) + (-1) + 3 + (-3) + (-3) + (-3) +$
$(-2) + (-3) + 5(-1) + (-2)(-1) + 3 + (-3) + (-1)(-1) + (-1) + (-2) +$
$(-3)(-1) + 1 + (-1)(-1) + (-1)(-1) + 2 + 3(-1) + 2(-1) + (-2) + (-1)$
$+ 2 + 2 + (-3) + (-1) + 2(-1) + 1 + (-3)(-1) + (-4) + 2(-1) + (-1) +$
$(-5)(-1) + (-1)(-1)]/128 = 0$

As expected, reducing the amplitude of the original signal by half did not affect the result of detecting information in the channels. The direct M-sequence was indeed transmitted via channel 95, the inverse M-sequence was transmitted via channel 156, no information was transmitted via channel 62, therefore, as a result of multiplication by the address sequence U_{62}, zero was obtained.

4   Conclusion

Code constructions based on linear sums of recurrent sequences of maximum length are multilayer orthogonal structures that are processed quite efficiently by the dual basis of the Galois field. The result of the sequential decomposition of their sections on a dual basis are double energy layers that contain the same elements of the field. Thus, the sum of the field vectors of even and odd levels is the same and equal to the field element, which, in turn, is the sum of the phases of the M-sequences that make up the multilayer structure.

The properties of the maximum length sequences considered in this paper are capable of providing constant control over the transmission of a compacted signal. A phase failure or a jump in the signal level will lead to a break in a number of field elements obtained as a result of processing the code structure with a dual basis. At the same time, orthogonality at the reception will allow you to unambiguously divide the signal into components.

The paper also suggests the idea of increasing energy efficiency by reducing the signal level by half. This procedure is designed to eliminate code redundancy. At the same time, it will not affect the result of determining the phase of the signal during reception.

Acknowledgments. The study was financially supported by the Russian Science Foundation within of scientific project No. 22-49-02023 "Development and study of methods for obtaining the reliability of tethered high-altitude unmanned telecommunication platforms of a new generation".

References

1. Baghani, M., Parsaeefard, S., Derakhshani, M., Saad, W.: Dynamic non-orthogonal multiple access and orthogonal multiple access in 5G wireless networks. IEEE Trans. Commun. **67**(9), 6360–6373 (2019). https://doi.org/10.1109/TCOMM. 2019.2919547
2. Sadkhan, S.-B.: Proposed development of scattering problem solution based on Walsh function. In: 4th Scientific International Conference Najaf (SICN), Najaf, pp. 54–58 (2019). https://doi.org/10.1109/SICN47020.2019.9019368
3. Torrieri, D.: Principles of Spread-Spectrum Communication Systems, 4th edn. Springer, Cham (2018). https://doi.org/10.1007/978-3-319-70569-9
4. Khudhair, A.-Y., Abd Khalid, R.-A.: Reduction of the noise effect to detect the DSSS signal using the artificial neural network. In: 1st Babylon International Conference on Information Technology and Science (BICITS), Babil, pp. 185–188 (2021). https://doi.org/10.1109/BICITS51482.2021.9509880
5. Qiang, X., Zhang, T.: Estimation of spreading code in non-periodic long-code DSSS signal. In: 6th International Conference on Wireless Communications, Signal Processing and Networking (WiSPNET), Chennai, pp. 162–165 (2021). https://doi.org/10.1109/WiSPNET51692.2021.9419440
6. Shehzadi, S., Sheikh, S.-A., Kulsoom, F., Zeeshan, M., Khan, Q.-U.: A robust timing and phase offset estimation technique for CPM-DSSS-based secured communication link. IEEE Access **9**, 111143–111151 (2021). https://doi.org/10.1109/ACCESS.2021.3102308
7. Qiu, Z., Peng, H., Li, T.: A blind despreading and demodulation method for QPSK-DSSS signal with unknown carrier offset based on matrix subspace analysis. IEEE Access **7**, 125700–125710 (2019). https://doi.org/10.1109/ACCESS.2019.2938785

8. Kukunin, D., Berezkin, A., Zadorozhnyaya, A., Karelin, E., Shestakov, A.: Model of adaptive data transmission system. In: 13th International Congress on Ultra Modern Telecommunications and Control Systems and Workshops (ICUMT), Brno, pp. 200–205 (2021). https://doi.org/10.1109/ICUMT54235.2021.9631637

9. Kukunin, D., Berezkin, A., Kirichek, R., Panteleimonov, I. : Phasing in asynchronous data transmission system using M-sequences. In: 5th International Conference on Future Networks & Distributed Systems (ICFNDS), Dubai, pp. 516–521 (2021). https://doi.org/10.1145/3508072.3508178

10. Leetang, K., Hachiya, H., Hirata, S.: Evaluation of ultrasonic target detection by alternate transmission of different codes in M-sequence pulse compression. In: IEEE International Ultrasonics Symposium (IUS), Xi'an, pp. 1–4 (2020). https://doi.org/10.1109/IUS46767.2020.9251455

11. Du, P., Shen, Y., Zeng, Y.: RFID multi-channel design method based on CDMA. In: International Conference on Artificial Intelligence and Advanced Manufacturing (AIAM), Dublin, pp. 92–95 (2019). https://doi.org/10.1109/AIAM48774.2019.00025

12. Kaiyang, H., Tingting, T., Yiao, Z.: Underwater small moving target detection using maximum length sequences. In: OES China Ocean Acoustics (COA), Harbin, pp. 727–730 (2021). https://doi.org/10.1109/COA50123.2021.9519954

Research and Development of Data Compression Methods Based on Neural Networks

A. Berezkin$^{(\boxtimes)}$ (ID), D. Kukunin (ID), and R. Kirichek (ID)

Bonch-Bruevich Saint-Petersburg State University of Telecommunications,
Bolshevikov Ave. 22 build. 1, 193232 St. Petersburg, Russia
aa.berezkin@mail.ru , kirichek@sut.ru

Abstract. This article discusses a neural network-based compression algorithm using error-correcting codes. The use of this algorithm has a number of advantages, on the one hand, the noise-resistant code allows to get rid of potentially high overheads provoked by the use of a neural network, lowering the required value of model accuracy to the value determined by the correctability of the used code. On the other hand, a trained neural network allows data compression without prior transformations.

Keywords: Autoencoder · Data compression · Neural networks

1 Introduction

Nowadays, the emergence of new types of communication networks, particularly, networks using tethered and autonomous UAVs, has led to the development of different data transmission methods [1, 7–9]. As expected in the future networks, it is required to consider having an ultra-low latency and high bandwidth for some applications such as augmented/virtual reality or tactile internet [6, 13]. The modern stage of telecommunications development imposes stringent requirements for speed, delay and reliability in transmitting and processing information.

One of the approaches to increase the speed of data transmission, and hence to eliminate delays, is the compression of information in communication channels on the fly. This approach requires the development of new information processing architectures that allow data processing in parallel. Neural networks by virtue of their structure can fulfill the set requirements.

Neural networks are actively used in image and other compression tasks [2, 10–12], since the generalization ability of this method allows to achieve lossy compression. Theoretically, it is possible to achieve full data recovery [3], but in practice developers face high overhead costs associated with an increase in the occupied space of the network itself, as well as, with an increase in the computational complexity of the algorithm. The proposed codec structure is shown in Fig. 1, where the neural network part of the codec is highlighted by a rectangle.

V. M. Vishnevskiy et al. (Eds.): DCCN 2022, LNCS 13766, pp. 103–116, 2022.
https://doi.org/10.1007/978-3-031-23207-7_9

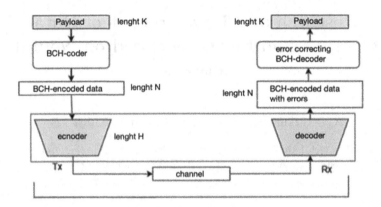

Fig. 1. Neural network codec architecture

This architecture consists of several independent processing units:

- Interference code encoder and decoder;
- A compression algorithm based on a neural network;

The use of this architecture has several advantages, on the one hand, the error-correcting code [3,4] allows to get rid of the potentially high overheads induced by the use of the neural network, lowering the required value of the final model accuracy to the value determined by the corrective capability of the used noise-correcting code. On the other hand, a trained autoencoder neural network allows data compression without prior transformations.

2   Development of a Research Stand

In this paper we investigated the operation of the autoencoder using several error-correcting codes of different multiplicity. The following error-correcting codes were used: (31, 21), (31, 16) and (31, 5).

These codes have different correction ability, as well as different number of information bits, which will allow to better explore the capabilities of the neural network and determine the dependence of the hidden state size not only on the number of guaranteed correctable errors, but also to determine the influence of the code distance on the generalization ability of the autoencoder neural network.

The criterion for comparing different architectures is the fact of lossless data recovery from the hidden state. This metric can be described as the accuracy of the neural network and defined as the fraction of correctly recovered symbols relative to the whole set. For a binary dataset, the accuracy formula is as follows:

$$accuracy = \frac{TP + TN}{TP + TN + FP + FN'}, \tag{1}$$

where $accuracy$ is the accuracy of the autoencoder, TP is the number of correctly recovered units, TN is the number of correctly recovered zeros, FP is the number of incorrectly recovered units, FN is the number of incorrectly recovered zeros.

For lossless data recovery, the accuracy should be equal to one. The accuracy shows the quality of the neural network with respect to the whole data set, not paying attention to errors in specific code combinations. In practice, there may be situations in which the neural network will perform well relative to the full data set, but some of the code combinations cannot be recovered in their original form. To control the accuracy of the autoencoder at the level of a particular code combination, an additional metric was introduced, defined as the minimum accuracy of the neural network relative to the code combination.

$$CodeWiseAcc = min(accuracy(X_{i=1..n})), \qquad (2)$$

where $CodeWiseAcc$ is minimum accuracy with respect to code combination, $accuracy$ is accuracy metric, $X_{i=1...n}$ is recovered element of original sequence. This metric allows to trace the maximum value of possible errors in a separate code combination and gives a better idea about the restoring capability of autoencoder.

The following formula was used to calculate the lower bound of autoencoder accuracy:

$$CodeWiseAcc_{m}in = \frac{n-t}{n}, \qquad (3)$$

where $CodeWiseAcc_{min}$ is the lower bound of the required accuracy, n is the number of elements of the code sequence containing check and information symbols, t is the number of guaranteed error correction by the error code.

The boundaries of acceptable accuracy of the autoencoder when using anti-interference codes to provide lossless compression are presented in the Table 1.

Table 1. Limits of acceptable neural network accuracy

Interference-resistant code	Lower limit of tolerable accuracy of autoencoder
BCH (31, 21)	0,9355
BCH (31, 16)	0,9032
BCH (31, 5)	0,7580

3   Autoencoder Architecture

The neural autoencoder consists of the following layers:

1. The input layer that takes in a number of values equal to the length of the code and outputs a vector of length 16
2. Intermediate activation function

3. Hidden layer, taking a vector of length 16 at the input and giving out a vector of size equal to the length of the vector of the hidden state
4. Intermediate activation function

The neural decoder consists of the following layers:

1. The input layer that takes as input a number of values equal to the given length of the hidden state vector, and outputs a vector of length 31
2. Intermediate activation function
3. Hidden layer, taking a vector of length 31 at the input and outputting a vector of length 31 which outputs a vector of length 775
4. activation function
5. Hidden layer, taking a length vector of length 775 and which outputs a vector of size equal to the length of the code being used
6. Final activation function

The hyperbolic tangent was chosen as the intermediate activation function. A sigmoidal function was chosen as the final activation function.

For this study, three data sets were generated containing all possible code combinations of length equal to the number of information characters of the corresponding BCH code. Based on the information bits, noise-resistant code combinations were generated using the BCH coder. Since a complete data set fully describing all possible allowed code combinations is known, the division into validation and test part is not required. Binary Cross-entropy was chosen as the error function of the neural network. Training occurred until one of the conditions was met:

1. A metric value equal to one was achieved, indicating lossless data recovery
2. The neural network error value has stopped decreasing for a significant number of epochs, thereby the network has reached its recovery limit

By changing the hidden state of the neural network, there was a determination of the maximum compression ratio, at which the possibility of lossless data recovery was achieved using the appropriate noise-correcting code.

Determination of the possible advantage of using a noise-correcting code was performed by comparing the results with an autoencoder trained to recover only information bits, without the use of a noise-correcting code.

4   Experimental Results

4.1   BCH-Based Codec (31, 5)

During the study it was found that the use of a noise-correcting code made it possible to represent the initial combination in the form of a vector of length 2, while the model without the mechanism used showed itself worse, achieving lossless compression with a hidden state length equal to 3. The results are given in the Table 2.

Table 2. Results based on the BCH codec (31, 5)

Model reconstruction accuracy	The size of the hidden state vector		
Using anti-jamming code	1	2	3
False	0,78125	0,899999976	1
True	0,637499988	1	1

Thus, in the course of simulation with the use of interference-free coding the compression coefficient was achieved equal to 0.4. On the other hand, without the use of noise-correcting code compression factor is equal to 0.6. Thus, increase of code redundancy by 83.

The importance of the introduced metric *CodeWideAcc* is clearly shown in Fig. 2. In the course of the experiment there was a contradiction: accuracy of the neural autoencoder lies in an acceptable zone, corresponding to lossless compression, but after the error correction algorithm the data recovery accuracy was 0.79 and 0.91, which does not correspond to the expected value of 1.

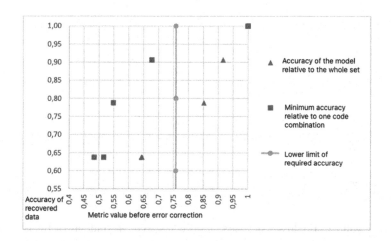

Fig. 2. Matching metrics before and after the error correction procedure

Figure 3 shows a graph of the change in the *CodeWiseAcc* metric as a function of the number of training epochs with the hidden state equal to one. It shows that the accuracy of the approach without the use of interference-free code is higher over the whole interval, but it never reaches the value equal to unity. This difference between the metrics is due primarily to the fact that the neural autoencoder based on the noise-correcting code recovers a vector of length 31, while the model without the use of noise-correcting coding only 5 initial information bits.

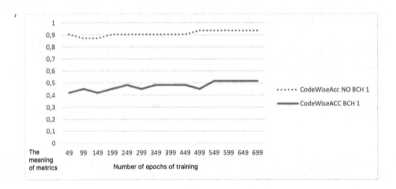

Fig. 3. The value of the CodeWiseAcc metric when the length of the hidden vector is equal to one

Figure 4 shows a graph of the change in the value of the loss function depending on the number of epochs of training with the length of the hidden vector equal to one.

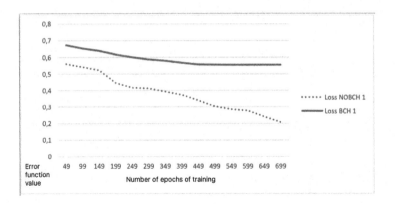

Fig. 4. The value of the loss function when the length of the hidden vector is equal to one

Figure 5 shows the graph of change in the metric *CodeWiseAcc* depending on the number of epochs of training with the hidden state equal to two. It can be seen that starting from 649 epochs the neural network autoencoder, built using noise-free code, achieves accuracy equal to one, which indicates lossless data compression.

Figure 6 shows a graph of the change in the value of the loss function as a function of the number of training epochs with the length of the hidden vector equal to two.

Despite the fact that the neural network autoencoder using the noise-free coding algorithm achieved a lossless compression effect, the error function values

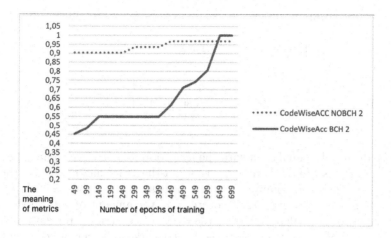

Fig. 5. The CodeWiseAcc metric value for a hidden vector length equal to two

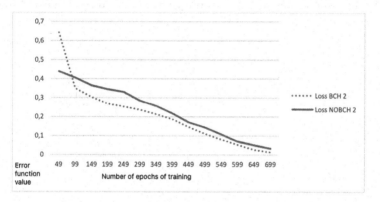

Fig. 6. The value of the loss function when the length of the hidden vector is equal to two

differ slightly, which confirms the theory of achieving full recovery of the original data by error correction.

4.2 BCH-Based Codec (31, 16)

During the study it was found that the use of a noise-correcting code made it possible to represent the initial combination in the form of a vector of length 7, while the model without the mechanism used showed itself worse, achieving lossless compression with a hidden state length equal to 8. The results are given in Table 3.

Table 3. Results based on the BCH codec (31, 16)

Model reconstruction accuracy	The size of the hidden state vector		
Using anti-jamming code	1	2	3
False	0,793	0,778	1
True	0,886	1	1

Thus, during the simulation with the use of interference-free coding the compression ratio was achieved equal to 0.43. On the other hand, without the use of noise-correcting code compression factor is equal to 0.5. Thus, a 52% increase in code redundancy entails a 12.5% decrease in the amount of data in the link.

Figure 7 shows the dependence of the obtained accuracy after data recovery by neural network autoencoder and the final accuracy after error correction. As can be seen, all the *CodeWiseAcc* values lying on the interval included in the lossless compression interval have accuracy after error correction equal to one.

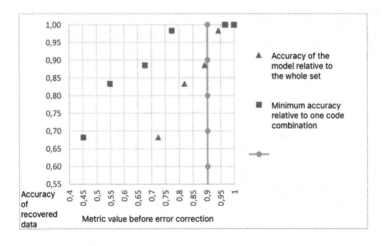

Fig. 7. Correspondence of metrics after code correction (31, 16)

Figure 8 shows the graph of change in the *CodeWiseAcc* metric depending on the number of training epochs with the hidden state equal to 7. You can see that the accuracy is increasing despite regular drawdowns. The solution is to keep the intermediate model, which has the highest accuracy.

Figure 9 shows a graph of the change in the value of the loss function as a function of the number of training epochs with the length of the hidden vector equal to 7.

On this graph you can see a strong difference between the approaches. This fact is explained by the low accuracy of the model without the use of interference-resistant codes. Figure 10 shows a graph of the change in the *CodeWiseAcc*

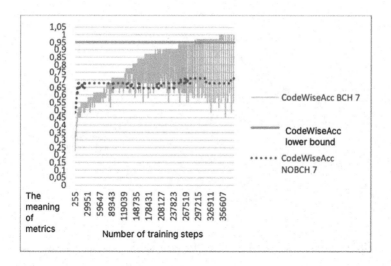

Fig. 8. The CodeWiseAcc metric value for a hidden vector length of seven

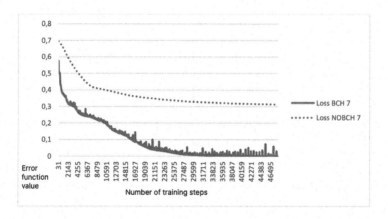

Fig. 9. The value of the loss function when the length of the hidden vector is equal to seven

metric as a function of the number of training epochs with the hidden state equal to 8.

Starting from 37567 epoch the neural network autoencoder constructed with the use of the interference-resistant code reaches the accuracy equal to one, which indicates lossless data compression. Neural network autoencoder without use of interference-free coding achieves this value at step 8863, which is much smaller. This fact is explained by the fact that the neural network autoencoder without the use of noise-correcting code recovers only 16 information symbols against 31.

Figure 11 shows a graph of the change in the value of the loss function depending on the number of training epochs with the length of the hidden vector equal to eight.

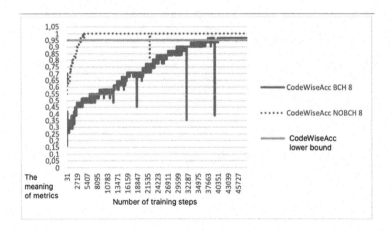

Fig. 10. The CodeWiseAcc metric value for a hidden vector length of eight

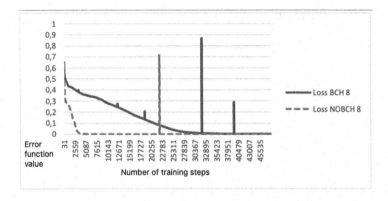

Fig. 11. The value of the loss function when the length of the hidden vector is equal to eight

Thus, the proposed implementation of the neural network codec allows to get rid of less redundancy in the channel while using the code with less correcting power. The reason of this fact may be the polynomial growth of the training set with the increase of information symbols, which requires to change the number of training parameters of the neural network in the higher side.

4.3   Code Based on BCH (31, 21)

During the study of the neural network autoencoder using the interference-resistant BSH code (31, 21), a contradiction was identified. A neural network autoencoder without the use of a noise-correcting code, having a hidden vector size equal to the length of the original sequence could not achieve the necessary accuracy for lossless compression. Table 4 shows the accuracy values depending on the hidden layer length and the condition of using the interference-resistant code.

Table 4. The results obtained based on the BCH codec (31, 21)

Model reconstruction accuracy	The size of the hidden state vector						
Using anti-jamming code	7	9	11	12	13	15	21
False	0,67	0,74	0,71	–	0,77	–	0,72
True	0,39	0,52	0,52	0,68	–	0,716	0,77

Figure 12 shows a graph of the dependence of the metric *CodeWiseAcc* on the size of the hidden state of the neural network autoenumerator using the interference-resistant code. As can be seen from the graph during the experiment it was not possible to achieve the required accuracy, which guarantees lossless data compression. However, as shown in [3] it is possible to achieve this by controlling the power of the training set and filtering elements that do not affect the final generalization ability.

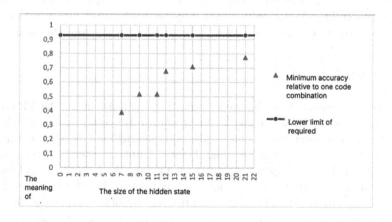

Fig. 12. CodeWiseAcc metric value depending on the hidden state

Figure 13 shows a graph of the change in the loss function value depending on the number of training epochs with the hidden vector length equal to 21 corresponding to the input combination length and using the neural network autoencoder model without using the noise coding mechanism. This graph shows that the values of the error function have reached a plateau, but the value of the function is not small enough to provide lossless compression.

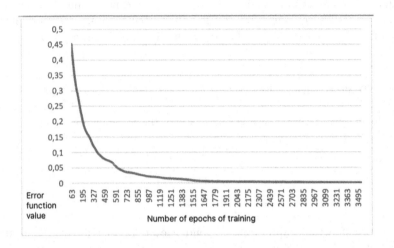

Fig. 13. The value of the loss function when the length of the hidden vector is equal to 21

Thus we can conclude that the structure of the neural network autoencoder, does not have sufficient complexity to provide lossless data compression for the given number of information symbols and the size of the training set. Therefore to study this code and codes with larger length of information bits it is necessary to use more complex architectures with a larger number of trainable parameters.

5   Conclusion

The following results were obtained during the analysis: using the FFT code (31, 5) it was possible to achieve a compression ratio equal to 0.4, which will reduce the data volume in the communication channel by 33% when modifying the architecture based only on the neural network autoencoder. When using the BCH code (31, 16), a compression ratio of 0.43 was achieved, which will reduce the amount of data in the communication channel by 12.5% when modifying the architecture based only on the neural network autoencoder. Also, the study revealed disadvantages that must be taken into account in the system design. During the comparison of the obtained accuracy when using codes with different minimum code distances, it was determined that the data structure formed by cyclic codes allows the neural network to better determine the weighting coefficients and restore the input sequence. The results of this comparison showed that a larger minimum code distance of the input sequence leads to greater relative accuracy of the model. This means that when the minimum code distance increases, the recovery potential of the neural network autoencoder increases, which can be used to solve problems in related areas.

Acknowledgments. The study was financially supported by the Russian Science Foundation within of scientific project No. 22-49-02023 "Development and study of methods for obtaining the reliability of tethered high-altitude unmanned telecommunication platforms of a new generation".

References

1. Koucheryavy, A., Vladyko, A., Kirichek, R.: State of the art and research challenges for public flying ubiquitous sensor networks. In: Balandin, S., Andreev, S., Koucheryavy, Y. (eds.) ruSMART 2015. LNCS, vol. 9247, pp. 299–308. Springer, Cham (2015). https://doi.org/10.1007/978-3-319-23126-6_27
2. Wu, Y., et al.: Deep image compression with latent optimization and piece-wise quantization approximation. In: IEEE/CVF Conference on Computer Vision and Pattern Recognition, Nashville, pp. 1926–1930 (2021). https://doi.org/10.1109/CVPR46437.2021
3. Berezkin, A., Zadorozhnyaya, A., Kukunin, D., Matveev, D., Kraeva, E.: Models and methods for decoding of error-correcting codes based on a neural network. In: 13th International Congress on Ultra Modern Telecommunications and Control Systems and Workshops (ICUMT), Brno, pp. 230–235 (2021). https://doi.org/10.1109/ICUMT54235.2021.9631637
4. Ahmed, S.: Linear block code decoder using neural network. In: IEEE International Joint Conference on Neural Networks (IJCNN 2008), Hong Kong, pp. 1111–1114 (2008). https://doi.org/10.1109/IJCNN13382.2008
5. Claytus Vaz, A., Nayak, G., Nayak, D.: Hamming code performance evaluation using artificial neural network decoder. In: 15th International Conference on Engineering of Modern Electric Systems (EMES), Oradea, pp. 37–40 (2019). https://doi.org/10.1109/EMES.2019.8795208
6. Al-Gaashani, M., A Muthanna, M.S., Abdukodir, K., Muthanna, A., Kirichek, R.: Intelligent system architecture for smart city and its applications based edge computing. In: 12th International Congress on Ultra Modern Telecommunications and Control Systems and Workshops (ICUMT), Brno, pp. 269–274 (2020). https://doi.org/10.1109/ICUMT51630.2020
7. Vladimirov, S., Kirichek, R., Vishnevsky, V.: Network coding for the interaction of unmanned flying platforms in data acquisition networks. In: The 4th International Conference on Future Networks and Distributed Systems (ICFNDS), St. Petersburg, pp. 1–7 (2020)
8. Vladimirov, S., Vishnevsky, V., Larionov, A., Kirichek, R.: The model of WBAN data acquisition network based on UFP. In: Vishnevskiy, V.M., Samouylov, K.E., Kozyrev, D.V. (eds.) DCCN 2020. LNCS, vol. 12563, pp. 220–231. Springer, Cham (2020). https://doi.org/10.1007/978-3-030-66471-8_18
9. Vishnevsky, V.M., Tereschenko, B.N., Tumchenok, D.A., Shirvanyan, A.M., Sokolov, A.: Principles of building a power transmission system for tethered unmanned telecommunication platforms. In: Vishnevskiy, V.M., Samouylov, K.E., Kozyrev, D.V. (eds.) DCCN 2019. LNCS, vol. 11965, pp. 94–110. Springer, Cham (2019). https://doi.org/10.1007/978-3-030-36614-8_8

10. Zhang, M., Zhang, H., Yuan, D., Zhang, M.: Compressive sensing and autoencoder based compressed data aggregation for green IoT networks. In: Global Communications Conference (GLOBECOM), Waikoloa, pp. 11732–11742 (2019). https://doi.org/10.1109/GLOBECOM38437.2019.9013373
11. Zhou, Y., Wang, C., Zhou, X.: DCT-based color image compression algorithm using an efficient lossless encoder. In: 14th IEEE International Conference on Signal Processing (ICSP), Beijing, pp. 450–454 (2018). https://doi.org/10.1109/ICSP43694.2018
12. Xu, T., Darwazeh, I.: Design and prototyping of neural network compression for non-orthogonal IoT signals. In: Wireless Communications and Networking Conference (WCNC), Marrakesh, pp. 1–6 (2019). https://doi.org/10.1109/WCNC44850.2019
13. Yastrebova, A., Kirichek, R., Koucheryavy, Y., Borodin, A., Koucheryavy, A.: Future networks 2030: architecture & requirements. In: 10th International Congress on Ultra Modern Telecommunications and Control Systems and Workshops (ICUMT), Moscow, pp. 1–8 (2018). https://doi.org/10.1109/ICUMT45195.2018

Anomaly Electrocardiograms Automatic Detection with Unsupervised Deep Learning Methods

Eugene Yu. Shchetinin[1] , Anastasia G. Glushkova[2] ,
and Leonid A. Sevastianov[3] (✉)

[1] Financial University under the Government of the Russian Federation,
Moscow, Russia
riviera-molto@mail.ru
[2] Oxford University, Oxford, UK
aglushkova@endeavorco.com
[3] Peoples Friendship University of Russia (RUDN), Moscow, Russia
sevastianov-la@rudn.ru

Abstract. Anomaly detection is an important problem in various fields of technology and industry, such as malicious intrusions into computer systems, finance and banking, health monitoring, etc. Currently, deep learning methods have achieved significant success in anomaly detection. Methods of detecting anomalies in a set of electrocardiograms containing normal ECG signals and ECG signals with various cardiovascular diseases have been investigated. To detect abnormal electrocardiograms, an autoencoder model in the form of a deep neural network with several fully connected layers was developed. Also, a method of selecting a threshold to separate abnormal ECG signals from normal ones was proposed, which consists in optimizing the ratio of performance indicators of the autoencoder model. A comparative computer analysis of the effectiveness of applying the proposed autoencoder model and other machine learning models, such as the support vector method, isolation forest, and random forest, to solve the problem of detecting abnormal ECG signals was carried out. For this purpose, metrics such as accuracy, recall, completeness, and f-score were used. The results showed that the proposed model surpassed other models with accuracy of 98.8%, precision of 95.75%, recall of 99.12%, f1-score of 98.75%.

Keywords: Anomaly detection · Electrocardiogram · Unsupervised deep learning · Autoencoder

1 Introduction

Anomaly detection is an important area of application of artificial intelligence in big data analysis, such as computer system security, fraud detection in bank

L. A. Sevastianov—This paper has been supported by the RUDN University Strategic Academic Leadership Program.

transfers, reliability of computer vision systems and others [2]. The detection of anomalies is also a key task of the analysis of biomedical information, since the violation of the stability of the recognition systems of dangerous diseases based on the analysis of biomedical signals and MRI, CT images, for example, can lead to erroneous screening of patients.

One of the main problems in machine learning and data analysis tasks is data points true labelling. In anomaly detection correct labelling is almost impossible due to both the unpredictability of anomalies' occurrence and the various forms of their existence. In addition, since the number of anomalies is too small and the methods of balancing classes still require improvement, the use of classification and supervised machine learning methods is problematic [4]. Therefore, one of the relevant approaches to solve this problem is the use of unsupervised machine learning methods [12], since in this case preliminary labelling of the source data into abnormal and normal data sets is not required. There are popular methods for solving the anomaly detection problem, which include the Isolation Forest algorithm, methods of nonparametric statistics, cluster analysis, and others [5,9,13]. The main class of classical anomaly detection methods is based on calculating distance to the nearest neighbours, thus forming clusters in the data, to assess whether the data are anomalous. Such methods are based on the use of a certain metric of the distance between observations. One-class classification approaches trained only on normal data are also widely used, for example, one-Class SVM [1,3]. Another group of methods uses the accuracy of the observation reconstruction to determine whether a data point is anomalous or not. Principal components analysis and its variants are examples of such methods [21]. The problems of using these methods to detect anomalies have been thoroughly investigated. For example, they include the low accuracy of detecting anomalous observations, the sensitivity of these methods to noise and distortions in the data, as well as a number of other shortcomings that require improvements and development of new methods for analysing anomalies [11,17].

At the present stage of development of anomaly detection methods, the use of unsupervised deep learning methods is becoming more and more effective [15]. Various autoencoder models designed to solve data reconstruction problems using reconstruction error to identify anomalies are popular as well. Approaches based on variational and generative autoencoders first learn the model to reconstruct normal data, and then identify anomalies as samples with high reconstruction errors [14,20,22]. In this paper, a model of a deep autoencoder for the detection of abnormal electrocardiograms in a data set studied in [8,16] is constructed and investigated. A method for selecting the optimal threshold for the separation of abnormal and normal ECG signals is proposed. A comparative analysis of the effectiveness of various machine learning models and the presented autoencoder was carried out using various performance indicators, such as accuracy, recall, precision and f1-score metrics, which showed the superiority of the proposed model for detecting anomalies in electrocardiograms.

2 Unsupervised Deep Learning Methods for Anomaly Detection

The autoencoder model consists of an encoder, a hidden layer of input data representation (latent representation) and a decoder. High-dimensional input data are transformed by the encoder into hidden representations of low-dimensional data. The dimension of the hidden representations is smaller than the one of the incoming source data. The task of the decoder is to recover the input data. The autoencoder accepts high-dimensional input data, compresses it to a representation in the space of a hidden layer. The decoder then takes the hidden representation of the data as input to restore the original data. In the end, the autoencoder outputs the recovered image or signal. An example of autoencoder model is presented on Fig. 1.

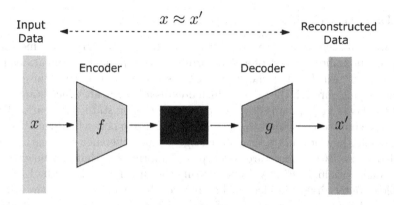

Fig. 1. Autoencoder model

The use of autoencoders makes it possible to generate data similar to the original, to train the encoder network, to highlight the most important features of the input image, to extract these features from the hidden space of the autoencoder and reconstruct images. The use of autoencoders for anomaly detection has a number of advantages, since it is not required to know the labels of all samples. However, hidden representations can be distorted by the presence of outliers in the training data. The objective function of data recovery is designed to reduce the size or compression of data, and not to perform a classification task. In this regard, it is necessary to formulate criteria for separating normal data and anomalies in dataset. There are various approaches to do this, for example, the Mahalanobis metric, the Local Outlier Factor method, etc., both which have advantages and disadvantages, which can be found in more detail in [4].

The paper proposed the following algorithm for separating normal and abnormal observations in dataset. For each observation and its reconstructed form the reconstruction error function (REF) is calculated as follows:

$$REF = \sqrt{\frac{1}{M} \sum_{i=1}^{M} (x_i - \check{x}_i)^2}, \qquad (1)$$

where x_i, $\check{x}_i$ are the characteristics of real and reconstructed observations, respectively, M is the number of analysed observations. An abnormal observation differs from a normal one in that the error of its recovery is greater. So, the images are sorted by REF (1) and are assigned to anomalies group based on a predefined threshold using maximum of REF values. The problem of optimal choice of the threshold was also investigated.

3   Performance Study of the Autoencoder Model for Anomalies Detection in the Electrocardiograms

3.1   Data Description

Computational experiments were carried out to test the proposed method of detecting anomalies on a set of electrocardiograms of patients with various heart diseases [8,16]. An electrocardiogram (ECG) is a tool for visualizing the electrical current passing through the heart, which creates a heartbeat, which starts at the top of the heart and spreads downwards. In the resting state, the heart cells are negatively charged compared to the external environment, while when they are depolarized, they become positively charged. The difference in polarization is recorded by an ECG. There are two types of information that can be extracted by ECG analysis [6,7]. Firstly, by measuring the time intervals on the ECG, it is possible to detect irregular electrical activity of the heart. Secondly, the strength of electrical activity gives an idea of areas of the heart that are overloaded or stressed. The data set under study was created and balanced in such a way that it contains 5,000 electrocardiogram records, of which the proportion of normal signals is 60%, the proportion of abnormal signals is 40%. Each line corresponds to one complete ECG record of a patient. Each electrocardiogram consists of 140 measurements and has a label indicating whether the ECG signal is normal or abnormal. This is a categorical variable with a value of 0 or 1 (normal ECG corresponds to label 0, abnormal ECG corresponds to label 1). During the preprocessing, the ECG signals were normalized as follows:

$$data = \frac{(data - data_{min})}{(data_{max} - data_{min})}, \qquad (2)$$

where $data_{min}$ is the minimum value of the observations, $data_{max}$ is the maximum value of the observations.

Further, all data paints were splitted into two sets: training $data\_train$ and test $data\_test$. The size of the training set was 4,000 ECG-signals, the test set was 1,000 ECG-signals. An example of a normal ECG signal is shown in Fig. 2. An example of an abnormal ECG signal is shown in Fig. 3.

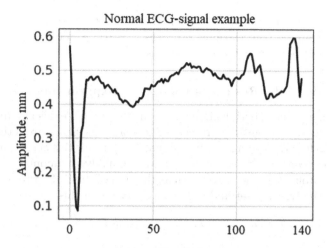

Fig. 2. Normal ECG-signal.

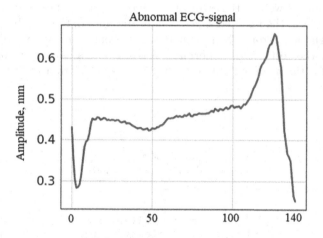

Fig. 3. Anomalous ECG-signal.

3.2   Deep Autoencoder Model

To detect abnormal ECG signals an autoencoder model based on deep neural networks was proposed. The autoencoder model was implemented in the Python programming language using the Keras framework [17]. The encoder consists of 5 fully connected layers: $Dense(128)$, $Dense(64)$, $Dense(32)$, $Dense(16)$, $Dense(8)$, each has ReLU activation function. The decoder consists of five fully connected layers: $Dense(8)$, $Dense(16)$, $Dense(32)$, $Dense(64)$, $Dense(128)$ with ReLU activation function and one fully connected layer $Dense(140)$ with sigmoid activation function. The loss function of signal reconstruction is given by RMS error between the original image and the image processed by the neural network (1). The Adam optimization method and the MSE loss function were used during training, the learning rate was $1E-04$. A total of 500 epochs of model training were conducted with *batch_size* parameter of 8.

3.3   Analysis of the Results of Computer Experiments

During the computer experiments for autoencoder model efficiency testing, the autoencoder model was initially trained on normal ECG signals from the training set in order to find the average error of signal recovery for this class and establish a threshold for cutting off signals that are, presumably, an anomaly. The initial value of the cutoff threshold was chosen as the sum of the RMS error of restoring signals from the training set and the RMS deviation of this error:

$$h = MSE(data\_train) + std(data\_train).$$

As a result of calculations, the average recovery error for signals from the training set was 0.00758. At the same time, the threshold was equal to h of 0.0116. Figure 4 shows the plot of normal ECG signals from the training set, the reconstructed signal, as well as the error of its reconstruction. The original signal is indicated in black, the restored signal is indicated in red, the area between them is a reconstruction error. Then, similar calculations were performed for ECG signals from the test set of abnormal electrocardiograms. The reconstruction error value for them turned out to be 0.035. The calculation results are shown in Fig. 4.

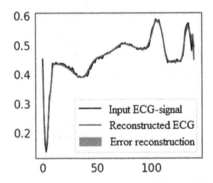

Fig. 4. Example of recovered ECG signals from the training set of normal ECG signals. (Color figure online)

If the recovery error for the tested signal exceeded one standard deviation from the normal signal, then such a signal was classified as abnormal. Comparing the results, we can see that the error of restoring abnormal signals exceeds the error of restoring normal signals. Figure 6 shows empirical density plots of the distribution of the reconstruction error (I) and its nonparametric kernel estimation by KDE method for normal ECG signals from the training dataset (II). It

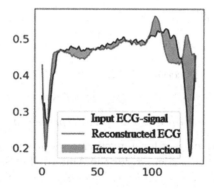

Fig. 5. Example of recovered ECG signals from the test set of abnormal ECG signals.

also shows the empirical density of the distribution of the reconstruction error (III) and its nonparametric kernel evaluation by the KDE method for abnormal ECG signals from the test dataset (IV). As can be seen from Fig. 6, when the initially set threshold h is 0.0116, a significant number of normal signals are mistakenly classified as abnormal, which allows us to use the selected low value of the threshold for cutting off abnormal signals.

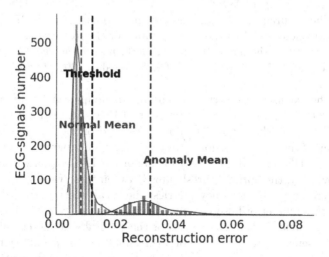

Fig. 6. Empirical distribution of the reconstruction error for normal signals and abnormal signals from the test set at a threshold value h of 0.0116.

To compare the results of implementation of the proposed autoencoder model with different threshold values and build an algorithm for its optimal selection, let's use machine learning performance metrics such as accuracy, precision, recall, f1-score defined below [9]:

$$accuracy = (TP + TN)/(TP + TN + FP + FN), \qquad (3)$$

$$precision = TP/(TP + FP), \qquad (4)$$

$$recall = TP/(TP + FN), \qquad (5)$$

$$f1 - score = 2 * precision * recall/(precision + recall), \qquad (6)$$

where TP – true positive cases for a given class (Normal), FN – false negative cases for a given class (in particular, for the Abnormal class, this is the number of anomalies mistakenly predicted by the model as non-anomalies), TN – true negative cases (Abnormal class), FP – false positive cases for this class (in particular, for the Normal class, this is the number of anomalies falsely predicted by the model as Normal). To calculate the metrics (3)–(6), the confusion matrix is used [20]. An example of its structure is shown in Fig. 9. Diagonal cells denote true positive values (TP, %) and true negative values (TN, %). When defined, the percentage values in the cells indicate the values of the accuracy indicator for the corresponding class. The remaining cells of the matrix contain the proportion of images that are evaluated by the model as false positive (vertical elements, FP, %) and false negative (horizontal elements, FN, %) images.

With the initially set threshold h of 0.0116, the performance metrics are the following: accuracy is 93.8%, precision is 99.6%, recall is 89.28%, f1-score is 94.16%. The accuracy of classification of the anomaly class is 87.86%. When choosing a threshold equal to the average value of the loss function on the test set of abnormal ECG-signals as 0.035, the performance metrics (3)–(6) are the following: accuracy is 75.2%, precision is 69.3%, recall is 100%, f1-score is 81.87%, AUC is 71.73%.

Plots of the reconstruction error (green) and its nonparametric kernel estimation using KDE method (green line) for normal ECG-signals from the training dataset with the threshold h of 0.035 are presented on Fig. 7. We can also see the histogram of the reconstruction error (red) and its nonparametric kernel estimation using KDE method (red line) for abnormal ECG signals from the test dataset. Obviously, the optimal threshold is between these values, because along with the decrease of the accuracy metrics, the recall metrics increased, which means that the number of false negative cases increased.

Thus, it is obvious that the metrics of the accuracy of the classification of anomalies are sensitive to the choice of a threshold, so it is necessary to develop an algorithm for its selection. To automate the selection of the optimal value of the anomaly clipping threshold, it is proposed to use methods of fine-tuning hyperparameters of the autoencoder model. The Keras-Tuner library in Tensorflow environment was used for software implementation [10]. Fine-tuning of the autoencoder parameters showed that the optimal cut-off threshold was reached at the following values of the hyperparameters: *batch_size* of 8, *epochs* of 500,

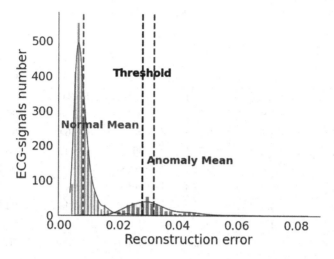

Fig. 7. Empirical distribution of the reconstruction error for normal signals and abnormal signals from the test set at the threshold value h of 0.035. (Color figure online)

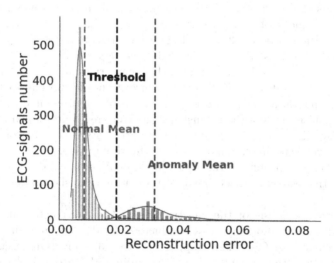

Fig. 8. The results of classification of normal and abnormal ECG signals from the test set of ECG signals at the optimal threshold value h of 0.023. (Color figure online)

learning rate lr of 1e−04. In computer experiments, the corresponding metrics (3)–(6) were calculated for the values of the cut-off threshold from the interval [Normal Mean, Anomaly Mean]. Their values are given in Table 1.

As follows from the Table 1, with an increase in h, precision decreases, but at the same time recall increases, due to False Negative values (4). Analysis of Table 1 shows that at a low threshold, the autoencoder model revealed a large

Table 1. .

Threshold, h	Accuracy, %	Precision, %	Recall, %	f1-score, %	Accuracy_Anomaly, %
0.0116	93.8	99.6	89.285	94.16	94.44
0.02	97.1	99.44	95.35	97.245	97
0.022	97.8	97.35	97.14	97.35	97.35
0.0225	98.7	98.75	98.92	98.64	98.66
0.023	98.8	98.7544	99.1	98.93	98.75
0.024	98.6	98.23	99.28	98.5	98.23
0.025	97.8	96.86	99.28	98	95.03
0.026	95.8	93.75	99.33	96.45	93.75
0.028	94.4	91.58	99.48	95.36	91.58
0.035	75	69.268	99.821	81.784	71.728

number of false anomalies (False Negative, FN), which in reality are normal ECG-signals. At a high threshold, the model revealed a large number of false normal signals (False Positive, FP). The balance of these values is reached at the threshold value h of 0.023, where the corresponding metrics are: accuracy of 98.8%, precision of 98.7544%, recall of 99.1%, f1-score of 98.93%. It can also be noted that with this value of threshold, *accuracy_anomaly* also reached its optimal value. Graphs of the histogram of the reconstruction error (green) and its nonparametric estimation by KDE method (green line) for normal ECG-signals from the training dataset when threshold h is 0.023. are presented on Fig. 8. It also plots the histogram of the reconstruction error (red) and its nonparametric estimation by KDE method (red line) for abnormal ECG signals from the test dataset. The vertical blue line indicates the threshold value.

On Fig. 9 the confusion matrix of classification of normal and abnormal ECG signals from the test set of ECG signals at the optimal threshold value h of 0.023 is shown.

Comparative analysis of the effectiveness of the constructed autoencoder model for detecting anomalies was conducted. To do this, machine learning methods such as the One-Class Support Vector Machine (One Class SVM) (with hyperparameters $kernel =' rbf'$, $gamma =' auto'$), Isolated Forest (IF) algorithm (with hyperparameters $max\_features = 100$, $max\_samples =' auto'$, $n\_estimators = 100$) and Random Forest algorithm (with parameters $n\_estimators = 100$, $criterion =' gini'$, $max\_depth =' None'$) were used. Additionally, an autoencoder based on recurrent neural network LSTM was constructed [19]. Table 2 shows the main results of these computations. For the RF method, the anomaly detection accuracy was 81.4%, for One-class SVM accuracy is 78.47%, the recurrent autoencoder provided accuracy of 92%. The results of the analysis allow us to assert that the proposed autoencoder model is supe-

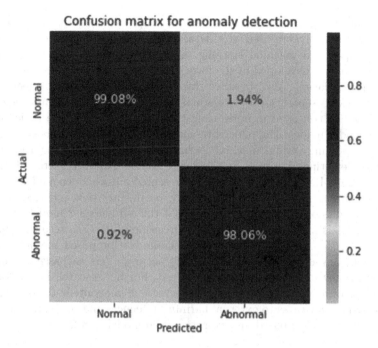

Fig. 9. Confusion matrix of classification of normal and abnormal ECG signals from the test set of ECG signals at the optimal threshold value h of 0.023.

rior to other models. In addition, it works much faster in comparison with the LSTM-based autoencoder.

Table 2. Metrics of statistical anomalies, where U (unsupervised) – unsupervised machine learning, S (supervised) – supervised machine learning.

Model	U/S	Accuracy, %	Precision, %	Recall, %	F1-score, %
Isolation Forest (IF)	U	57.3	64.72		
Oneclass SVM	U	42.26	41.32	91.3	56.3
Random Forest (RF)	S	81.4			
Deep autoencoder	U	98.8	95.75	99.12	98.75
LSTM-AE	U	91.36	94.72	84.57	93.16

4   Conclusion

Methods of anomaly detection using methods of unsupervised machine and deep learning were investigated in the work. A set of electrocardiograms containing both normal ECG signals and ECG signals of people with various cardiovascular diseases (anomalies) were used as a dataset. To detect abnormal electrocardiograms, unsupervised machine learning methods were used and an automatic encoder model was developed in the form of a deep neural network with several fully connected layers. Also, to solve this problem, a method for selecting the threshold for the separation of abnormal ECG signals from normal ones was proposed, which consisted in optimizing the ratio of performance indicators of the autoencoder model. The paper presents a comparative analysis of the effectiveness of applying various machine learning models to find the solution of the problem of detecting abnormal ECG signals, such as Support Vector Machines, Isolation Forest, Random Forest and the presented autoencoder model. For this purpose, accuracy metrics such as accuracy, recall, precision and f1-score were used. The results of the analysis showed that the autoencoder model surpassed the other models with accuracy of 98.8% precision of 95.75%, recall of 99.12%, f1-score of 98.75%. If class labels are available, the proposed method can be used as a binary classification method. Its accuracy has surpassed results of classification methods such as Random Forest. Further research in the form of development of new methods of anomaly detection can contribute to the study of the model's robustness to external harmful influences and to the development of autoencoder models based on generative deep models [18, 20].

5 Program Code

Autoencoder model in Keras

```
class AnomalyDetector(Model):
  def __init__(self):
    super(AnomalyDetector, self).__init__()
    self.encoder = tf.keras.Sequential([
      layers.Dense(128, activation='relu'),
      layers.Dense(64, activation='relu'),
      layers.Dense(32, activation="relu"),
      layers.Dense(16, activation="relu"),
      layers.Dense(8, activation="relu")])

    self.decoder = tf.keras.Sequential([
      layers.Dense(8, activation='relu'),
      layers.Dense(16, activation="relu"),
      layers.Dense(32, activation="relu"),
      layers.Dense(64, activation='relu'),
      layers.Dense(128, activation='relu'),
      layers.Dense(140, activation="sigmoid")])

  def call(self, x):
    encoded = self.encoder(x)
```

```
        decoded = self.decoder(encoded)
        return decoded

autoencoder = AnomalyDetector()
```

Prediction anomaly class with autoencoder model

```
reconstructions = autoencoder.predict(anomalous_test_data)
test_loss = tf.keras.losses.mae(reconstructions, anomalous_test_data)

def predict(model, data, threshold):
  reconstructions = model(data)
  loss = tf.keras.losses.mae(reconstructions, data)
  return tf.math.less(loss, threshold)

def print_stats(predictions, labels):
  print("Accuracy = {}".format(accuracy_score(labels, preds)))
  print("Precision = {}".format(precision_score(labels, preds)))
  print("Recall = {}".format(recall_score(labels, preds)))
  print("F1-score = {}".format(f1_score(labels, preds)))
  print("AUC-score= {}".format(roc_auc_score(labels, preds)))
  CLASSES=['Normal','Abnormal']
  print(classification_report(labels, preds, target_names=CLASSES))

preds = predict(autoencoder, test_data, threshold)
print_stats(preds, test_labels)
```

LSTM-based autoencoder modell

```
timesteps = 140
input_dim = 1
inputs = Input(shape=(timesteps, input_dim))
encoded = LSTM(8)(inputs)
#encoded = LSTM(4)(encoded)
#this RepeatVector is the only that is needed
# to restore the sequence length
encoded = RepeatVector(timesteps)(encoded)
encoded = LSTM(2)(encoded)
encoded = RepeatVector(timesteps)(encoded)
decoded = LSTM(4,return_sequences = True)(encoded)
decoded = LSTM(input_dim,return_sequences = True)(decoded)

LSTM_autoencoder = Model(inputs, decoded)
```

References

1. Beam, A.L., Kohane, I.S.: Big data and machine learning in health care. JAMA **319**(13), 1317–1318 (2018). https://doi.org/10.1001/jama.2017.18391
2. Boukerche, A., Zheng, L., Alfandi, O.: Outlier detection: methods, models, and classification. ACM Comput. Surv. **53**(3) (2020). https://doi.org/10.1145/3381028

3. Candès, E.J., Li, X., Ma, Y., Wright, J.: Robust principal component analysis? J. ACM **58**(3) (2011). https://doi.org/10.1145/1970392.1970395
4. Chalapathy, R., Chawla, S.: Deep learning for anomaly detection: a survey (2019). https://doi.org/10.48550/arxiv.1901.03407. https://arxiv.org/abs/1901.03407
5. Chandola, V., Banerjee, A., Kumar, V.: Anomaly detection: a survey. ACM Comput. Surv. **41**(3) (2009). https://doi.org/10.1145/1541880.1541882
6. Clifford, G.D., et al.: Classification of normal/abnormal heart sound recordings: the PhysioNet/computing in cardiology challenge 2016. In: 2016 Computing in Cardiology Conference (CinC), pp. 609–612 (2016)
7. Ebrahimi, Z., Loni, M., Daneshtalab, M., Gharehbaghi, A.: A review on deep learning methods for ECG arrhythmia classification. Expert Syst. Appl.: X **7**, 100033 (2020). https://doi.org/10.1016/j.eswax.2020.100033. https://www.sciencedirect.com/science/article/pii/S2590188520300123
8. Goldberger, A., et al.: Components of a new research resource for complex physiologic signals. Circulation **101**(23), E215–E220 (2000)
9. Liu, F.T., Ting, K.M., Zhou, Z.H.: Isolation forest. In: 2008 Eighth IEEE International Conference on Data Mining, pp. 413–422 (2008). https://doi.org/10.1109/ICDM.2008.17
10. O'Malley, T., et al.: KerasTuner (2019). https://github.com/keras-team/keras-tuner
11. Pang, G., Shen, C., Cao, L., Hengel, A.V.D.: Deep learning for anomaly detection: a review. ACM Comput. Surv. **54**(2) (2021). https://doi.org/10.1145/3439950
12. Patel, A.: Hands-On Unsupervised Learning Using Python: How to Build Applied Machine Learning Solutions from Unlabeled Data. O'Reilly Media (2019)
13. Pimentel, M.A., Clifton, D.A., Clifton, L., Tarassenko, L.: A review of novelty detection. Sig. Process. **99**, 215–249 (2014). https://doi.org/10.1016/j.sigpro.2013.12.026. https://www.sciencedirect.com/science/article/pii/S016516841300515X
14. Pol, A.A., Berger, V., Germain, C., Cerminara, G., Pierini, M.: Anomaly detection with conditional variational autoencoders. In: 2019 18th IEEE International Conference On Machine Learning And Applications (ICMLA), pp. 1651–1657 (2019). https://doi.org/10.1109/ICMLA.2019.00270
15. Ruff, L., et al.: A unifying review of deep and shallow anomaly detection. Proc. IEEE **109**(5), 756–795 (2021). https://doi.org/10.1109/JPROC.2021.3052449
16. Shchetinin, E.Y., Sevastianov, L.A., Demidova, A.V., Glushkova, A.G.: Cardiac arrhythmia disorders detection with deep learning models. In: Vishnevskiy, V.M., Samouylov, K.E., Kozyrev, D.V. (eds.) DCCN 2021. CCIS, vol. 1552, pp. 371–384. Springer, Cham (2022). https://doi.org/10.1007/978-3-030-97110-6_29
17. Thudumu, S., Branch, P., Jin, J., Singh, J.J.: A comprehensive survey of anomaly detection techniques for high dimensional big data. J. Big Data **7**(1), 1–30 (2020). https://doi.org/10.1186/s40537-020-00320-x
18. Yang, Z., Bozchalooi, I.S., Darve, E.: Regularized cycle consistent generative adversarial network for anomaly detection (2020). https://doi.org/10.48550/ARXIV.2001.06591. https://arxiv.org/abs/2001.06591
19. Yu, Y., Si, X., Hu, C., Zhang, J.: A review of recurrent neural networks: LSTM cells and network architectures. Neural Comput. **31**(7), 1235–1270 (2019). https://doi.org/10.1162/neco_a_01199
20. Zenati, H., Foo, C.S., Lecouat, B., Manek, G., Chandrasekhar, V.R.: Efficient GAN-Based Anomaly Detection (2018). https://doi.org/10.48550/arxiv.1802.06222. https://arxiv.org/abs/1802.06222

21. Zhao, Z., et al.: Robust anomaly detection on unreliable data. In: 2019 49th Annual IEEE/IFIP International Conference on Dependable Systems and Networks (DSN), pp. 630–637 (2019). https://doi.org/10.1109/DSN.2019.00068
22. Zhou, C., Paffenroth, R.C.: Anomaly detection with robust deep autoencoders. In: Proceedings of the 23rd ACM SIGKDD International Conference on Knowledge Discovery and Data Mining, KDD 2017, pp. 665–674. Association for Computing Machinery, New York (2017). https://doi.org/10.1145/3097983.3098052

Using Neural Networks for Channel Quality Prediction in Wireless 5G Networks

Ekaterina Bobrikova[1], Anna Platonova[1(✉)], Ekaterina Medvedeva[1,2], Yu. V. Gaidamaka[1,2], and Sergey Shorgin[2]

[1] Peoples' Friendship University of Russia (RUDN University), 6 Miklukho-Maklaya Street, Moscow 117198, Russian Federation
platonova_aa@pfur.ru
[2] Federal Research Center "Computer Science and Control" of the Russian Academy of Sciences (FRC CSC RAS), 44-2 Vavilov Street, Moscow 119333, Russian Federation
https://www.rudn.ru, https://www.frccsc.ru

Abstract. The article proposes a method for assigning a modulation coding scheme (MCS) by a base station (BS) scheduler on an unmanned aerial vehicle (UAV), based on predicting the value of the signal-to-interference-to-noise ratio (SINR) on the mobile user equipment (UE) at the next time slot from a sequence of known values of this ratio in the past. Prediction is performed using machine learning. For this, a neural network was built and applied to solve the problem of multi-parameter optimization using the stochastic gradient method. The trained neural network for the predicted SINR value allows the scheduler to select the modulation-code scheme correctly, thereby ensuring the level of data transmission quality in the radio channel necessary to provide the service.

Keywords: SINR · Machine learning · Neural network

1 Introduction

One of the main tasks that the scheduler of the BS of UAVs of LTE and LTE-A generation 5G wireless communication networks should solve is the task of selecting the MCS when transmitting data in the radio channel between the base station/access point and the user. The correct assignment of the MCS allows avoiding data retransmission, which leads to inefficient use of the resources of the receiving-transmitting equipment and reduces the quality of the user's service, in particular, increases the delay in providing the service. The task of predicting SINR becomes especially relevant due to the limited capacity of the UAV battery, when repeated requests due to incorrect MCS assignment at the next cycle reduce

The reported study was funded by RSF, project number 22-29-00694, https://rscf.ru/en/project/22-29-00694.

the UAV lifetime. In the case of stationary users, the task of selecting the MCS is not difficult, since the MCS is determined by the channel quality indicator (CQI) value, which the BS scheduler periodically receives from the UE [1]. CQI depends on the SINR value on the user's equipment, which is affected by the radio access technology, the distance between transceivers, the radio signal propagation environment, including the presence of obstacles and signal blockers, the power of interfering transmitters, etc. Due to the mobile user's movement, its geolocation of UE, the distance to the BS and the radio channel parameters change. The receiver regularly captures and sends the BS a new CQI value, according to which the BS scheduler sets up the corresponding MCS for transmission from the BS. This should allow the scheduler to avoid the inefficient default MCS selection scheme when after receiving an unsatisfactory CQI, the BS scheduler traverses successively MCSs until it receives a CQI value from the UE that is acceptable for data transmission with adequate quality. The predicting correct MCS on the BS side instead of traversing step by step consequent MCSs can substantially decrease the delay and overhead due signaling. The arised problem of predicting the CQI value of a mobile user can be solved in two stages. First, prediction of the SINR value on the mobile user's equipment based on the known values of this ratio in the past, and second, the MCS assignment according to the SINR obtained at the first stage. The paper proposes an approach to prediction of the SINR using the apparatus of neural networks [2].

The paper is organized as follows. In Sect. 2 the overview on the related works is presented. In Sect. 3 the system model is described and the mathematical apparatus is presented. In Sect. 4 the results of the numerical experiment are discussed. The conclusions and further research tasks are presented in the last section.

2   Related Work

Let's consider several works [3–8] on integrating machine learning with wireless networks.

The aim of [3] is to use machine learning algorithms for prediction the signal-to-noise ratio (SNR). For SNR prediction for downlink channel estimation in 5G system, the authors proposed the following methods: support vector machine (SVM), stochastic gradient descent (SGD), and multilayer perceptron (MLP). The authors used time and SNR as features while the target they considered was CQI, and for part of the experiment they also added SNR as a target. The machine learning performance measure they considered is improving the accuracy of SNR predictions, and network performance is measured in terms of reducing errors in SNR calculation.

The paper [4] discusses the problems and possibilities of radio resource management (RRM) based on machine learning. The authors considered two different approaches to machine learning, namely reinforcement learning (RL) and artificial neural networks (ANN). An architecture for training and studying machine learning models for 5G networks has also been proposed. The authors focused on predicting various RRM parameters such as power control, cell, bandwidth, etc.

The work [5] provides a detailed guide on the use of machine learning based on artificial neural networks for the implementation of various applications in wireless networks. In particular, the authors presented an overview of a number of key types of neural networks, such as recurrent, peaking, and deep neural networks. The authors also investigated the issue of spectrum management and multiple access technologies. The authors used the ANN model for resource management. This model also has the ability to switch between different frequency bands, which is a key feature of 5G networks.

The work [6] considers the functionality of the upper layers of the Open System International (OSI) model. The authors proposed a machine learning method based on double state-action-reward-state-action (SARSA) to improve the quality of experience (QoE) of a user in an IP network. With this model, they were able to improve the peak signal-to-noise ratio by up to 7% and improve video quality scores by up to 25%. The performance of the model was tested in a live video streaming session.

The paper [7] compares various machine learning models for analyzing cellular network traffic. It also addresses two different and highly relevant issues related to end users and applications running on their smartphones: anomaly detection generated by smartphone applications and QoE prediction for popular applications. The Ensemble Learning (EL) model, which combines six different algorithms, is preferred. The algorithms in the ensemble are decision trees, Naive Bayes, MLP, SVM, random forests (RF), and K nearest neighbors (K-NN). The proposed models are evaluated using real measurements of cellular traffic obtained in live networks and end devices. The results show that decision tree-based models are the most accurate for solving these problems, and joint models, in particular overlay models, can significantly improve performance and reliability.

Reference [8] explores an artificial intelligence implementation paradigm to reduce resource usage at the sounding reference signal (SRS) physical layer. The authors proposed a new uplink SINR prediction method on SRS using an ANN based scheme. This research has the potential value to improve the most valuable aspects of RRM.

3   System Model and Mathematical Problem Statement

One of the most popular machine learning methods is the apparatus of neural network. The simplest single-layer neural network - a single-layer perceptron [9] - will be used in the paper for predicting the SINR value of a mobile user based on known SINR values in the past. A system model is built for one cell of a wireless communication network, divided into squares, each with its own indicator, one of cells houses a BS. We consider the user's movement model to be known - the data set of movement trajectories is either specified from observations, or generated using an analytical movement model, for example, in the form of a

Grid Random Walk model in a discrete time. The user's trajectory is the final sequence of square cell indicators in which the user was on the corresponding time step.

We believe that the square cell indicator uniquely determines the distance between the BS and the UE. Then the user's movement trajectory can be set by a sequence of MCS values, or by a sequence of CQI values for the UE at adjacent time steps, or even by the SINR value, which uniquely determines the CQI value. In machine learning terminology, we can say that the approach using a neural network allows for a given training set - the user's movement trajectory - to predict the answer, i.e. the square cell indicator at the next time step. As a mathematical model for such a prediction, a single-layer neural network was chosen, which, as a result of solving the problem of multi-parameter optimization based on the data set given at the input, is trained to output an approximate answer that meets the specified criteria.

In the course of the research, it was found that greater accuracy of predictions is achieved when choosing as precedents of the training set not the indicators of square cells, but the SINR values. This is illustrated in the next sections when constructing a mathematical model, as well as in a numerical experiment.

Let the i-th object $\mathbf{x}_i$ be a vector with n sequential timeslots as components, l be the number of objects, $i = 1, ..., l$. Let's introduce the features of the object $\mathbf{x}_i$: $f_1(\mathbf{x}_i), ..., f_n(\mathbf{x}_i)$ are known SINR values corresponding to n sequential timeslots of the object $\mathbf{x}_i$, i.e. n is the number of features. Let's denote $\mathbf{f}(\mathbf{x}_i) = (f_1(\mathbf{x}_i), ..., f_n(\mathbf{x}_i))$. Let's consider the known SINR value in the next fifth timeslot as the answer y_i. Note that the number of answers is equal to the number of objects, the pair object and the answer $(\mathbf{x}_i, y_i)$ is called a precedent, the set of pairs $X^l = (\mathbf{x}_i, y_i)_{i=1}^l$ - training set.

As an approximating algorithm, we choose a linear model

$$a(\mathbf{x}, \mathbf{w}) = \langle \mathbf{w}, \mathbf{f}(\mathbf{x}) \rangle = \sum_{j=1}^{n} w_j f_j(\mathbf{x}), \tag{1}$$

where $\mathbf{x}$ is an object, n is the number of features, w_j are unknown feature weights, $j = 1, ..., n$.

The process of selecting the optimal parameter $\mathbf{w}$ based on the training set X^l is called the learning the a algorithm. The optimal parameter $\mathbf{w}$ of the model is the parameter that provides the minimum value to the quality functional $Q(a, X^l)$ of the a algorithm on the training set X^l, i.e.

$$Q(a, X^l) = \frac{1}{l} \sum_{i=1}^{n} \mathcal{L}(a, y_i) \to \min_{\mathbf{w}}. \tag{2}$$

Here the function

$$\mathcal{L}(a, y_i) = (a(\mathbf{x}_i, \mathbf{w}) - y_i)^2 \tag{3}$$

is called the loss function and reflects the accuracy of the approximation on the object $\mathbf{x}_i$.

For the numerical optimization of functional (2), the work uses the stochastic gradient method [2, 10].

The described process of selecting the optimal parameter $\mathbf{w}$ based on the training set $X^l = (\mathbf{x}_i, y_i)_{i=1}^l$ is the operation of the simplest single-layer neural network. For the object $\mathbf{x}_i$, the result of applying the linear function $a(\mathbf{x}_i, \mathbf{w})$ is called the predicted SINR value on the object $\mathbf{x}_i$ or simply the prediction on the object $\mathbf{x}_i$. Note that $a(\mathbf{x}_i, \mathbf{w})$ is an approximation of the known value y_i of the unknown function $y = y(\mathbf{x})$ on object $\mathbf{x}_i$.

4 An Example of Using the Neural Network Apparatus for Predicting SINR

An illustration of a neural network's operation for predicting SINR in a cell of an LTE network was carried out for an example of sampling trajectories of user's movement. We considered 2 cases: when there is 1 BS in the coverage area and when there are 3 BSs. For the BS coverage area, the Monte Carlo method generated 106 trajectories according to the Grid Random Walk motion model [11] in the form of square cell indicator's sequences of different lengths. Further, for the LTE technology, according to [1, 11], from the trajectories, the corresponding sequences of the values SINR were obtained, and from them - "fives", i.e. precedents which constituted a training set of length $l = 29669$ for the first case and $l = 60872$ for the second case. The precedents that make up the training set are "object-answer" pairs, where the object is a set of four sequential timeslots, a vector of features of length $n = 4$ is given for each object, and the features and answers are the SINR values.

For example, a sequence of the SINR values

$$\dots 11.02 \to 10.93 \to 10.85 \to 10.44 \to 10.49 \to 10.14 \to 10.11 \to 9.98 \dots$$

gives a training set with the following "object - answer" pairs:

$(11.02, 10.93, 10.85, 10.44) \to 10.49;$ $(10.93, 10.85, 10.44, 10.49) \to 10.14;$
$(10.85, 10.44, 10.49, 10.14) \to 10.11;$ $(10.44, 10.49, 10.14, 10.11) \to 9.98$ etc.

The initial values of the weights of features for the linear model (1) are chosen $\mathbf{w} = (0.25,\ 0.25,\ 0.25,\ 0.25)$, the gradient step (learning rate) $h = 10^{-4}$ [12].

Let us consider the prediction results from information about the SINR values of the user in 4 consecutive timeslots of the SINR value on the user's equipment in the next 5th timeslot using the constructed neural network.

Case 1. Consider case with 1 BS in the coverage area. Figure 1 and Fig. 2 show the black dotted line the predictions $a(\mathbf{x}_i, \mathbf{w})$, the gray solid line is the answers y_i. For clarity, in Fig. 2 shows the predictions and answers for the first 200 objects in the training set.

Figure 3 and Fig. 4 show the change in the loss function $\mathcal{L}(a, y_i)$ for all objects of the training set and for the first 200 objects, respectively.

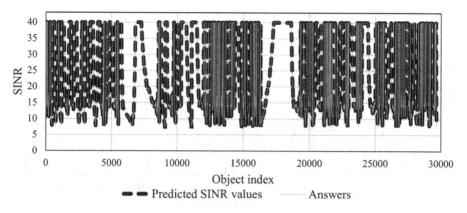

Fig. 1. Case 1: predicted SINR values and answers for all objects in the training set.

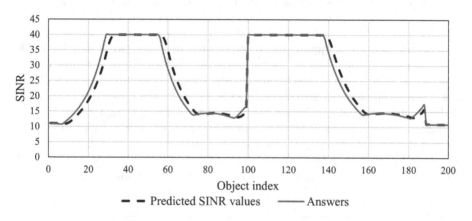

Fig. 2. Case 1: predicted SINR values and answers for the first 200 object in the training set.

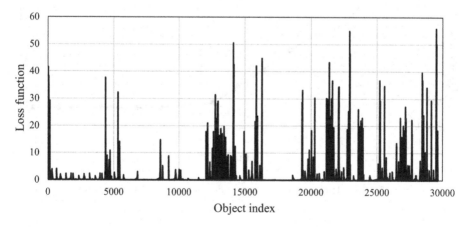

Fig. 3. Case 1: loss function for all objects in the training set.

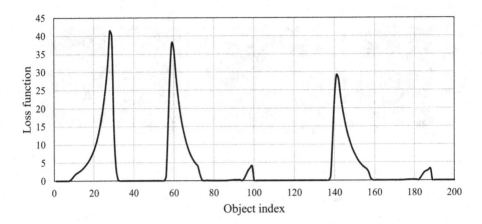

Fig. 4. Case 1: loss function for the first 200 objects in the training set.

The criterion for stopping the training of a neural network can be stabilization of one of the following model parameters - weights, quality functional, accuracy. It took 15 epochs to train, and the maximum accuracy equal to 0.866 was achieved already at the 10th epoch (Fig. 6). Here, the epoch is one iteration of the stochastic gradient method for all objects of the training set.

The coincidence of the predicted values and answers in Fig. 1 and Fig. 2 also confirm the prediction quality indicators: the quality functional (2) (Fig. 5) and the accuracy (Fig. 6), defined as the ratio of the number of coincident predicted values and answers to the total number of objects in the training set.

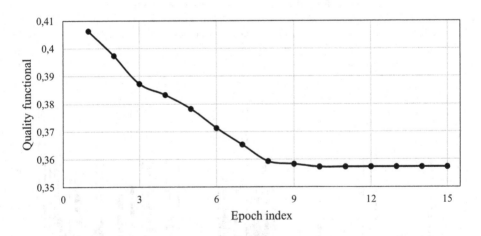

Fig. 5. Case 1: changing the assessment of the quality functional.

On Fig. 6 we considered for the precision option: 1) $\mathcal{L}\left(a, y_i\right) < 0.1$ - gray line. 2) $\mathcal{L}\left(a, y_i\right) < 1$ - black line.

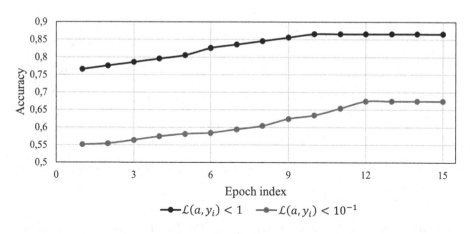

Fig. 6. Case 1: accuracy.

Case 2. Consider case with 3 BSs in the coverage area. Figure 7 and Fig. 8 show the black dotted line the predictions $a\left(\mathbf{x}_i, \mathbf{w}\right)$, the gray solid line is the answers y_i. For clarity, in Fig. 8 shows the predictions and answers for the first 250 objects in the training set. If we compare Fig. 1 and Fig. 7, it can be seen that in the second case, the SINR values reach a maximum of about 20 dB. This is due to the fact that with an increase in the number of BSs in the coverage area, interference occurs. Thus, the SINR values cannot reach the maximum, as in the first case.

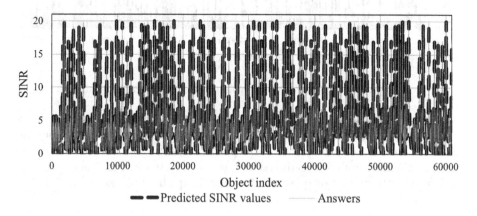

Fig. 7. Case 2: predicted SINR values and answers for all objects in the training set.

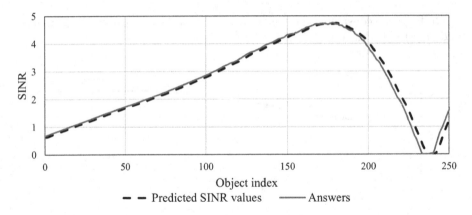

Fig. 8. Case 2: predicted SINR values and answers for the first 250 objects in the training set.

Figure 9 and Fig. 10 show the change in the loss function $\mathcal{L}(a, y_i)$ for all objects of the training set and for the first 250 objects, respectively.

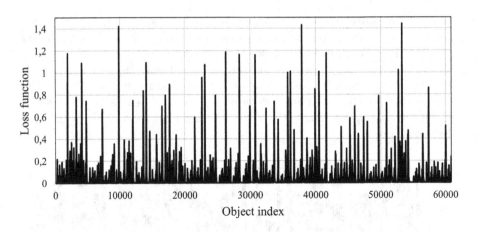

Fig. 9. Case 2: loss function for all objects in the training set.

In this case, training also took 15 epochs, and the maximum accuracy of 0.998 was already achieved at the 11th epoch (Fig. 12).

The coincidence of the predicted values and answers in Fig. 7 and Fig. 8 is confirmed by the prediction quality indicators: the quality functional (2) (Fig. 11) and the accuracy (Fig. 12), defined as the ratio of the number of matching predicted values and the answers to the total number of objects in the training set.

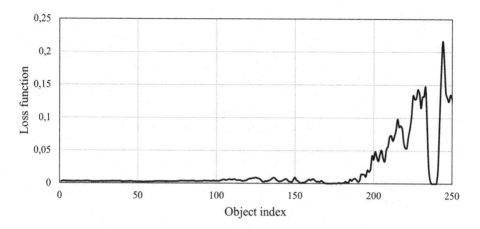

Fig. 10. Case 2: loss function for the first 250 objects in the training set.

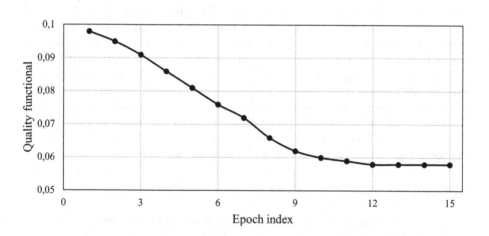

Fig. 11. Case 2: changing the assessment of the quality functional.

On Fig. 12, we considered the accuracy option: 1) $\mathcal{L}(a, y_i) < 0.1$ - gray line; 2) $\mathcal{L}(a, y_i) < 1$ is the black line.

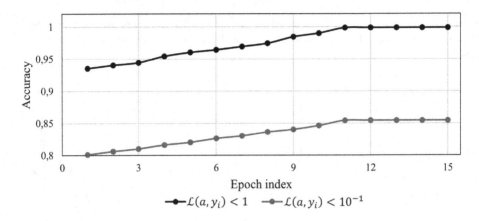

Fig. 12. Case 2: accuracy.

5   Conclusion

The proposed method for assigning a MCS by the BS scheduler, based on predicting the SINR value using the neural network apparatus, consists of two stages. At the first stage, for a given user movement model, a single-layer neural network is built and trained to predict the SINR value on the mobile UE based on the known values of this ratio in the past. At the second stage, according to the predicted the SINR value, the MCS is determined, which is required to transmit data to the user when providing a service with an appropriate quality level. Note that the optimization problem solved when training the network is multivariable, while the parameters of the neural network depend significantly on the user's movement model. In the future, to solve the problem of predicting SINR, it is planned to develop both supervised learning methods, for example, the use of a multilayer neural network, and the use of a reinforcement learning method for prediction in the absence of a training set.

References

1. Ghosh, A., Ratasuk, R.: Essentials of LTE and LTE-A. Cambridge University Press, Cambridge (2011)
2. Vorontsov, K.: Matematicheskie metody obucheniya po precedentam (teoriya obucheniya mashin) [Mathematical teaching methods by precedents (machine learning theory)] (2011). http://www.machinelearning.ru/wiki/images/6/6d/Voron-ML-1.pdf
3. Saija, K., Nethi, S., Chaudhuri, S., Karthik, R. M.: A machine learning approach for SNR prediction in 5G systems. In: 2019 IEEE International Conference on Advanced Networks and Telecommunications Systems (ANTS), pp. 1–6. IEEE (2019)
4. Calabrese, F.D., Wang, L., Ghadimi, E., Peters, G., Hanzo, L., Soldati, P.: Learning radio resource management in RANs: framework, opportunities, and challenges. IEEE Commun. Mag. **56**(9), 138–145 (2018)

5. Chen, M., Challita, U., Saad, W., Yin, C., Debbah, M.: Artificial neural networks-based machine learning for wireless networks: a tutorial. IEEE Commun. Surv. Tutor. **21**(4), 3039–3071 (2019)
6. Kumar, D., Logganathan, N., Kafle, V.P.: Double sarsa based machine learning to improve quality of video streaming over HTTP through wireless networks. In: 2018 ITU Kaleidoscope: Machine Learning for a 5G Future (ITU K), pp. 1–8. IEEE (2018)
7. Casas, P.: Machine learning models for wireless network monitoring and analysis. In: 2018 IEEE Wireless Communications and Networking Conference Workshops (WCNCW), pp. 242–247. IEEE (2018)
8. Ullah, R., et al.: A machine learning approach for 5G SINR prediction. Electronics **9**(10), 1660 (2020)
9. Averkin, A., Gaaze-Rapoport, M.G., Pospelov, D.A.: Tolkovyj slovar' po iskusstvennomu intellektu [Dictionary of Artificial Intelligence]. Radio i svyaz' [Radio and communication] (1992)
10. Trask, A.W.: Grokking Deep Learning. Simon and Schuster (2019)
11. Gaidamaka, Y., Samouylov, K., Shorgin, S.: Method of modeling interference characteristics in heterogeneous fifth generation wireless networks with device-to-device communications. Inform. Primen. **11**, 2–9 (2017)
12. Bobrikova, E., Platonova, A., Yartseva, I., Khairov, E.: K zadache predskazaniya sinr v besprovodnoj seti s podvizhnymi pol'zovatelyami s pomoshch'yu apparatanejronnyh setej [To the problem of predicting sinr in a wireless network with mobile users using a neural network apparatus]. In: Informatsionno-telekommunikatsionnye tekhnologii i matematicheskoe modelirovanie vysokotekhnologichnykh sistem [Information and Telecommunication Technologies and Mathematical Modeling of High-Tech Systems 2021], pp. 30–32. RUDN, Moscow (2021)

Analysis and Formalization of Requirements of URLLC, mMTC, eMBB Scenarios for the Physical and Data Link Layers of a 5G Mobile Transport Network

Dmitry Aminev[1], Evgenia Bogdanova[3], and Dmitry Kozyrev[1,2(✉)] (ORCID)

[1] V. A. Trapeznikov Institute of Control Sciences of Russian Academy of Sciences, 65 Profsoyuznaya Street, Moscow 117997, Russia
aminev.d.a@ya.ru
[2] Peoples' Friendship University of Russia (RUDN University), 6 Miklukho-Maklaya Street, Moscow 117198, Russian Federation
kozyrev-dv@rudn.ru
[3] T8 LLC, 44-1, ul. Krasnobogatyrskaya, Moscow 107076, Russia
bogdanova@t8.ru

Abstract. The evolution of mobile communication networks from 2G to 5G is briefly described. The assignment of scenarios for 5G mobile networks is noted. The architecture of the physical and data link layers of a 5G transport network is disclosed. The analysis of the requirements of the scenarios eMBB, URLLC, mMTC to the physical and data link layers of a transport network, their approximate numerical evaluation were carried out. The requirements that have a dominant influence in the construction of the network are identified, and their formalization is introduced. A mathematical model of requirements based on a matrix of weighting coefficients is proposed.

Keywords: Transport network · Formalization of requirements · Data transmission path · Mobile communications · 5G · Mathematical model · Matrix of weight coefficients

1 Introduction

The transport telecommunications network is the major part of the infrastructure of a mobile operator, and the quality of the services provided largely depends on its characteristics. Transport networks have evolved along with mobile communication standards from 2G/3G to 4G LTE and 5G (Fig. 1).

Traditionally, starting from 2G, the transport equipment occupied only the Backhaul segment, performing traffic aggregation and relaying between the radio

This paper has been supported by the RUDN University Strategic Academic Leadership Program and funded by RFBR according to the research project number 19-29-06043.

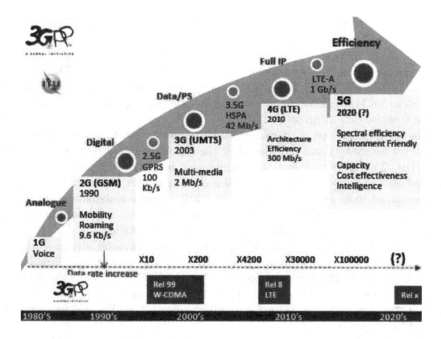

Fig. 1. Evolution of transport networks along with mobile communication standards from 2G/3G to 4G LTE and 5G

access means (RAN) at the periphery and the network core (Core). With the development of mobile technologies in connection with the growth of traffic in radio access and the transition to packet data, the role of a universal, flexible, high-performance transport infrastructure has increased. It is expanding towards the radio access network. The first open standard for digital mobile communications 2G (GSM technology - Global System for Mobile Communication) provided information transfer at a speed of 9.6–14.4 kbit/s, which made it possible to offer subscribers, in addition to voice calls, the service of exchanging short test SMS messages.

In the 1990s, GSM was improved with the advent of digital encryption. GPRS and EDGE technologies, which are often referred to as the 2.5G intermediate stage, allowed the user to access the Internet from a mobile device, and the user's peak speed was 474 kbps.

The 2G/2.5G network architecture is a wireless RAN section between user mobile terminals and base stations, a section of aggregation of base station traffic to base station controllers, and a backbone section between base station controllers and the network core (mobile switches, GPRS support nodes). The 2G operators transport network (aggregation sections and backbones) is characterized by the use of plesiochronous digital hierarchy (PDH, ITU-T G.703) and synchronous digital hierarchy (SDH, ITU-T G.707) technologies.

By the early 2000s, the third generation (3G) of mobile communication technologies offered the user a set of services that combined both a high-speed mobile Internet access and a radio technology that creates a data (voice, messages, etc.) transmission channel. There are several stages of development here. The 3G generation includes UMTS, CDMA2000 technologies, which provide data transfer rates up to 3.6 Mbps. The 3.5G generation includes HSPA/HPSA + technologies with a transfer rate of up to 14.4 Mbps.

The third generation is characterized by packet traffic, therefore, the Ethernet packet technology was introduced on the transport infrastructure over PDH and SDH. The transition met the requirement for increased transport capacity, and also provided the network with a mechanism for rapid traffic recovery and enhanced tools for operation, administration, and maintenance. Thus, NG SDH (Next Generation SDH) solutions were used on 3G transport networks, combining the advantages of channel and packet technologies.

In 2008, the international association 3GPP that develops advanced mobile communication standards, approved 4G (LTE) as the next mobile broadband network standard. The standard provides data transfer rates up to 326.4 Mbps from the base station to the user and up to 172.8 Mbps in the opposite direction. In connection with the increased subscriber speeds, the demand for network bandwidth is growing.

The principle of providing voice communication services has changed—instead of circuit switching, packet switching (IP-telephony) is used, and services are divided by priority. In 4G, compared to previous generations, there is a delay requirement. So an uncongested 3G network is characterized by delays of about 100 ms. Such a time interval is quite large, so the contribution to it from the transport network is considered insignificant - there are no separate requirements for transport. In the LTE network, delays are reduced by 4–6 times, the impact on the size of this value from the side of the transport infrastructure increases, and there are requirements for delay in transport for high-priority services. Thus, one of the components of the LTE network is a transport network based on the IP/MPLS protocol stack, which provides high-speed data transmission.

Starting from the 4G generation, the integration of IP/MPLS and optical transport levels on the Backhaul segment becomes typical, since the bandwidth of the existing SDH network turned out to be insufficient. The industry has come to the conclusion that it is necessary to create an OTN/WDM network [11] in order to transmit data in the format traditional for data networks. This meant using fixed frame sizes instead of the fixed frame rates used in SDH. This fundamental change made it possible to coordinate IP traffic with OTN more efficiently than with SDH. The integration of Internet Protocol (IP) and OTN is much more suitable for today's multi-service 4G LTE networks. On the mobile Backhaul, the base stations and the core of the network are connected by transport channels of more than 100 Gbps.

The transition to the 5G standard leads to changes in all components of mobile networks: the radio access network, the core, the single control plane and the transport network that connects all components into a single integrated solution. So in 5G networks, optical transport solutions penetrate the RAN - at the level of standards, new segments Fronthaul, Midhaul, Backhaul come forward [3,5,14]. In world practice, this process is called "Fiberization". Let us consider in more detail the evolution of mobile technologies, services and corresponding requirements for transport.

The problems of network planning and optimization for various 5G user applications have been discussed in a number of sources, however, standards and specifications regarding the requirements for different network levels are under development [13]. It is noted that the correct approach of operators in the implementation of 5G is to plan a strategy for providing 5G services at the first stage, determine the requirements for all network layers at the second stage (taking into account standards as they develop), and, finally, design the network taking into account the requirements model. According to the authors' opinion, this sequence will provide the operator with a better return on investment in the transition to next-generation mobile communications.

Paper [13] also provides an example of a L1/L2/L3 network layer requirements model for 5G video services (for VR, 4K, streaming broadcasting applications). The model takes into account different components of the 5G network: radio, network core, transport network. The main parameters are bandwidth (including at the time of peak load), packet loss (reliability), delay (processing delay and traffic latency), coverage quality. The article describes the experience of network planning based on the assessment of user experience, but there is no mathematical apparatus with formalized requirements and an optimization method. In addition, the input parameters of the model are not formalized.

2 Statement of the Problem

Figure 2 shows the architecture of a transport IMT-2020/5G network with the decomposition of the functional elements of the base station in accordance with 3GPP recommendations [1,3]. The following sections are distinguished between the main components RU (Remote Unit), DU (Distributed Unit), CU (Central Unit): Fronthaul, Midhaul and Backhaul. The access domain in the terminology of the classical transport network corresponds to the Fronthaul segment; the metro aggregation domain corresponds to the Midhaul segment; the domain of the backbone transport network corresponds to the Backhaul segment [2–4]. The presence or absence of Fronthaul and Midhaul segments is determined by the deployment scenario of the radio access network.

The IMT-2020/5G network is considered by the standards as a packet network, where IP/Ethernet packet switches and routers organize the connectivity of the main components of the 5G network at the L2/L3 level of the OSI model [5]. These connections are organized over the physical L0 infrastructure, for example, in a dark fiber or using one of the physical layer technologies - PON, xWDM, wireless.

At the L1 level, a transport technology such as OTN (Optical Transport Network) can be used between packet and physical networks, which provides

Decomposition of traditional Base Station functionality

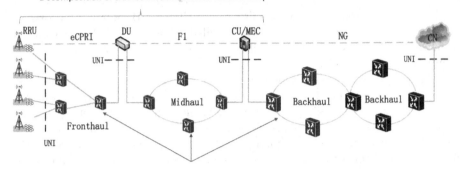

Fig. 2. Transport network architecture for IMT-2020/5G

aggregation and transparent transmission of large amounts of traffic, and also provides a number of advantages when organizing end-to-end 5G connections. Figure 2 shows the logical model of the transport network, which combines technologies of different levels.

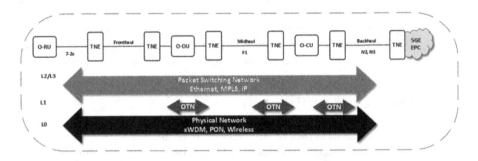

Fig. 3. Logical model of the transport network (general case)

Within the framework of the current paper, the requirements for the two lower levels L0, L1, namely, the xWDM physical layer equipment and the OTN transport equipment, are considered. L2/L3 switching/routing devices that implement the transmission of IP/Ethernet packet traffic are client equipment and are considered in the paper only from the point of view of requirements for interaction interfaces.

The three categories of use cases defined by the 3GPP standard for 5G/IMT-2020 networks have a significant impact on the requirements for all these components and the transport network in particular:

– enhanced Mobile Broadband (eMBB);
– Ultra-Reliable Low Latency Communication (URLLC);
– Massive Machine-Type Communications (mMTC).

The eMBB scenario assumes a large network bandwidth required to support the high data transfer rates of mobile users. The URLLC scenario imposes significant restrictions on delays and imposes increased requirements on the throughput and reliability of network equipment. The mMTC scenario assumes the connection of a large number of moving objects and places demands on network flexibility and connection availability.

Each specific scenario of the three categories provides a number of requirements, fulfillment of which will determine the optimal network architecture. At the same time, the architecture must be scalable in order to support user scenarios not only at the initial stage of technology implementation, but also in the future, with the development of eMBB, URLLC, mMTC scenarios. Carrying out such optimization by mathematical methods is impossible without preliminary formalization of scenario requirements.

Thus, it is necessary to solve several problems:

- identify the main requirements for each of the 3 scenarios;
- carry out formalization of the main requirements, taking into account their subsequent normalization, that is, formalization by reducing the absolute values of quantities to relative dimensionless coefficients;
- compose a mathematical model of requirements, presented in the form of relations and connections of matrices of dimensionless coefficients, on the basis of which the network should then be optimized for each of the scenarios.

3   Analysis and Formalization of Requirements

Based on the above-noted categories of user scenarios, it is possible to identify the general requirements for the transport network: high bandwidth with the ability to scale, low latency and transmission of synchronization signals, reliability, as well as network flexibility and availability of connections at any time. The fulfillment of each of these requirements can be ensured by the implementation of various software and hardware functions.

3.1   Bandwidth Scaling

Analytical studies [5, 6] predict a tenfold increase in throughput and the number of user devices in 5G networks compared to the previous generation of mobile communications. In this regard, the transport equipment on each segment must provide a high throughput of the IMT-2020/5G network and the ability to scale it as the user traffic increases.

In the general case, the throughput of Fronthaul depends on the number of antenna sectors, the throughput of the radio channel on each carrier, and the implementation of the MIMO functionality for each carrier. The O-RAN specification [7] provides an estimate of the Fronthaul traffic for different scenarios, for the 7.2x base station functional separation option:

- small cells: one sector, carriers in either mmWave or Sub6 band with low order MIMO;

- medium cells: many sectors, carriers in both Sub6 and mmWave bands with medium order MIMO;
- large cells: many sectors, carriers in both bands, mmWave- and Sub 6-, using Massive MIMO.

Option 7.2x assumes that the radio access equipment on the Fronthaul segment, after processing traffic between base stations DU/CU and processing units RU, generates eCPRI flows that are transmitted through standard 10GE/25GE L2/L3 interfaces. This means that optical transport equipment must support 10GE/25GE client signals.

The required capacity on the Fronthaul segment can be achieved in the following ways:

1. *Organization of dark fibers for RU and DU connections according to the point-to-point principle (L0)*—each fiber can transmit the traffic of one antenna at a speed of 10 Gb/s or 25 Gb/s, depending on the interface. There is no transport layer L1.
2. *Channel multiplexing at the physical layer (L0)*—the efficiency of fiber use is increased through the use of optical xWDM multiplexers. More than 80 channels of 10 or 25 Gbps can be transmitted in one pair of fibers. As in the first case, there is no L1 transport technology.
3. *OTN aggregation*—it's possible to aggregate a large number of user traffic flows using OTN optical transport technology (at the L1 level). For even more efficient use of fiber, OTN aggregators (muxponders) can be connected to xWDM multiplexer ports.

Further we formalize the requirements for throughput and determine the numerical criteria for scaling the capacity of the transport infrastructure for three scenarios. Define the following variables:

N_{fiber}—number of fiber pairs available at the operator's site,
N_{10G}/N_{25G}—required number of connections with a speed of 10Gb/s/25Gb/s,
N_{mux}—number of available optical multiplexer ports,
N_{OTN10G}, N_{OTN25G}[1]—number of available user interfaces of 10GE and 25GE OTN aggregators.

For scenarios 1–3, the bandwidth scalability of the Fronthaul segment B_{scal} depends on the availability of transport resources and on the potential to increase the capacity to the maximum value on a particular Fronthaul segment.

The availability of resources will be determined by the coefficients k_{fiber}, k_{mux} or k_{OTN} for scenarios 1–3, respectively. The possibility of increasing the

---

[1] The number of available N_{OTN10G}, N_{OTN25G} user interfaces is given for the scenario of the best use of OTN muxponder ports, when the maximum possible number of ports is used for 25GE users, and the remaining device capacity is occupied by 10GE users. In this case, the conditions $N_{OTN10G} \geq N_{10G}$, $N_{OTN25G} \geq N_{25G}$ are met.

capacity to the maximum value B_{max} for a particular Fronthaul segment is determined by the coefficient b_{free}:

$$b_{free} = \frac{(B_{max} - N_{10G} * 10 - N_{25G} * 25)}{B_{max}}$$

The scalability of capacity B_{scal} for different architectures can be estimated as follows:

1) When using P2P connections:

$$B_{scal} = k_{fiber} * b_{free}, \ k_{fiber} = \frac{N_{fiber}}{N_{10G} + N_{25G}},$$

where k_{fiber}—is the fiber availability factor. 2) When using optical multiplexers:

$$B_{scal} = k_{mux} * b_{free}, \ k_{mux} = \frac{N_{mux} * N_{fiber}}{N_{10G} + N_{25G}},$$

where k_{mux}—is the availability factor of optical multiplexer ports. 3) When using OTN aggregation:

$$B_{scal} = k_{OTN} * b_{free}, \ k_{OTN} = \frac{(N_{OTN_{10G}} + N_{OTN_{25G}}) * N_{mux} * N_{fiber}}{N_{10G} + N_{25G}},$$

where k_{OTN}—is the availability factor of user interfaces of the OTN muxponder.

The absolute values of the maximum throughput of Backhaul and Midhaul are not currently defined in the standards and specifications. Assuming that Midhaul and Backhaul are implemented using OTN/DWDM technology, the requirement for the throughput of transport equipment in these sections will always be met.

3.2 Network Temporal Characteristics: Delays and Signal Synchronization Support

The key differences between the 5G transport network and classical transport networks are the stringent requirements for delays and support for the transmission of synchronization signals. The 5G network is packet-oriented and cannot meet these requirements without additional mechanisms. Delay and jitter values are not deterministic for packet switches and routers. Only the use of special TSN mechanisms (TSN—Time Sensitive Networking) delivers the predictable temporal characteristics.

The delay at the optical layer is deterministic. It is caused, first of all, by the signal propagation delay in the optical fiber (5 microsecond/km). In OTN/DWDM, the largest latency contributor is the FEC (Forward Error Correction) procedure, as well as the presence of buffers for processing OTN headers in processors. Delays TL0/L1 and TL2/L3 at different levels will be further considered separately. End-to-end latency is currently specified for several user scenarios. Designate this value as TE2E. Figure 3 shows the delay distribution by network segments for URLLC and eMBB according to ITU-T [8,9] and 3GPP [10][2] recommendations. The delay values for user plane (U) and control plane (C) traffic do not always match, since these flows can be terminated in different elements of the 5G network.

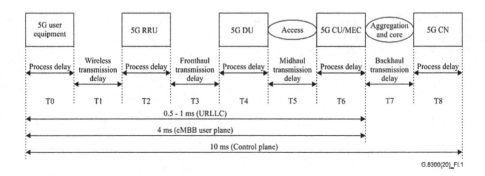

Fig. 4. Distribution of required delays on a 5G network for URLLC and eMBB scenarios for user plane and control plane

The segment most critical to delays is Fronthaul. The distance between RU and BBU should not exceed 10–20 km, which corresponds to a delay of 100–200 μs for a bidirectional transmission and is determined by the limitations of the CPRI/eCPRI protocols that must be supported by the transport equipment. The exact values of the allowable delay between the termination points of the CPRI and eCPRI interfaces are given in Table 1 and are designated as T_{E2E_FH}.

Table 1. Requirements of the CPRI and eCPRI protocols for the transport network

Requirement for transport equipment FH	Acceptable values when using CPRI	Acceptable values when using eCPRI
Acceptable end-to-end delay T_{E2E_FH}	50–250 μs, depending on the mobile application	100 μs–100 ms, depending on the traffic class
Permissible time error (TE)	± 16 ns	± 10 ns … ±1,36 μs depending on the traffic class
Frequency setting accuracy (FE)	50 ppb	16 ppb
Phase setting accuracy (PE)	1,5 ns	1,5 ns

---

[2] At present, the delay requirements for the mMTC scenario are not defined by the standards.

In addition to the acceptable absolute values of delays on the Fronthaul segment, it is necessary to take into account the asymmetry of delays in the transmission directions (upwards - from RU to DU, downwards - from DU to RU). The limitations are related to phase and time synchronization between base station elements. The CPRI and eCPRI protocols have different accuracy requirements. In general, the delay asymmetry at the Fronthaul should not exceed the TE (Time Error) value (Table 1). Delay asymmetry in DWDM equipment is determined by the following factors:

- different optical length of fibers in bidirectional transmission;
- presence of dispersion compensators in the DWDM-network with incoherent transponders;
- OTN/DWDM transponders containing FIFO buffers.

Additionally, network time error occurs in IP routers and Ethernet switches.

In transport equipment, it is possible to compensate for delay asymmetry both at the electrical level (L1) in signal processing and at the physical level (L0). It is possible to eliminate delay asymmetry using architectural solutions, for example, by organizing bidirectional transmission over a single optical fiber (the so-called BiDi solutions).

5G systems have stringent requirements for phase/time synchronization between the radio equipment units. Synchronization signals are delivered from the primary synchronization source to the RU radio units via the transport network. Network synchronization protocols operate at the L2 level—this is the so-called synchronous Ethernet (SyncE) and IEEE 1588-2008 Precision Time Protocol (PTP). The xWDM optical network provides a physical layer for transparent transmission of Ethernet traffic, which includes 1588 and SyncE packets, and must provide the required accuracy setting of frequency, time, and phase.

We formalize the requirements for temporal characteristics and determine the numerical criteria necessary to include this target parameter into the model of the optimal transport network. Introduce the following variables:

$T_{L0/L1}$—the total value of delays (T_{FIBER}, T_{FEC}, T_{OTN}) in all L0/L1 equipment elements when the signal passes from RU to the core of the 5G CN network (Fig. 4, is fixed for the given network configuration).

$T_{L2/L3}$—the total value of delays in all elements of L2/L3 equipment when the signal passes from RU to the core of the 5G CN network (Fig. 3, is not a constant for this network configuration and depends on the load).

T_{E2E_FH}—the total value of delays in all L0/L1 equipment elements on the Fronthaul segment;

T_{E2Emax}—the maximum allowable delay for the entire network (Fig. 4), includes all sources of delay in the end-to-end signal transmission.

$T_{E2E\_FH\_max}$—the maximum allowable delay on the Fronthaul segment.

TE, FE, PE—errors of time, frequency and phase settings, respectively, on the Fronthaul segment.

TE_{max}—the maximum allowable time error on the Fronthaul segment;

FE_{max}—the maximum allowable frequency setting error on the Fronthaul segment;

PE_{max}—the maximum allowable phase setting error on the Fronthaul segment.

The target parameter k_{time} characterizing the temporal characteristics of the network, consists of the terms:

$$k_{E2E} = \frac{T_{E2E\_max} - T_{L2/L3} - T_{L0/L1}}{T_{E2E\_max}};\ k_{FH} = \frac{T_{E2E\_FH\_max} - T_{E2E\_FH}}{T_{E2E\_FH\_max}};$$

$$k_{TE} = \frac{TE_{max} - TE}{TE_{max}};\ k_{FE} = \frac{FE_{max} - FE}{FE_{max}};\ k_{PE} = \frac{PE_{max} - PE}{PE_{max}}.$$

Then, $k_{time} = k_{E2E} + k_{FH} + k_{TE} + k_{FE} + k_{PE}$. We believe that in the formation of the target "delay" parameter, in the general case, each of the terms has the same weight, since they are all equally important for the transport layer to meet the requirements for temporal characteristics.

3.3 Network Flexibility and Connectivity

Transport network flexibility refers to the connectivity of nodes at the transport layer. Flexibility provides the ability to establish a connection between any two users and is determined by the support for traffic switching in IP routers (L3), Ethernet switches (L2), OTN cross-switches (L1), optical wavelength switching devices (L0).

Connections on the Fronthaul segment are mostly point-to-point, and there is no need for cross-connect. The requirement for flexibility and, accordingly, support for cross-connect is imposed on segments with a more complex topology, Midhaul and Backhaul.

The network implemented on the IP/MPLS technology stack provides highly granular switching at the level of individual packets. Data packets are directed to the correct port corresponding to the required direction by processing IP headers or special MPLS labels. However, these operations require high performance equipment and, as a result, high power consumption. In addition, delays in L2/L3 devices are load dependent and non-deterministic.

In a complex Midhail/Backhaul topology, for some L2/L3 devices, most of the traffic is transit, so to increase the efficiency of using network resources, it is advisable to organize the so-called bypass of intermediate switches and routers. Then the traffic, bypassing L2/L3 processing, will go along the route formed from OTN or photon switches (Fig. 5a).

Thus, the advantages of switching L1/L0 are the reduction of power consumption of routers, the unloading of L2/L3 resources, as well as the reduction of traffic delay along the route and its determinancy.

To organize transparent end-to-end connections and manage traffic at the level of individual client services, OTN cross-connect equipment with a fully available switching matrix is used.

If traffic control (redirection of flows in intermediate nodes) at the level of individual wavelengths is necessary, transport equipment must include photonic switches—optical input-output multiplexers ROADM (Reconfigurable Optical Add-Drop Multiplexer). ROADMs do not provide the same granularity as OTN switches, but they allow more efficient management of large amounts of traffic without optoelectronic conversion (Fig. 5b). Flexibility in this case will be determined by the number of directions in a ROADM node.

The maximum efficiency of optical (photonic) switching is provided by the "CDC" (Colorless, Contentionless, Directionless) configuration of the ROADM device, which allows switching traffic from any input port, i.e. from any direction, to any output port, regardless of the incoming wavelength. Same as for OTN switches, we call this property full availability.

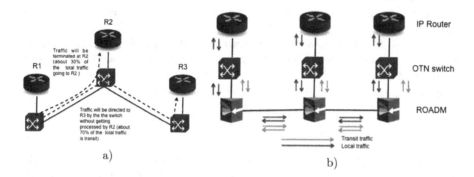

Fig. 5. Bypass of intermediate L2/L3 switches with switching in OTN/DWDM nodes (a); switching levels in the transport network (b) – electronic OTN switching (L1) and wavelength switching in ROADM (L0)

Within the current article, the capabilities and characteristics of switching equipment are not considered in detail. Note that the possibility of bypassing L2/L3 nodes over the transport network is provided by fully available switching at the transport layer, i.e. the ability at any network node to associate any input port corresponding to a certain direction with any output port/direction.

For the model of an optimal transport 5G/IMT-2020 network, we introduce a characteristic showing the ability to bypass switches/routers at the level of transport equipment, determined by the presence of a connection at the transport level between two nodes and the capacity that this connection provides.

Connectivity is determined by whether the node has an OTN cross-connect or a full-switched ROADM. The capacity limit at the transport layer may be due to the fact that the number of wavelengths (per L0) or the number of OTN channels (per L1) is less than the number of client streams at peak load. The peak load on the network occurs when clients from all directions simultaneously switch to a specific node along one of the linear sections.

Introduce coefficients k_{L0} and k_{L1} for each pair of adjacent network nodes. They show the ratio of available connections at the transport layer to the maximum possible number of client flows from all directions between these nodes:

$$k_{L0} = \frac{N_{L0}}{N_{L2/L3}}, \ k_{L1} = \frac{N_{L1}}{N_{L2/L3}},$$

where N_{L0}—number of connections at the photon level (wavelengths) for a given pair of adjacent nodes, N_{L1}—number of OTN channels for a given pair of adjacent nodes, $N_{L2/L3}$—maximum number of client flows from all directions for a given pair of adjacent nodes.

For simplicity, assume that the bandwidth of each transport channel is not less than the bandwidth required by the user, therefore only the number of channels appears in the formula, since OTN provides capacity scalability, and modern equipment aggregates and transmits user streams at speeds up to 400 Gbps.

In the model, it is necessary to take into account k_{L0} and k_{L1} for each pair of adjacent nodes and evaluate the flexibility of the network as a whole, so we sum the obtained coefficients and normalize them to the number of connections/pairs of adjacent nodes in the Midhaul or Backhaul network:

$$K_{L0} = \frac{\sum_1^M k_{L0}}{M}, \ K_{L1} = \frac{\sum_1^M k_{L1}}{M}.$$

If the ports of IP routers are associated with an optical wavelength and the connection between the source and the receiver is established through ROADM nodes, the network is characterized only by the coefficient K_{L0}. If more granular switching is used and the L2/L3 ports are connected to the user ports of the OTN switches, the K_{L1} connectivity factor must be used. The physical connectivity of the nodes must be supported by the control plane. Traffic routes can be built by the operator in the transport equipment control system or automatically with the use of the ASON/GMPLS protocols (ASON—Automatic Switch Optical Networks, GMPLS—Generalized Multiprotocol Label Switching). Route calculation in a simple case occurs in a distributed way in the network nodes or centrally in the SDN controller if the network is program-controlled.

3.4   Connection Reliability

Network reliability is determined by the mechanism for implementing recovery switching and the availability of these connections at any time [15, 16]. In transport networks, there are several mechanisms for organizing the connection protection.

Standard hardware protection switching mechanisms include the following schemes: 1+1 OLP (Optical Line Protection), 1+1 OMSP (Optical Multiplex Section Protection) or E-SNCP (Electrical SubNetwork Connection Protection) [11, 12]. Each of the "1 + 1" schemes assumes that the traffic is transmitted simultaneously along the main and backup routes. If a defect or failure condition in the primary direction is detected at the receiver, a guaranteed switchover to the backup route occurs in less than 50 ms. Difference of OLP, OMSP, E-SNCP is in the object of protection. So in OLP, the optical linear path (amplifiers, fiber) is protected, in OMSP the multiplex section (linear path and input-output multiplexers) is protected, the SNCP class involves the protection of channel-forming equipment (OTN transponders).

The control plane allows to expand the capabilities of hardware redundancy if it is implemented using the ASON/GMPLS (Automatically Switched Optical Network/Generalized Multiprotocol Label Switching) protocol group. Special protocols of the ASON stack automatically determine the topology, detect available network resources and reserve them in case of failover.

There are several types of connection protection switches in ASON networks. In the event of a break on the working route, if "1+1" protection is not provided, or in case a double failure occurs, the traffic is switched to a backup route, network resources for which are pre-reserved. Before a failure, these same resources can be used to carry low priority traffic to other clients. This scenario is efficient in the use of network resources and ensures guaranteed delivery of traffic even in the event of double failures.

The control plane of ASON networks allows the search for an alternative route even if resources were not reserved in advance. The route is calculated in real time based on the availability of network resources. This scenario provides the best resource efficiency, but does not guarantee that there is a free route on the network at a given time. Unlike "1+1" protection scenarios or pre-redundant protection connections, where the switching time is guaranteed to be less than 50 ms, real-time calculation of an alternative route and switching to it can take several to tens of seconds. Thus, the latter mechanism is not suitable for the 5G URLLC scenario and can only be used as a complement to the first two protection methods. The highest network reliability is ensured by the simultaneous use of the three mentioned mechanisms, when they complement each other. This approach allows achieving high reliability even with multiple failures.

To define the reliability quantitatively, introduce the target parameter P, which will be determined by the sum of p for each pair of network nodes. Consider the reliability p for a specific connection, it will be determined by the implemented switching mechanisms to a protective route and the availability of the found route. Let the maximum value of p be 1 and depend on the coefficients p_{1+1}, $p_{reserve}$, $p_{compute}$, in the presence of the appropriate type of protection, multiplied by the availability of the backup route (a_{1+1}, $a_{reserve}$, $a_{compute}$). For the implemented "1+1" scenario, the backup route is always up, so availability $a_{1+1} = 1$. For the ASON scenario with resource reservation, $a_{reserve} = 1$ if no higher priority traffic is transmitted on the backup route. For an ASON scenario with route calculation $a_{compute} = k_{L0}$, $a_{compute} = k_{L1}$, where k_{L0} and k_{L1} are the coefficients of connection availability for each pair of adjacent network nodes introduced earlier. Thus, when implementing all protective schemes, the resulting reliability parameter P will look as follows:

$$P = f\left(|A|, p_{1+1} \times a_{1+1}, p_{reserve} \times a_{reserve}, p_{compute} \times a_{compute}\right),$$

where $A = \begin{bmatrix} a_1^1 & a_1^2 & ... & a_1^n \\ ... & ... & ... & ... \\ a_m^1 & a_m^2 & ... & a_m^n \end{bmatrix}$ —network connectivity matrix on Midhaul and Back-

haul segments.

4 Model of Requirements of URLLC, mMTC, eMBB Scenarios for the Physical and Data Link Layers of the Transport Network

When formalizing requirements for the 5G network, the following parameters were introduced that characterize the transport layer of the 5G network. Let's write them as a vector $(B_{scal}, P, K_{L0}/K_{L1}, k_{time})$. For eMBB, URLLC, mMTC scenarios , these parameters have different priority, so they should be taken into account with different weight coefficients w. Introduce the network weight matrix W, where the subscript of each element corresponds to the parameter, and the superscript corresponds to the user scenario eMBB, URLLC, mMTC. The *scal* index corresponds to B_{scal} , *prot*—to the P parameter, *flex*—to the K_{L0}/K_{L1} parameters, *time*—to the k_{time} parameter.

$$W = \begin{bmatrix} w_{scal}^{eMBB} & w_{prot}^{eMBB} & w_{flex}^{eMBB} & w_{time}^{eMBB} \\ w_{scal}^{URLLC} & w_{prot}^{URLLC} & w_{flex}^{URLLC} & w_{time}^{URLLC} \\ w_{scal}^{mMTC} & w_{prot}^{mMTC} & w_{flex}^{mMTC} & w_{time}^{mMTC} \end{bmatrix},$$

where the sum of elements in each row is 1.

When multiplying the weight matrix by the parameter vector, we get a vector that characterizes any 5G transport network:

$$(a) = \begin{bmatrix} w_{scal}^{eMBB} & w_{prot}^{eMBB} & w_{flex}^{eMBB} & w_{time}^{eMBB} \\ w_{scal}^{URLLC} & w_{prot}^{URLLC} & w_{flex}^{URLLC} & w_{time}^{URLLC} \\ w_{scal}^{mMTC} & w_{prot}^{mMTC} & w_{flex}^{mMTC} & w_{time}^{mMTC} \end{bmatrix} \times \begin{pmatrix} B_{scal} \\ P \\ K_{L0}/K_{L1} \\ k_{time} \end{pmatrix} =$$

$$= \begin{pmatrix} w_{scal}^{eMBB} \cdot B_{scal} & w_{prot}^{eMBB} \cdot P & w_{flex}^{eMBB} \cdot K_{L0}/K_{L1} & w_{time}^{eMBB} \cdot k_{time} \\ w_{scal}^{URLLC} \cdot B_{scal} & w_{prot}^{URLLC} \cdot P & w_{flex}^{URLLC} \cdot K_{L0}/K_{L1} & w_{time}^{URLLC} \cdot k_{time} \\ w_{scal}^{mMTC} \cdot B_{scal} & w_{prot}^{mMTC} \cdot P & w_{flex}^{mMTC} \cdot K_{L0}/K_{L1} & w_{time}^{mMTC} \cdot k_{time} \end{pmatrix}.$$

This vector will be the target parameter for optimizing a specific network topology for a custom scenario.

Requirements from three types of services to the 5G transport network were determined. Network modeling based on the initial data of the requirements model can be carried out in the AnyLogic environment, which allows one to design complex systems using several methods: discrete-event (process) and agent-based modeling.

The first approach involves the approximation of the real process by discrete events, i.e. modeling only "important events" in the functioning system. The second, agent-based, approach involves modeling the behavior of decentralized agents within the entire system as a whole.

To model the 5G transport network, a combined method was used, where user services are represented as agents with their own parameters. Agents (users) are placed in the environment—a network represented by the discrete-event method. When interacting with the environment, agents can, in turn, change the parameters of the system.

When formalizing the requirements, a matrix is formed that characterizes any service passing through any network. In fact, the coefficient matrix W defines a specific service, and the parameter vector (B_{scal}, P, K_{L0}/K_{L1}, k_{time}) defines the network through which this service passes.

Then, under the agent we will understand the established end-to-end connection or service, for which the matrix of coefficients W is defined. The coefficients of the matrix will be fixed parameters of the agent. Another parameter of the service (agent) will be the required bandwidth B. When passing through the simulated network, which is equivalent to establishing a connection for this service, its resources will be occupied (for example, a DWDM channel or a part of it equal to the value B). In addition to fixed values, we assign to agents a set of variables that will change when passing through the network. These variables correspond to the previously introduced elements (B_{scal}, P, K_{L0}/K_{L1}, k_{time}), and the calculated result vector (Factor eMBB, Factor URLLC, Factor mMTC) will determine the suitability of the network for a given service.

5   Conlusions

Based on the analysis of the general requirements for the IMT-2020/5G transport network, the main requirements for the physical and data link layers were identified: high bandwidth throughput with the ability to scale, low delays and transmission of synchronization signals, reliability, as well as network flexibility and connection availability at any time.

As a result of formalization, a vector of parameters (B_{scal}, P, K_{L0}/K_{L1}, k_{time}) was introduced, which helps to solve the problem of constructing an optimal architecture at the stage of transport network design, as well as in constructing optimal traffic transmission routes on the implemented network, depending on the scenario (eMBB, URLLC, mMTC). When optimizing the architecture, depending on the chosen scenario, the formalized requirements are to be taken into account with the appropriate weighting coefficients w. So, for example, for the URLLC scenario, the critical parameter is the $T_{L0/L1}$ delay, so the k_{time} parameter will have the highest weight. For the eMBB scenario, bandwidth is important, which means that the B_{scal} parameter will have the maximum weight in the model. The mMTC scenario assumes multiple dynamic connections of moving objects, and the weight K_{L0}/K_{L1} will be the most significant.

When modeling a network of arbitrary topology in the AnyLogic environment, it is proposed to use a combined agent-process approach, where services with a given set of parameters and variables act as agents, and the network itself is modeled using built-in elements. The suitability of the network for a given service is determined by the resulting vector *(Factor eMBB, Factor URLLC, Factor mMTC)*.

References

1. 3GPP Technical specification. NG-RAN; Architecture description (2022)
2. Bogdanova, E.: Optical route in the transport IMT-2020/5G network. Last mile, no. 7, pp. 40–47 (2019). (in Russian)
3. ITU-T Recommendation G.8300: Characteristics of transport networks to support IMT-2020/5G (2020)
4. ITU-T Technical Report. Transport network support of IMT-2020/5G (2018)
5. 3GPP Technical specification. NG-RAN; NG data transport (2022)
6. Light Reading. 5G Network & Service Strategies. Operator Survey (2021)
7. O-RAN.WG9.XTRP-REQ-v01.00. Technical specification. O-RAN Open X-haul Transport Working Group 9. Xhaul Transport Requirements (2021)
8. ITU-T Recommendation G.8271.1: Network limits for time synchronization in packet networks with full timing support from the network (2022)
9. ITU-T Recommendation G.8273.2: Timing characteristics of telecom boundary clocks and telecom time slave clocks for use with full timing support from the network (2022)
10. 3GPP Technical report. Study on scenarios and requirements for next generation access technologies (2022)
11. Recommendation ITU-T G.709/Y.1331. Interfaces for optical transport network (2022)
12. Recommendation ITU-T G.831.1. Optical transport network: Linear Protection (2017)
13. Whitepaper: Ovum. 5G Service Experience-Based Network Planning Criteria (2019). https://carrier.huawei.com/~/media/CNBGV2/download/products/servies/5G-Planning-Criteria-White-Paper.pdf
14. 3GPP Technical Specification 38.401: NG-RAN. Architecture description (2020)
15. Aminev, D., Golovinov, E., Kozyrev, D., Larionov, A., Sokolov, A.: Reliability evaluation of a distributed communication network of weather stations. In: Vishnevskiy, V.M., Samouylov, K.E., Kozyrev, D.V. (eds.) DCCN 2019. LNCS, vol. 11965, pp. 591–606. Springer, Cham (2019). https://doi.org/10.1007/978-3-030-36614-8_45
16. Golovinov, E., Aminev, D., Tatunov, S., Polesskiy, S., Kozyrev, D.: Optimization of SPTA acquisition for a distributed communication network of weather stations. In: Vishnevskiy, V.M., Samouylov, K.E., Kozyrev, D.V. (eds.) DCCN 2020. LNCS, vol. 12563, pp. 666–679. Springer, Cham (2020). https://doi.org/10.1007/978-3-030-66471-8_51

Traffic Arrival Model for Millimeter Wave 5G NR Systems

E. M. Khayrov[1]($\boxtimes$) (ID), V. A. Prosvirov[1] (ID), and Anna Platonova[2] (ID)

[1] Higher School of Economics (HSE University), 20 Myasnitskaya Street,
Moscow 101000, Russian Federation
emil.khayrov@gmail.com
[2] Peoples' Friendship University of Russia (RUDN University),
6 Miklukho-Maklaya Street, Moscow 117198, Russian Federation
platonova-aa@rudn.ru

Abstract. The introduction of integrated access and backhaul (IAB) technology in beyond 5G systems will result in multiple wireless backhaul links between the user equipment (UE) and the base station (BS). This configuration motivates the research into traffic at the packet level to evaluate the quality of service (QoS) provided to users. At the same time, specifics of directional communications in millimeter wave (mmWave) and terahertz (THz) frequencies with frequent outages caused by blockage and micromobility lead to periods of silence and further rate compensation. In this paper, we consider the traffic as the process of receiving requirements for resource blocks, i.e. the number of resource blocks required to allocate on the BS to provide a constant bit rate at the UE. We propose a Markov model for the specified traffic system that explicitly accounts for such mmWave/THz-specific features as micromobility and blockage effects. Our numerical results demonstrate that the Poisson arrival process assumption underestimates the actual queuing delay at BS by approximately 10%.

Keywords: Packet traffic · mmWave · 5G · Blockage · Micromobility · Markov chain

1 Introduction

The evolution of 5G networks is growing rapidly, bringing new advantages for next-generation wireless networks such as higher data rates and lower latency. This is achieved by using a millimeter-wave spectrum, where the length of the radio wave has the millimeter values with a frequency of 30–300 GHz. However, the mmWave brings new challenges compared to 4G LTE: for example, high signal attenuation, which leads to high sensitivity for any type of blockages, such as buildings, vehicles, or even humans or trees. One of the possible solutions

The research was funded by the Russian Science Foundation, project №21-79-00142 (https://rscf.ru/en/project/21-79-00142/).

to address this issue is to deploy base stations closer to each other, covering more areas to eliminate the gaps, but it will impose significant costs on network operators [3,8].

To reduce such expenditures, the 3GPP standardization organization proposed the integrated access and backhaul (IAB) technology. IAB relies on relay nodes (IAB nodes) to serve as an intermediate traffic transmitter to deliver traffic between UE and BS in a multi-hop way. The additional complexity of such configuration requires analyzing additional metrics of latency along with conventional performance metrics, such as spectral efficiency or amount of time in outage conditions. At this point, the research of traffic at the packet level becomes crucial [14].

The traffic rate is highly affected by blockages, which are often caused by humans. The use of directional antennas brings another issue to traffic. Any small movement of the UE may lead to connection loss for a short time, required to reallocate beams, with the following rate compensation [13]. Such variability affects the traffic transmission to BS drastically and should be carefully analyzed.

In this paper, we characterize the traffic structure in the 5G NR scenario, where the UE communicates with the BS (or a specific IAB node) through the wireless channel. We assume that the UE requires some constant bit rate for its application, and the BS allocates required resource blocks to sustain that bit rate. The simulation of such traffic includes the analysis of human blockages, which implies variability of required resource blocks, and the micromobility effects with temporal rate discontinuation and subsequent rate compensation. After that, we build a mathematical model of the process of receiving requirements for resource blocks (the number of resource blocks required in a certain state) from the UE to the BS as a special case of the Markovian Arrival Process (MAP) - Markov Modulated Poisson Process (MMPP) and provide its comparison with the standard $M/M/1$ model. The main idea behind that lies in the possibility of approximating such a complex system of traffic arrival by a simple Poisson process.

This paper is organized as follows. In Sect. 2 we provide a short review of communication link modeling in mmWave. Section 3 describes our system model. Section 4 contains detailed description of our traffic model, which is shown as a MMPP process in Sect. 5, and in Sect. 6 we provide some results of model evaluation. Section 7 concludes the paper.

2   Related Work

Millimeter waves provide a wide range of benefits such as enhanced Mobile Broadband (eMBB), Ultra Reliable Low Latency Communications (URLLC), and massive Machine Type Communications (mMTC). However, they also have fairly high signal attenuation, and the problem of blockage in such systems should be considered separately. Some researchers develop a model with performance analysis of blockages caused by humans [16]. Since blockage itself represents some kind of binary operation (blocked/non-blocked), the analytical model is often represented as a renewal process [8,9,12].

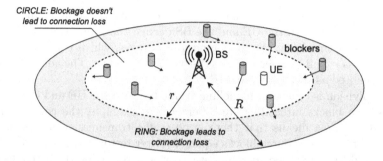

Fig. 1. System model: the communication link between the UE and the BS may be blocked by blockers - other humans.

In [2], a different approach to analyze blockage is provided. This approach uses deep learning techniques to predict a blockage arrival in the next time slot. [6] provides a model with unmanned aerial vehicles (UAVs) performing as base stations. This approach considers dynamic human blockages in mmWave-drone-BS networks. One of the key goals of this work is to define the optimal height of the access point located at the UAV to serve the maximum number of users, taking into account human blockages as well as the optimal signal-to-noise ratio (SNR).

Some researchers analyzed blockages with respect to the proportional fair scheduler [5,10]. In [5], the human blockage is predicted by detecting changes in the received power signal when the blockage is barely starting to occur. [10] proposes a modification of the standard proportional fair (SPF) scheduler that not only improves the priority of UEs in degraded conditions but also speeds up the response time to any kinds of fluctuations, primarily blockages.

It is fair to notice that there are a lot of papers that consider scenarios with static users/blockers. Although the implementation of movement increases the difficulty of such a model, it conducts to more realistic scenarios with practical use, which is why we are interested in dynamic blockages.

3 System Model

3.1 Deployment

We consider a single mmWave/THz BS mounted at height h_A having coverage of a circular shape with radius r, see Fig. 1. We assume that r is such that no UEs inside it experience outage conditions, which means that UEs at the cell edge do not lose connection in 95% of the cases.

Since UEs can have a connection beyond that area when the connection link is clear and non-blocked, we take into account an extra area beyond coverage r, which will be defined as a ring-shaped area with an outer radius R. In this area, if the blockage between the UE and the BS occurs, the connection is considered to be lost. For simplicity, the model without outages, which is a circular-shape area of radius r, will be referred to as the 'CIRCLE' model. The additional space will be denoted as the 'RING' model.

This division is crucial in modelling traffic between the UE and the BS since the resource blocks intensities are different. For example, the connection loss in the RING area means that there is no traffic transmission at all, and the number of required resource blocks equals 0. After restoring the connection, the traffic should be increased to compensate for lost resources. Thus, it allows us to carefully analyze models both independently and in combination.

3.2   Blockages

Pedestrians serve as blockers for mmWave/THz propagation. We assume the density of blockers to be λ_B bl/m^2. Each blocker introduces a loss of 20 dB. Blockers are assumed to move in $\mathbb{R}^2$ according to the Random Direction Mobility model (RDM, [11]). According to RDM, a human chooses a direction randomly in $[0, 2\pi)$ and then moves in this direction for an exponentially distributed time with parameter μ_B at a speed v.

We consider the blocking area as a rectangle with a width of d_m and length l (see Fig. 2), where d_m is the diameter of blocker and $l = d(h_B - h_U)/(h_A - h_U)$, where d is the distance between the UE and the BS and h_B, h_U are the heights of blockers and UEs respectively ($h_U < h_B$). A communication channel is considered blocked if a blocker enters the rectangle and therefore blocks the line-of-sight (LoS) channel.

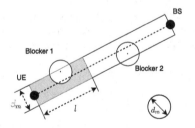
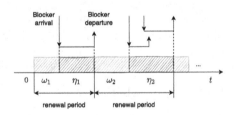

Fig. 2. Blockage area: blockers break the LoS if they enter the gray area.

Fig. 3. Illustration of the blockage renewal process.

The blockage process can be described as a renewal process with two states: the LoS period, which occurs when the communication channel is not blocked by any blocker and is described with random value ω, and the blocking (nLoS) period when the communication channel is disrupted, described with random value η (see Fig. 3). The LoS period is assumed exponentially with parameter

$\gamma(x)$ as a function of the distance between the UE and the BS. The average time of LoS period is $E[\omega(x)] = 1/\gamma(x)$.

Since nLoS period may be caused by one or more blockers following one another, the nLoS period can be approximated with $M/M/\infty$ model with parameters $\gamma(x)$ and $\varepsilon(x)$, where $\gamma(x)$ means the blocker arrival rate as a function of the distance and ε corresponds to intensity of blockers' departure. The average time of nLoS period can be described as a congestion period of $M/M/\infty$ (the time that the process spends above a fixed level, in our case, when the number of blockers before the UE is more than 0): $E[\eta(x)] = \frac{1}{\gamma(x)} \sum_{i>0} \frac{1}{i!} \left(\frac{\gamma(x)}{\varepsilon(x)} \right)^i$.

3.3   Micromobility Effects

Micromobility (MM) provides another level of difficulty to our traffic transmission model. To simplify it, we assume that the time required to restore communication during MM coincides with the exhaustive beam search time [9]: $T_{MM} = N_U N_T \delta$, where N_U and N_T are the numbers of BS and UE antenna configurations and δ is the antenna array switching time. Since micromobility leads to connection loss, we define a short period after MM when the lost resources are compensated. For simplicity, we assume that it takes the same amount of time as the MM, and the requirements for resource blocks are multiplied by two. The time interval between two consecutive micromobility effects can be approximated with exponential distribution with parameter β: $f_m \sim Exp(\beta)$ [7]. Parameter β depends on the application running on the UE (watching movies, talking on the phone, playing games, etc.) since each type of action implies different angular and positioning motion.

3.4   Wireless Channel

The traffic transmission process depends on many factors. Both the distance between the BS and the UE and the communication link state (blocked/non-blocked) strongly influence the received Quality-of-Service (QoS). In our scenario, we assume that user requires some constant bit rate C for its application (e.g. VR or high-active game).

The signal-to-noise ratio (SNR) at the UE can be written as

$$S(d) = \frac{P_{\text{BS}} G_{\text{BS}} G_{\text{UE}}}{L(d) N_0 R_b \chi_{\sigma_{\text{SF}}} X}, \tag{1}$$

where P_{BS} is the transmit power, G_{BS} and G_{UE} are the antenna gains at the BS and the UE, N_0 is the thermal noise, R_b is the physical resource block size, $\chi_{\sigma_{\text{SF}}}$ is the shadow fading, which is normally distributed in dB with zero mean and standard deviation σ_{SF}, X is the aggregated losses coefficient, which accumulates the losses of interference, noise figure and cable losses, and

$$L_{[\text{dB}]}(d) = 10\gamma_c \log_{10} d + \beta_b + 20 \log_{10} f_c \tag{2}$$

is the path loss function of distance d from the BS to the UE, f_c is the carrier frequency (GHz), β_b is a blockage coefficient (32.4 dB when the UE is not blocked by human blockers and 52.4 dB otherwise), and γ_c is the path loss exponent.

We define the requirements for resource blocks that needs to be allocated at the BS to provide the UE a constant bit rate per second as the number of resource blocks per second, which highly depends on the received signal power (1) on the antenna. Thus we define these requirements as a function of SNR (S) as follows:

$$R_p(S) = e_u \frac{B}{N_s R_{PRB}}, \tag{3}$$

where e_u is the spectral efficiency (based on MCS from Table 1), N_s is the number of channel subscribers and B is the bandwidth, calculated to provide a stable bit rate for the UE's application from the Shannon channel capacity expression [15]

$$B = \frac{C}{log_2(1 + S(d))}. \tag{4}$$

Table 1. CQI, MCS, AND SNR mapping for 5G NR.

CQI	MCS	Spectral efficiency	SNR in dB
0	outage		
1	QPSK, 78/1024	0.15237	−9.478
2	QPSK, 120/1024	0.2344	−6.658
3	QPSK, 193/1024	0.377	−4.098
4	QPSK, 308/1024	0.6016	−1.798
5	QPSK, 449/1024	0.877	0.399
6	QPSK, 602/1024	1.1758	2.424
7	16QAM, 378/1024	1.4766	4.489
8	16QAM, 490/1024	1.9141	6.367
9	16QAM, 616/1024	2.4063	8.456
10	64QAM, 466/1024	2.7305	10.266
11	64QAM, 567/1024	3.3223	12.218
12	64QAM, 666/1024	3.9023	14.122
13	64QAM, 772/1024	4.5234	15.849
14	64QAM, 873/1024	5.1152	17.786
15	64QAM, 948/1024	5.5547	19.809

4   Traffic Model

In this section, we analyze the intensities required for our traffic transmission model. We define transition intensity as the average number of transitions per second from one state to another. We obtain the transition intensities from the average time in a corresponding state.

The arrival intensity denotes the traffic that arrives under certain requirements for resource blocks to provide a constant bit rate. For simplicity, we assume the arrival intensity to be in resource blocks per second.

We build a tractable model of traffic transmission in the form of a Continuous Time Markov Chain (CTMC), where each state corresponds to certain channel conditions (blockage, micromobility). We define the arrival intensities based on (1).

4.1 Model Without Outage (CIRCLE Model)

Formally, the arrival intensity when there is no blockage can be denoted as $\lambda_1^{(0)} = R_p(S_1)$, where S_1 is the SNR value in LoS. Taking into account the micromobility problem, we define $\lambda_2^{(0)}$ to be the arrival intensity when MM is happening and $\lambda_3^{(0)}$ to be the arrival intensity when MM is compensated during non-blockage time. As a matter of MM nature, $\lambda_2^{(0)} = 0$. Since we assume that MM compensation has the same time as MM, we define $\lambda_3^{(0)} = 2\lambda_1^{(0)}$. In case of nLoS, blockage will decrease the SNR value, which results in lower value of e_u. In this case, the requirements for resource blocks increase, meaning that the arrival intensity in the blockage state increases as well: $\lambda_4^{(0)} = R_p(S2)$, where S_2 is the SNR value in nLoS ($S_2 < S_1$). However, micromobility can be happening during blockage state, since the UE can still have some signal reception while in blockage. We define the MM and its compensation states in the same way: $\lambda_5^{(0)} = 0$ and $\lambda_6^{(0)} = 2\lambda_4^{(0)}$.

To model the traffic transmission we define a new process with two components: $\psi^{(0)}(t) = (\psi_1^{(0)}(t), \psi_2^{(0)}(t))$. The first component, $\psi_1^{(0)}(t)$, describes the blockage state at some time t, which will have one of two possible values: non-blockage (or LoS state, 0) and blockage (nLoS state, 1). Second component $\psi_2^{(0)}(t)$ describes the micromobility process: no micromobility (or 0), micromobility happened (or 1) and micromobility compensation time (or 2). State space for this process will be denoted as $\chi_0 = \{0,1\} \times \{0,1,2\}$. The corresponding transition intensity diagram for this process is shown in Fig. 4.

The transition intensities (i.e. average number of transitions per unit time) of this process can be described as follows. Let $\tau_b^{(0)}$ be the transition intensity from non-blocked to blocked state. Since we define blockages as renewal process with ω (LoS) and η (nLoS) periods, this transition intensity can be denoted by $\tau_b^{(0)} = 1/E[T_\omega]$, where $E[T_\omega]$ is the average time in LoS state. Similarly, the reverse transition from blocked to non-blocked state is denoted as $\tau_{br}^{(0)} = 1/E[T_\eta]$, where $E[T_\eta]$ is the average time in nLoS state. Since micromobility effects are independent of current state (blocked/non-blocked), the transition intensity in both cases will be the same and denoted as $\tau_m^{(0)} = E[f_m] = 1/\beta$. Taking into account that MM takes the time T_{MM}, the transition intensities from MM to its

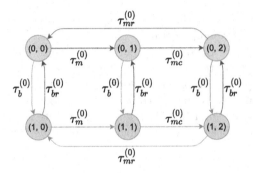

Fig. 4. Transition intensity diagram for CIRCLE model

compensation and from compensation to blocked/non-blocked state is denoted as $\tau_{mc}^{(0)} = 1/T_{MM}$ and $\tau_{mr}^{(0)} = 1/T_{MM}$ correspondingly.

4.2 Model with Outages (RING Model)

Micromobility effects in this model are largely the same as in the previous model. The only difference here is the fact that when the UE is in the blocked state, he loses connection and there is no micromobility. In this blockage state, the arrival intensity drops to zero and a new compensation period is allocated.

Similarly to CIRCLE model, the arrival intensity when blockage didn't happened can be denoted as $\lambda_1^{(1)} = R_p(S)$. Micromobility intensities are similar to previous model. We define $\lambda_2^{(1)} = 0$ and $\lambda_3^{(1)} = 2\lambda_1^{(1)}$ to be the arrival intensities when MM is happening and when MM is compensated correspondingly during non-blockage time. In case of nLoS, blockage leads to connection loss, which means that its arrival intensity $\lambda_4^{(1)} = 0$. Since micromobility can't be happening during blockage state (because of connection loss), we define $\lambda_5^{(1)} = 0$ and $\lambda_6^{(1)} = 0$. The compensation period arrival intensity depends on the blockers density: with more blockers the chance that compensation can be interrupted with another blockage also increases, which results in inequality of periods. In that case we define $\lambda_7^{(1)} = k\lambda_1^{(1)}$, where $k = (E[T_\eta] + E[T_\theta])/E[T_\theta]$ - compensation power coefficient. During compensation period the micromobility arrival intensities are defined as $\lambda_8^{(1)} = 0$ and $\lambda_9^{(1)} = 2\lambda_7^{(1)}$.

To model the traffic transmission we define similar process with two components: $\psi^{(1)}(t) = (\psi_1^{(1)}(t), \psi_2^{(1)}(t))$. The first component, $\psi_1^{(1)}(t)$, describes the blockage state at some time t, which will have one of three possible values: non-blockage (or LoS state, 0), blockage (LoS interruption, 1) and compensation (LoS state with increased intensity, 2). Second component $\psi_2^{(1)}(t)$ describes the micromobility process: no micromobility (or 0), micromobility happened (or 1) and micromobility compensation time (or 2). State space for this process will

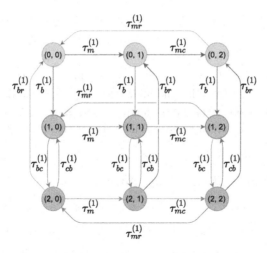

Fig. 5. Transition intensity diagram for RING model

be denoted as $\chi_1 = \{0, 1, 2\} \times \{0, 1, 2\}$. The corresponding transition intensity diagram for this process shown in Fig. 5.

The transition intensities of this process are described similarly to the previous model. The only difference is the compensation period related states. Let $\tau_{bc}^{(1)}$ be the transition intensity from blocked to compensation state. Since in this state the channel can be disrupted again, we define $\tau_{cb}^{(1)}$ as the transition from compensated to blocked state. Thus, the transitions from compensated state to non-blocked and blocked depends on each other, so we define their intensities as a weighted sum: $\tau_{br}^{(1)} + \tau_{bc}^{(1)} = 1/E[T_\theta]$, where $E[T_\theta]$ is the average time in compensation period state. Their individual intensities can be described as:

$$\tau_{br}^{(1)} = \frac{1}{E[T_\theta]} \frac{E[T_\theta]}{E[T_\omega]}, \tau_{cb}^{(1)} = \frac{1}{E[T_\theta]} \frac{(E[T_\omega] - E[T_\theta])}{E[T_\omega]} \tag{5}$$

This transition intensity can be denoted as $\tau_{bc}^{(1)} = 1/E[T_\theta]$. All previously defined transitions are the same:

$$\tau_b^{(1)} = \frac{1}{E[T_\omega]}, \tau_{br}^{(1)} = \frac{1}{E[T_\eta]}, \tau_m^{(1)} = E[f_m] = \frac{1}{\beta}, \tau_{mc}^{(1)} = \frac{1}{T_{MM}}, \tau_{mr}^{(1)} = \frac{1}{T_{MM}}. \tag{6}$$

4.3 Models Superposition

Two previously defined models represent the traffic transmission models depending on the distance from the BS locally for different outage conditions. Here we define the function of distance and merge two defined models into one, which can be accomplished by introducing third component of distance to previously defined two-component processes. We define this new process as

$\psi^{(2)}(t) = (\psi_0^{(2)}(t), \psi_1^{(2)}(t), \psi_2^{(2)}(t))$, where $\psi_0^{(2)}(t)$ is defined as the model compo-nent and will have one of two possible values: CIRCLE model (or 0, it will rep-resent the process $\psi^{(0)}(t)$) and RING model (or 1, $\psi^{(1)}(t)$). Sub-process $\psi_1^{(2)}(t)$ describes the blockage state as a union of sub-processes $\psi_1^{(0)}(t)$ and $\psi_1^{(1)}(t)$, while $\psi_2^{(2)}(t)$ represents the micromobility state, again, as a union of $\psi_2^{(0)}(t)$ and $\psi_2^{(1)}(t)$. State space for $\psi^{(2)}(t)$ will be denoted as $\chi_1 = \{0,1\} \times \{0,1,2\} \times \{0,1,2\}$. The corresponding transition intensity diagram for this process shown in Fig. 6.

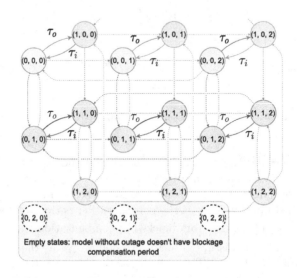

Fig. 6. Transition intensity diagram for model with superposition

In Fig. 6 we demonstrate a combination of two predefined models. The tran-sitions $\tau_i = 1/E[T_R]$ and $\tau_o = 1/E[T_C]$ here denotes the transition intensities from RING to CIRCLE and from CIRCLE to RING, where $E[T_C]$ and $E[T_R]$ are the average times of the UE being in corresponding areas.

5 Performance Comparison

In this section, we provide the analysis of our traffic in the form of queuing systems $M/M/1$ and $MMPP/M/1$. By building two different queuing models we want to find out if there will be any difference in serving the traffic in such systems. A minimal difference in behavior will allow us to approximate such a complex system with a simple model with Poisson distributed traffic arrival. Thus, the main metric of these models is the distribution of the average delay of queuing time (or average waiting time).

In $M/M/1$, the stationary distribution $W(x)$ of the queuing delay of the system is given by:

$$W(x) = 1 - \rho e^{(\lambda - \mu)x}. \tag{7}$$

The average time of customer waiting in queue can be computed as follows:

$$W = \int_0^\infty (1 - W(x))dx = \frac{1}{\mu - \lambda} - \frac{1}{\mu} = \frac{\rho}{\mu - \lambda}. \tag{8}$$

A MMPP is represented by matrices such as the arrival intensity matrix $\mathbf{Q_1} = diag\{\lambda_1, ..., \lambda_K\}$, the infinitesimal generator matrix $\mathbf{Q}$ of the underlying Markov chain that defines the transition intensity according to the states from Sect. 4 and $\mathbf{Q_0} = \mathbf{Q} - \mathbf{Q_1}$.

In fact, the process of serving the requirements for resource blocks inside our system can be described as Quasi-Birth-and-Death (QBD) process with the generator given by $\mathbf{A}$ [4]:

$$\mathbf{A} = \begin{bmatrix} \mathbf{Q_0} & \mathbf{Q_1} & \mathbf{0} & \mathbf{0} & \cdots \\ \mu\mathbf{I} & \mathbf{Q_0} - \mu\mathbf{I} & \mathbf{Q_1} & \mathbf{0} & \cdots \\ \mathbf{0} & \mu\mathbf{I} & \mathbf{Q_0} - \mu\mathbf{I} & \mathbf{Q_1} & \cdots \\ \vdots & \vdots & \vdots & \vdots & \ddots \end{bmatrix}. \tag{9}$$

The stationary distribution of this process is computed as follows:

$$\pi_i = \pi_0 \mathbf{R}^i, \quad i \geq 0, \tag{10}$$

where $\mathbf{R}$ is the minimal non-negative solution of the matrix equation

$$\mu\mathbf{R}^2\mathbf{I} + \mathbf{R}(\mathbf{Q_0} - \mu\mathbf{I}) + \mathbf{Q_1} = \mathbf{0}, \tag{11}$$

and the vector π_0 is the unique solution of the system of linear equations

$$\begin{aligned} \pi_0(\mathbf{Q_0} + \mu\mathbf{R}\mathbf{I}) &= \mathbf{0}, \\ \pi_0(\mathbf{I} - \mathbf{R})^{-1}\mathbf{e} &= 1, \end{aligned} \tag{12}$$

where $\mathbf{e}$ is the unit vector and $\mathbf{I}$ is the identity matrix.

In $MMPP/M/1$, the average time of waiting in queue can be found as follows:

$$w(s) = \pi_0(\mathbf{I} - \mathbf{R}\frac{\mu}{\mu + s})^{-1}\mathbf{e}. \tag{13}$$

6   Numerical Results

6.1   Arrival Process Analysis

The parameters utilized for the physical channel and the simulation process is provided in Table 2. We use the assumption that the UE requires a high-speed connection at the bit rate of 50 Mbps.

Figure 7 shows the time trace sample of resource blocks requirements for some UE, located in CIRCLE area from simulation. It is shown that when the

Table 2. Numerical results parameters

Symbol	Value	Description
f_c	28 GHz	Operating frequency
G_{BS}	11.57 dBm	BS antenna gain
G_{UE}	5.57 dBm	UE antenna gain
R_{PRB}	1.44 MHz	PRB size
h_B, h_A, h_U	1.7, 3, 1.3 m	Blocker, BS and UE heights
X	2, 3, 7 dB	Cable losses, interference margin, noise figure
N_0	−174 dBm/Hz	Noise power spectral density
P_{BS}	2 W (33 dBm)	Transmitting BS power
p_{out}	0.05	Outage probability
S_{thre}	−9.478 dB	SNR threshold
σ_{SFnLoS}	8.2 dB	nLoS shadow fading STD
σ_{SFLoS}	4 dB	LoS shadow fading STD
C	50 Mbps	Required bit rate at the UE
T	10000	Simulation time
Δt	0.1 s	Time slot length
Δt_2	0.001 s	Alternative time slot length (for MM)
N_s	30	Number of subscribers
v	1 m/s	Blockers speed
β	3	MM distribution parameter

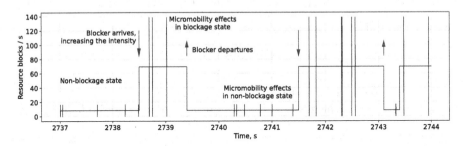

Fig. 7. Resource trace sample from simulation

blocker arrives before the UE, the requirements for resource blocks is increasing to provide a constant bit rate for a smooth user experience. Departure of blocker decreases the requirements correspondingly. Frequent short oscillations show the micromobility effects when resources drops to zero for a short amount of time and then compensated for the same period with twice the rate. It is fair to notice that although the effects of micromobility do not depend on blockages, the arrival intensity during their compensation depends on the current blocked/non-blocked state [1].

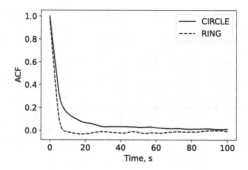

Fig. 8. Autocorrelation function (ACF) for CIRCLE and RING resource blocks traces

The autocorellation function shows that the process of allocating the required resource blocks to provide a constant bit rate is a short-term memory process, as shown in Fig. 8. That proves that this process of resource blocks allocation can be approximated by the Markov process.

6.2 Queuing Time Assessment

This section contains the numerical results for $M/M/1$ and $MMPP/M/1$ for CIRCLE and RING models and their superposition in terms of average waiting time (8) and (13).

The resource blocks service rate (i.e. the average number of processed resource blocks per unit time) can be assessed as follows. We assume that the time of one packet transmission is 3 ms (1 ms for transmission, 1 ms for decoding, and 1 ms for response). The probability of packet loss coincides with the block error rate (BLER) and equals 0.1. The average service time $1/\mu$ is then calculated as $3/(1 - 0.1) = 3.333$ ms and the resource blocks service rate equals μ.

Figure 9 compares the pdf of the average queuing delay (or waiting time) of $M/M/1$ model with the $MMPP/M/1$ model. It is shown that Poisson process (PP) behavior is similar to MMPP. The requirements for resource blocks in MMPP models experience more queuing delay, which is caused by the variations in arrival rate at different states of the communication channel. Table 3 provides the parameters for the distribution of average delay of the system.

It is shown that the difference between mean queuing delay in MMPP and PP models for every model is around 10%. The similarity of behavior of such models shows that the complexity of such model changes the structure not that drastically when load is not so high, allowing for approximating the model with simpler queuing systems.

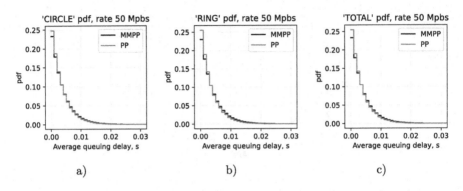

Fig. 9. The pdf of average queuing delay (waiting time) for a) CIRCLE model, b) RING model and c) their superposition TOTAL model.

Table 3. Average delay assessment

	Circle		Ring		Total	
	MMPP	PP	MMPP	PP	MMPP	PP
Mean, ms	3.7909	3.5247	3.8496	3.4194	3.7904	3.4284
Stddev, ms	3.7767	3.4988	3.8302	3.3895	3.7774	3.3978
Arrival rate	0.01187 res.bl/ms		0.03539 res.bl/ms		0.0346 res.bl/ms	

7   Conclusion

In this paper, we evaluate a complex model of traffic transmission in mmWave with blockages and micromobility effects using Markov-based models M/M/1 and MMPP/M/1. We build the tractable model of packet traffic between the UE and the BS with the necessity of providing a constant bit rate to the UE despite different outage conditions, which requires rate compensation. We show that the average queuing delay distribution in both models has a similar pattern, and for the considered set of input parameters we observed no more than 10% difference between the proposed model and Poisson approximation in terms of queuing delay at BS.

References

1. 3GPP: Study on channel model for frequencies from 0.5 to 100 GHz (Release 14). 3GPP TR 38.901 V14.1.1 (2017)
2. Alrabeiah, M., Alkhateeb, A.: Deep learning for mmWave beam and blockage prediction using sub-6 GHZ channels. IEEE Trans. Commun. **68**(9), 5504–5518 (2020). https://doi.org/10.1109/TCOMM.2020.3003670
3. Bai, T., Vaze, R., Heath, R.W.: Analysis of blockage effects on urban cellular networks. IEEE Trans. Wireless Commun. **13**(9), 5070–5083 (2014)

4. Dudin, A.N., Klimenok, V.I., Vishnevsky, V.M.: The Theory of Queuing Systems with Correlated Flows. Springer, Cham (2020). https://doi.org/10.1007/978-3-030-32072-0
5. Firyaguna, F., Bonfante, A., Kibilda, J., Marchetti, N.: Performance evaluation of scheduling in 5G-mmWave networks under human blockage. CoRR abs/2007.13112 (2020). https://arxiv.org/abs/2007.13112
6. Gapeyenko, M., Bor-Yaliniz, I., Andreev, S., Yanikomeroglu, H., Koucheryavy, Y.: Effects of blockage in deploying mmWave drone base stations for 5G networks and beyond. In: 2018 IEEE International Conference on Communications Workshops (ICC Workshops), pp. 1–6 (2018). https://doi.org/10.1109/ICCW.2018.8403671
7. Gapeyenko, M., et al.: On the temporal effects of mobile blockers in urban millimeter-wave cellular scenarios. IEEE Trans. Veh. Technol. **66**(11), 10124–10138 (2017)
8. Gapeyenko, M., et al.: Analysis of human-body blockage in urban millimeter-wave cellular communications. In: 2016 IEEE International Conference on Communications (ICC), pp. 1–7. IEEE (2016)
9. Gerasimenko, M., Moltchanov, D., Gapeyenko, M., Andreev, S., Koucheryavy, Y.: Capacity of multiconnectivity mmWave systems with dynamic blockage and directional antennas. IEEE Trans. Veh. Technol. **68**(4), 3534–3549 (2019)
10. Ma, J., Aijaz, A., Beach, M.: Recent results on proportional fair scheduling for mmWave-based industrial wireless networks. In: 2020 IEEE 92nd Vehicular Technology Conference (VTC2020-Fall), pp. 1–5 (2020). https://doi.org/10.1109/VTC2020-Fall49728.2020.9348753
11. Nain, P., Towsley, D., Liu, B., Liu, Z.: Properties of random direction models. In: IEEE 24th Annual Joint Conference of the IEEE Computer and Communications Societies, vol. 3, pp. 1897–1907 (2005)
12. Ometov, A., Moltchanov, D., Komarov, M., Volvenko, S.V., Koucheryavy, Y.: Packet level performance assessment of mmWave backhauling technology for 3G PP NR systems. IEEE Access **7**, 9860–9871 (2019). https://doi.org/10.1109/ACCESS.2018.2890558
13. Petrov, V., Moltchanov, D., Koucheryavy, Y., Jornet, J.M.: Capacity and outage of terahertz communications with user micro-mobility and beam misalignment. IEEE Trans. Veh. Technol. **69**(6), 6822–6827 (2020)
14. Samuylov, A., et al.: Characterizing resource allocation trade-offs in 5G NR serving multicast and unicast traffic. IEEE Trans. Wireless Commun. **19**(5), 3421–3434 (2020)
15. Shannon, C.: Communication in the presence of noise. Proc. IRE **37**(1), 10–21 (1949). https://doi.org/10.1109/JRPROC.1949.232969
16. Wang, M., Liu, J., Chen, W., Ephremides, A.: Joint queue-aware and channel-aware delay optimal scheduling of arbitrarily bursty traffic over multi-state time-varying channels. IEEE Trans. Commun. **67**(1), 503–517 (2018)

Model for Analyzing Impact of Path Loss on eMBB Bit Rate Degradation Under Priority URLLC Transmission in 5G Network

Irina Kochetkova[1,2]([envelope]) [ORCID], Elena Makeeva[1] [ORCID], Anastasia Ageeva[1], and Andrey Gorshenin[2] [ORCID]

[1] Peoples' Friendship University of Russia (RUDN University), 6 Miklukho-Maklaya Street, Moscow, Russian Federation
kochetkova-ia@rudn.ru

[2] Federal Research Center "Computer Science and Control" of the Russian Academy of Sciences, 44-2 Vavilova Street, Moscow, Russian Federation

Abstract. One of the challenging tasks in 5G networks is to organize a joint URLLC (ultra-reliable and low-latency communication) and eMBB (enhanced mobile broadband) transmission in such a way that provides the priority to URLLC connections. The eMBB users suffer quality of service degradation, primarily bit rate degradation, as well as service interruption. In the paper, we provide a queuing model for analyzing this effect depending on several path loss models. The queuing system is of type resource queuing system, where the resource has three-dimensional structure – frequency bandwidth, radio frame length, and transmitted signal power. Due to different URLLC and eMBB bit rate requirements, we use weighted round robin (WRR) resource allocation scheme. The stationary probability distribution depends on the conditional probabilities of session acceptance. We provide the formulas for calculating eMBB metrics – average bit rate, interruption probability, and blocking probability. A numerical example illustrates the impact of two path loss models for macro- and microcells on eMBB metrics.

Keywords: 5G · URLLC · eMBB · Bit rate degradation · Service interruption · Path loss model · Weighted round robin · Queuing system · Resource queuing system · Interruption probability

1 Introduction

Fifth generation (5G) systems require a huge collective effort to specify, standardize, manufacture, design, and deploy cellular networks [1,2]. 5G systems

The research was supported by the Ministry of Science and Higher Education of the Russian Federation, project No. 075-15-2020-799. The research was carried out using the infrastructure of the Shared Research Facilities "High Performance Computing and Big Data" (CKP "Informatics") of the Federal Research Center "Computer Science and Control" of the Russian Academy of Sciences.

V. M. Vishnevskiy et al. (Eds.): DCCN 2022, LNCS 13766, pp. 176–189, 2022.
https://doi.org/10.1007/978-3-031-23207-7_14

are being developed for numerous different application scenarios, e.g. hotspot connectivity, wireless control of industrial manufacturing, production processes [1,3,4]. For example, industrial manufacturing requires the support for ultra-reliable low latency service (URLLC) and enhanced mobile broadband (eMBB). Therefore, 5G systems must support the joint transmission of URLLC and eMBB sessions.

The eMBB traffic needs to support Gbps with moderate latency, while URLLC traffic requires extremely low latency with very high reliability (99.999%) [5,6]. The joint operation of URLLC and eMBB services on the same radio resource leads to a user-scheduling problem that is quite difficult to solve due to the QoS requirements and location [7–10,12,13].

In this paper, we will consider the problem of joint URLLC and eMBB transmission within one base station in the form of a queuing system with a decrease in eMBB traffic bit rate, signal attenuation, and URLLC traffic priority. Under such a consideration, eMBB traffic is strongly affected and its speed suffers greatly.

The rest of the paper is organized as follows. In Sect. 2, we specify the system model and schemes of priority URLLC transmission and resource allocation. We mathematically describe the path loss model and derive the conditional probabilities of session acceptance in Sect. 3. In Sect. 4, we model the joint URLLC and eMBB transmission as a resource queuing system [11], as well as numerically illustrate some eMBB metrics. Conclusions are provided in the last section.

2   System Model

2.1   General Assumptions

Our system model represents one cell of radius R with one base station (BS) in the center and the uniformly distributed devices generating two traffic types – eMBB and URLLC, as shown in Fig. 1. The bandwidth of this cell is N physical resource blocks (PRB). Resource block is divided by 7 mini-slots. One PRB is equal to one time slot, which corresponds to 0.5 ms. We make assumptions that eMBB devices are distributed relative to slots, i.e. 1 PRB refers to one eMBB devices, and URLLC devices relative to mini-slots, i.e. 1 mini-slot – to one URLLC device. Thus, the maximum number of eMBB sessions that can be served simultaneously is N, and the maximum number of URLLC devices $C = 7N$.

The eMBB and URLLC sessions arrive according to the Poisson process with rates λ_m and λ_u, respectively. The eMBB sessions start their service at maximum bit rate b_1 occupying 1 PRB. When an URLLC sessions arrives, an eMBB session bit rate could be partially reduced to values $b_1 > b_2 > ... > b_K$ even up to 0 due to the priority of URLLC traffic, $b = (b_1, ..., b_K)$. The session duration is exponentially distributed with rates μ_m and μ_u respectively.

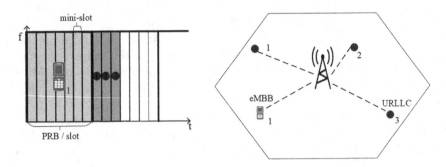

Fig. 1. Considered cell and radio frame allocation when accepting priority URLLC sessions

2.2   Priority URLLC Transmission

On an URLLC session arrival, four cases are possible:

- If there is at least 1 free mini-slot, then URLLC traffic will be transmitted.
- If there is no free mini-slot and at least 1 eMBB session is not at the minimum rate, then URLLC traffic will be transmitted due to the bit rate degradation of such eMBB session.
- If there is no free mini-slot and all eMBB sessions device are at the minimum rate, then URLLC traffic will be transmitted due to the interruption of an eMBB session.
- If there is no free mini-slot and no eMBB sessions, then URLLC traffic will be blocked.

When URLLC traffic is transmitted, the following cases are possible:

- If there is no one eMBB sessions in the system or eMBB sessions are serviced at maximum speed, then the resources occupied by URLLC traffic are released.
- If there is at least 1 eMBB session is not at the maximum speed, the mini-slots occupied by it are released and the reduced eMBB bit rate is restored.

When the eMBB session arrives, it can be accepted for service if there is at least 1 free PRB in the system, otherwise the session will be blocked. When eMBB traffic is transmitted, the PRB occupied by it are released.

2.3   Resource Requirement

We consider a three-dimensional structure of the resource as shown in Fig. 2, where $\widehat{F}$ is total frequency bandwidth, $\widehat{T} = 1$ is normalized radio frame length, $\widehat{P}$ is maximum transmitted signal power.

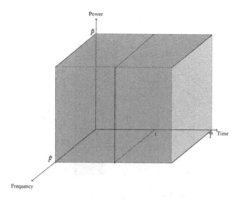

Fig. 2. Three-dimensional resource structure – frequency bandwidth, radio frame length, and transmitted signal power

The need for resources can be expressed as a vector $\boldsymbol{y} = (y_f, y_t, y_p)$ and the resource allocation policy is based on its division by frequency and time, i.e. given u sessions the set of possible resource occupation configurations looks like

$$\mathcal{Y}(u) = \{(\boldsymbol{y}_1, ..., \boldsymbol{y}_u) : 0 < y_f^i \leq \widehat{F}, 0 < y_t^i \leq 1, 0 < y_p^i \leq \widehat{P},$$
$$\sum_{i=1}^{u} y_f^i \leq \widehat{F}, \sum_{i=1}^{u} y_t^i \leq 1, y_p^i \leq \widehat{P}, i = 1, ..., u\}, u > 0.$$

The transmission bit rate b can be represented from the Shannon theorem as follows

$$b = y_f y_t log_2 \left(1 + \frac{y_p \cdot PL}{N_0 + I_0} \right),$$

where PL is variable path loss, N_0 is constant noise power, I_0 constant interference power.

2.4   RR Resource Allocation

One of the resource allocation schemes is round robin (RR). According to it, the time resource is divided into equal parts between all user sessions, which in order to ensure a guaranteed minimum speed, regulate the data transmission power themselves. So set $\mathcal{Y}_{RR}(u)$ of possible configurations of resource occupation have the form

$$\boldsymbol{y}_{RR}(u) = \left(y_f = \widehat{F}, y_t = \frac{1}{u}, y_p = \frac{N_0 + I_0}{PL} \left(2^{b \cdot u / \widehat{F}} - 1 \right) \right), u > 0;$$

$$\mathcal{Y}_{RR}(u) = \left\{ \begin{array}{c} (\boldsymbol{y}_1, ..., \boldsymbol{y}_u) : y_f(i) = \widehat{F}, y_t(i) = \frac{1}{u}, \\ y_p(i) = \frac{N_0 + I_0}{PL} \left(2^{b \cdot u / \widehat{F}} - 1 \right) \leq \widehat{P}, i = 1, ..., u, PL > 0 \end{array} \right\}, u > 0.$$

The example for this resource allocation is shown in Fig. 3.

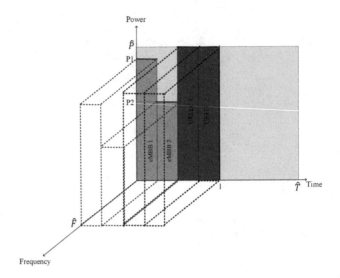

Fig. 3. Example of Round Robin (RR) resource allocation for 2 types of sessions with different bit rate requirements

2.5 WRR Resource Allocation

As we consider different bit rates for URLLC and eMBB sessions, we use weighted round robin (WRR) scheme. Let L be the number of bit rate levels and $b_l, l = \overline{1, L}$ be the bit rate level. Then the possible configurations of resource occupation can be represented as follows

$$\boldsymbol{y}_l^{WRR}(u_1, ..., u_L) = \left(\begin{array}{c} y_f = \widehat{F}, y_t = \dfrac{b_l}{\sum_{i=1}^{L} b_i u_i}, \\ y_p = \dfrac{N_0 + I_0}{PL} \left(2^{\sum_{i=1}^{L} b_i u_i / \widehat{F}} - 1 \right) \end{array} \right),$$

$$u_l \geq 0, l = 1, ..., L, \sum_{l=1}^{L} u_l > 0;$$

$$\mathcal{Y}_l^{WRR}(u_1, ..., u_L) = \left\{ \begin{array}{c} (\boldsymbol{y}_1^1, ..., \boldsymbol{y}_{u_1}^1, ..., \boldsymbol{y}_1^L, ..., \boldsymbol{y}_{u_L}^L) : y_f(l, i) = \widehat{F}, \\ y_t(l, i) = \dfrac{b_l}{\sum_{i=1}^{L} b_i u_i}, y_p(l, i) = \dfrac{N_0 + I_0}{PL} \cdot \\ \cdot \left(2^{\sum_{j=1}^{L} b_j u_j / \widehat{F}} - 1 \right) \leq \widehat{P}, \\ i = 1, ..., u_l, l = 1, ..., L, PL > 0 \end{array} \right\},$$

$$u_l \geq 0, l = 1, .., L, \sum_{l=1}^{L} u_l > 0.$$

2.6 Resource Allocation for URLLC & eMBB

Let us adopt WRR scheme for the considered system model. Let $\boldsymbol{x} = (m_1, ..., m_K, n)$ be the vector of eMBB and URLLC sessions number, $\sum_{k=1}^{K} m_k$ be the number of eMBB sessions in the system, n be the number of URLLC sessions, $\boldsymbol{x} \in \mathcal{X}$, where $\mathcal{X}$ is the set of possible combination of eMBB and URLLC sessions number in the system. Then the aggregated bit rate of URLLC and eMBB sessions is $c(\boldsymbol{x}) = n + \sum_{k=1}^{K} b_k m_k$.

We consider the following resource allocation scheme

$$\boldsymbol{y}^u(\boldsymbol{x}) = \left(y_f = \widehat{F}, y_t = \frac{1}{c(\boldsymbol{x})}, y_p = \widehat{P} \right), \boldsymbol{x} \in \mathcal{X};$$

$$\boldsymbol{y}_k^m(\boldsymbol{x}) = \left(y_f = \widehat{F}, y_t = \frac{b_k}{c(\boldsymbol{x})}, y_p = \frac{N_0 + I_0}{PL} \left[2^{c(\boldsymbol{x})/\widehat{F}} - 1 \right] \right), \boldsymbol{x} \in \mathcal{X}, k = \overline{1, K};$$

$$\mathcal{Y}(\boldsymbol{x}) = \left\{ \begin{array}{c} (\boldsymbol{y}_{11}^m, ..., \boldsymbol{y}_{1m_1}^m, ..., \boldsymbol{y}_{K1}^m, ..., \boldsymbol{y}_{Km_K}^m, \boldsymbol{y}_1^u, ..., \boldsymbol{y}_n^u) : \\ y_f^u(i) = \widehat{F}, y_t^u(i) = \frac{1}{c(\boldsymbol{x})}, y_p^u(i) = \widehat{P}, i = 1, ..., n, \\ y_f^m(k, j) = \widehat{F}, y_t^m(k, j) = \frac{b_k}{c(\boldsymbol{x})}, \\ y_p^m(k, j) = \frac{N_0 + I_0}{PL} \left[2^{c(\boldsymbol{x})/\widehat{F}} - 1 \right] \leq \widehat{P}, \\ j = 1, ..., m_k, k = 1, ..., K, PL > 0 \end{array} \right\}, \boldsymbol{x} \in \mathcal{X}.$$

3 Path Loss Based Admission Control

3.1 Path Loss Model

To study the effect of electromagnetic wave attenuation, we consider two scenarios – macrocell (UMa) and urban microcell street canyon (UMi-Street Canyon) with non-light of sight (see Table 1). Center frequency f_c has a range of $0.5\,\text{GHz} < f_c < 100\,\text{GHz}$, d_{3D} is the distance between the top of the base station (BS) and the height of the user terminal (UT) [14].

Based on the formulas from Table 1, we could express the PL formula as follows

$$PL(\text{dB}) = 10A \cdot \log_{10}(d) + 10B \log_{10}(f_c) + C;$$

$$PL(\text{dB}) = 10A \cdot \log_{10}(d) + B;$$

$$PL = d^A 10^B,$$

where d is the distance between BS and UT, A and B are constants.

For the considered scenarios, the PL formulas are calculated as following

$$UMaPL = 10^{3.24} d^3,$$

$$UMiPL = 10^{3.24} d^{3.19}.$$

Table 1. Parameters of path loss models

Scenario	Path Loss (dB)
UMa, NLOS	$PL = 32.4 + 20\log_{10}(f_c) + 30\log_{10}(d_{3D})$
UMi-Street Canyon, NLOS	$PL = 32.4 + 20\log_{10}(f_c) + 31.9\log_{10}(d_{3D})$

The PL cumulative distribution function (CDF) $F_{PL}(x)$ can be express in terms of CDF of distance between UT and BS. We assume that the devices are distributed uniformly around the circle R, then CDF $F_d(x) = \mathbf{P}\{\xi_d \le x\} = x^2/R^2$, and probability density function (PDF) $f_d(x) = 2x/R^2$. Therefore, $\xi_{PL} = A\xi_d^B$ and

$$F_{PL}(x) = \mathbf{P}\{\xi_{PL} \le x\} = \mathbf{P}\left\{\xi_d \le \left(\frac{x}{A}\right)^{-B}\right\} = F_d\left(\frac{x}{A}\right)^{-B} = \left(\frac{x}{A}\right)^{-2B} \cdot \frac{1}{R^2}.$$

3.2 Conditional Probability of Session Acceptance

Given the effect of random PL fading, the probability of accepting a request for the transmission of traffic is a conditional probability that $u+1$ sessions could be in the system given that u are in the system

$$p(0) = \mathrm{P}\{\mathbf{y}_1 \in \mathcal{Y}(1)\},$$
$$p(u) = \mathrm{P}\{(\mathbf{y}_1,...,\mathbf{y}_u,\mathbf{y}_{u+1}) \in \mathcal{Y}(u+1)|(\mathbf{y}_1,...,\mathbf{y}_u) \in \mathcal{Y}(u)\}$$
$$= \frac{\mathrm{P}\{(\mathbf{y}_1,...,\mathbf{y}_u,\mathbf{y}_{u+1}) \in \mathcal{Y}(u+1)\}}{\mathrm{P}\{(\mathbf{y}_1,...,\mathbf{y}_u) \in \mathcal{Y}(u)\}}, u = 1,2,...$$

where $A(u) = \mathrm{P}\{(\mathbf{y}_1,...,\mathbf{y}_u) \in \mathcal{Y}(u)\}$ is probability that u sessions could be in the system.

3.3 Conditional Probability for RR

Thereby, the probability $A_{RR}(u)$ for the RR resource allocation policy is calculated as follows

$$A_{RR}(u) = \mathrm{P}\{(\mathbf{y}_1,...,\mathbf{y}_u) \in \mathcal{Y}_{RR}(u)\}$$
$$= \mathrm{P}\left\{y_p^i = \frac{N_0+I_0}{PL}\left(2^{bu/\widehat{F}}-1\right) \le \widehat{P}, i=1,...,u\right\}$$
$$= \left[\mathrm{P}\left\{\frac{N_0+I_0}{PL}\left(2^{bu/\widehat{F}}-1\right) \le \widehat{P}\right\}\right]^u = \left[\mathrm{P}\left\{PL \ge \frac{N_0+I_0}{\widehat{P}}\left(2^{bu/\widehat{F}}-1\right)\right\}\right]^u$$
$$= \left[1 - F_{PL}\left(\frac{N_0+I_0}{\widehat{P}}\left(2^{bu/\widehat{F}}-1\right)\right)\right]^u, u=0,1,...$$

$$p_{RR}(u) = \frac{A_{RR}(u+1)}{A_{RR}(u)} = \frac{\left[1 - F_{PL}\left(\frac{N_0+I_0}{\widehat{P}}\left(2^{b(u+1)/\widehat{F}}-1\right)\right)\right]^{u+1}}{\left[1 - F_{PL}\left(\frac{N_0+I_0}{\widehat{P}}\left(2^{bu/\widehat{F}}-1\right)\right)\right]^u}, u=1,2,...$$

3.4   Conditional Probability for WRR

For the WRR scheme, the formulas are transformed to

$$A_{WRR}(u_1, ..., u_L) = P\left\{(\boldsymbol{y}_1^1, ..., \boldsymbol{y}_{u_1}^1, ..., \boldsymbol{y}_1^L, ..., \boldsymbol{y}_{u_L}^L) \in \mathcal{Y}_{WRR}(u_1, ..., u_L)\right\}$$

$$= P\left\{y_p(l, i) = \frac{N_0 + I_0}{PL}\left(2^{\sum_{j=1}^{L} b_j u_j / \widehat{F}} - 1\right) \le \widehat{P}, i = 1, ..., u_l, l = 1, ..., L\right\}$$

$$= \left[\frac{N_0 + I_0}{PL}\left(2^{\sum_{j=1}^{L} b_j u_j / \widehat{F}} - 1\right) \le \widehat{P}\right]^{\sum_{l=1}^{L} u_l}$$

$$= \left[P\left\{PL \ge \frac{N_0 + I_0}{\widehat{P}}\left(2^{\sum_{j=1}^{L} b_j u_j / \widehat{F}} - 1\right)\right\}\right]^{\sum_{l=1}^{L} u_l}$$

$$= \left[1 - F_{PL}\left(\frac{N_0 + I_0}{\widehat{P}}\left(2^{\sum_{j=1}^{L} b_j u_j / \widehat{F}} - 1\right)\right)\right]^{\sum_{l=1}^{L} u_l},$$

$$u_l \ge 0, l = 1, .., L, \sum_{l=1}^{L} u_l > 0.$$

$$p_{WRR}^l(u_1, ..., u_L) = \frac{A_{WRR}(u_1, ..., u_l + 1, ..., u_L)}{A_{WRR}(u_1, ..., u_L)}$$

$$= \frac{\left[1 - F_{PL}\left(\frac{N_0 + I_0}{\widehat{P}}\left(2^{\sum_{j=1}^{L}(b_j u_j + b_l)} - 1\right)\right)\right]^{\sum_{l=1}^{L} u_l + 1}}{\left[1 - F_{PL}\left(\frac{N_0 + I_0}{\widehat{P}}\left(2^{\sum_{j=1}^{L} b_j u_j} - 1\right)\right)\right]^{\sum_{l=1}^{L} u_l}}, u_l \ge 0, l = 1, .., L, \sum_{l=1}^{L} u_l > 0.$$

3.5   Conditional Probability for URLLC & eMBB

We apply the received formulas for WRR to the eMBB and URLLC case: $(p_m(\boldsymbol{x}), x \ge 0)$ and $(p_u(\boldsymbol{x}, x \ge 0)$

$$A(\boldsymbol{x}) = P\left\{(\boldsymbol{y}_{11}^m, ..., \boldsymbol{y}_{1m_1}^m, ..., \boldsymbol{y}_{K1}^m, ..., \boldsymbol{y}_{Km_K}^m, \boldsymbol{y}_1^u, ..., \boldsymbol{y}_n^u) \in \mathcal{X}\right\}$$

$$= \left(1 - F_{PL}\left(\frac{N_0 + I_0}{\widehat{P}}\left[2^{c(\boldsymbol{x})/\widehat{F}} - 1\right]\right)\right)^{\sum_{k=1}^{K} m_k}, \boldsymbol{x} \in \mathcal{X};$$

$$p_m(\boldsymbol{x}) = \frac{A(\boldsymbol{x} + \boldsymbol{e}_1)}{A(\boldsymbol{x})}$$

$$= \frac{\left[1 - \left(A^{-1}\left(\frac{N_0 + I_0}{\widehat{P}}\left[2^{c(\boldsymbol{x} + \boldsymbol{e}_1)/\widehat{F}} - 1\right]\right)^{-2B} \cdot \frac{1}{R^2}\right)\right]^{\sum_{k=1}^{K} m_k + 1}}{\left[1 - \left(A^{-1}\left(\frac{N_0 + I_0}{\widehat{P}}\left[2^{c(\boldsymbol{x})/\widehat{F}} - 1\right]\right)^{-2B} \cdot \frac{1}{R^2}\right)\right]^{\sum_{k=1}^{K} m_k}}, \boldsymbol{x} \in \mathcal{X}. \quad (1)$$

$$p_u(\boldsymbol{x}) = \frac{A(\boldsymbol{x} + \boldsymbol{e}_{K+1})}{A(\boldsymbol{x})}$$

$$= \frac{\left[1 - \left(A^{-1}\left(\frac{N_0 + I_0}{\widehat{P}}\left[2^{c(\boldsymbol{x} + \boldsymbol{e}_{K+1})/\widehat{F}} - 1\right]\right)^{-2B} \cdot \frac{1}{R^2}\right)\right]^{\sum_{k=1}^{K} m_k}}{\left[1 - \left(A^{-1}\left(\frac{N_0 + I_0}{\widehat{P}}\left[2^{c(\boldsymbol{x})/\widehat{F}} - 1\right]\right)^{-2B} \cdot \frac{1}{R^2}\right)\right]^{\sum_{k=1}^{K} m_k}}, \boldsymbol{x} \in \mathcal{X}. \tag{2}$$

4 Queuing Model Analysis

4.1 Continuous-Time Markov Chain

Let us describe the system using the Continuous-Time Markov Chain (CTMC) $\boldsymbol{X}(t)$ with states $\boldsymbol{x} = (\boldsymbol{m}, n) = (m_1, ..., m_K, n)$ and state space

$$\mathcal{X} = \left\{ \begin{array}{c} (\boldsymbol{m}, n) = (m_1, ..., m_K, n) : m_k \geq 0, k = \overline{1, K}, 0 \leq n \leq C, \\ \displaystyle\sum_{k=1}^{K} m_k \leq N, \sum_{k=1}^{N} m_k \cdot b_k + n \leq C. \end{array} \right\} \tag{3}$$

4.2 Infinitesimal Generator

Taking into account the conditional probabilities that a new eMBB/URLLC session could be accepted given that $\sum_{k=1}^{K} m_k + n$ sessions are in the system ($p_m(\boldsymbol{x}) / p_u(\boldsymbol{x})$, $\boldsymbol{x} \in \mathcal{X}$), which are calculated by (1) and (2), the elements of the infinitesimal generator look like as follows

$$a(\boldsymbol{m}, n)(\boldsymbol{m} + \boldsymbol{e}_1, n) = \lambda_m \cdot p_m(\boldsymbol{x}),$$
$$\sum_{k=1}^{K} m_k \cdot b_k + n + b_1 \leq C, \sum_{k=1}^{K} m_k < N;$$
$$a(\boldsymbol{m}, n)(\boldsymbol{m} - \boldsymbol{e}_k, n) = m_k \cdot \mu_m,$$
$$m_k > 0;$$
$$a(\boldsymbol{m}, n)(\boldsymbol{m}, n + 1) = \lambda_u \cdot p_u(\boldsymbol{x}),$$
$$\sum_{k=1}^{K} m_k \cdot b_k + n + 1 \leq C;$$
$$a(\boldsymbol{m}, n)(\boldsymbol{m} - \boldsymbol{e}_k + \boldsymbol{e}_{k+1}, n + 1) = \lambda_u \cdot p_u(\boldsymbol{x}),$$
$$\sum_{k=1}^{K} m_k \cdot b_k + n + 1 > C, \sum_{i=1}^{K-1} m_i = 0, m_k > 0, k = \overline{1, K - 1};$$
$$a(\boldsymbol{m}, n)(\boldsymbol{m} - \boldsymbol{e}_K, n + 1) = \lambda_u \cdot p_u(\boldsymbol{x}),$$
$$\sum_{k=1}^{K} m_k \cdot b_k + n + 1 > C, \sum_{k=1}^{K-1} m_k = 0, m_K > 0;$$

$$a(\boldsymbol{m}, n)(\boldsymbol{m}, n-1) = n \cdot \mu_u,$$

$$n > 0, \sum_{k=2}^{K} m_k = 0;$$

$$a(\boldsymbol{m}, n)(\boldsymbol{m} - \boldsymbol{e}_k + \boldsymbol{e}_{k-1}, n-1) = n \cdot \mu_u,$$

$$n > 0, \sum_{i=k+1}^{K} m_i = 0, m_k > 0, k = \overline{2, K}.$$

Solving the equilibrium equations, the stationary probabilities $\pi(\boldsymbol{x}), (\boldsymbol{x}) \in \mathcal{X}$ can be found.

4.3 Performance Metrics

After finding the stationary probabilities $\pi(\boldsymbol{x}), (\boldsymbol{x}) \in \mathcal{X}$ the following performance metrics can be calculated:

- Interruption probability of eMBB traffic

$$p_m^{\mathrm{in}} = \sum_{m_K=1}^{M} \frac{1}{m_K} \frac{\lambda_u \cdot p_u \left(0, ..., 0, m_K, C - b_K m_K\right)}{\lambda_u \cdot p_u \left(0, ..., 0, m_K, C - b_K m_K\right) + \left(C - b_K m_K\right) \mu_u + m_K \mu_m}.$$

$$\cdot \pi \left(0, ..., 0, m_K, C - b_K m_K\right).$$

- Blocking probability of eMBB traffic

$$p_m^{\mathrm{bl}} = \sum_{\boldsymbol{x} \in \mathcal{X}} \pi \left(\boldsymbol{x}\right) \left(1 - p_m(\boldsymbol{x})\right).$$

- Average eMBB bit rate

$$\bar{b} = \frac{\sum_{\boldsymbol{x} \in \mathcal{X} \setminus \{m=0\}} \sum_{k=1}^{K} b_k m_k \left(\sum_{k=1}^{K} m_k\right)^{-1} \pi(\boldsymbol{x})}{1 - \sum_{n=0}^{C} \pi(\boldsymbol{0}, n)}.$$

4.4 Numerical Example

For the numerical example, we considered the effect of signal attenuation functions on eMBB and URLLC sessions bit rate, taking into account the bit rate degradation. Let us consider a channel with a bandwidth (B) of 1.4 MHz of $N = 6$ PRBs. Then the total number of mini-slots is $C = 6 \cdot 7 = 42$. The eMBB sessions arrival rate is $\lambda_m = 1$ sessions per second and URLLC sessions is $\lambda_u = \lambda_m * i, i = 1, ..., 100000$ sessions per second. We assume that the maximum eMBB session rate is $b_1 = 7$ and when a URLLC session arrives, the eMBB rate could be reduced by 1 until it reaches the minimum rate $b_K = 5$, i.e. $\boldsymbol{b} = 7, 6, 5$. The URLLC rate is constant and equal to $d = 1$. The service time of one eMBB session is 60 s $(\mu_m = 1/60)$, and URLLC session is 1 ms, i.e. 0.001 s. $(\mu_u = 1000)$. Initial data are shown in the Table 2.

Table 2. Parameters for numerical example

Parameter	Definition	Value	
		UMa	UMi
B	Bandwidth	1.6 MHz	1.6 MHZ
f_c	Center frequency	1 GHz	1 GHz
d	Distance between BS and user terminal	35 m	10 m
R	Cell radius	5000 m	5000 m
N_0	Noise coefficient	9 dB	9 dB
I_0	Interference coefficient	–	–
$\widehat{F}$	Maximum frequency/Carrier frequency	6 GHz	6 GHz
$\widehat{P}$	Maximum power	49 dBm	44 dBm
A	PL coefficient	3	3.19
B	PL coefficient	3.24	3.24

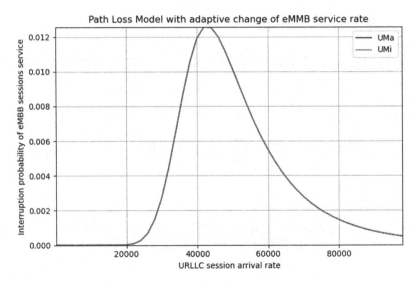

Fig. 4. Interruption probability of eMBB session for the macrocell (UMa) and urban microcell street canyon (UMi-Street Canyon) path loss models: absolute values

Let us illustrate the behavior of the performance metrics such as interruption probability of eMBB sessions and average rate of eMBB sessions for the UMa and UMi-street canyon scenarios. The Fig. 4 illustrates that the interruption probability depending on the increasing URLLC session arrival rate. The difference between two scenarios is insignificant, as shown in Fig. 5. Either the ambiguous behavior of the schedule can be explained by the fact that at first the number of eMBB sessions in the system grows, and then it will decrease due to the fact that URLLC sessions enter the system much more often than eMBB

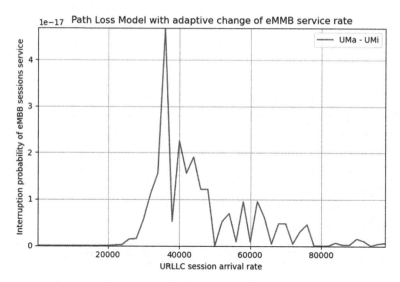

Fig. 5. Interruption probability of eMBB session for the macrocell (UMa) and urban microcell street canyon (UMi-Street Canyon) path loss models: differences

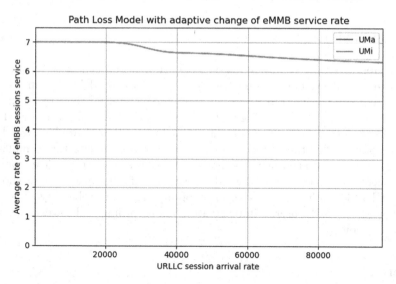

Fig. 6. Average eMBB session bit rate for the macrocell (UMa) and urban microcell street canyon (UMi-Street Canyon) path loss models: absolute values

sessions, and new eMBB sessions do not arrive. The Fig. 6 shows the behavior of the average eMBB session bit rate: the measure is decreasing depending on increasing the URLLC session arrival intensity for two scenarios with insignificant difference Fig. 7.

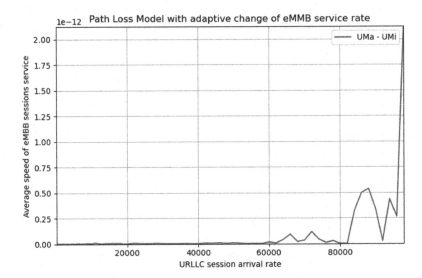

Fig. 7. Average eMBB session bit rate for the macrocell (UMa) and urban microcell street canyon (UMi-Street Canyon) path loss models: differences

5   Conclusion

In this paper, we have considered two signal attenuation scenarios (UMa, UMi-street-canyon). A system for joint transfer of eMBB and URLLC traffic with a possible reduction in the transmission rate of eMBB traffic for these two scenarios in the form of a resource queuing system was proposed. We also considered the principles of resource allocation for different schemes (Round Robbin, Weighted Round Robbin). For the introduced resource queuing system, the formulas of the performance measures for this model were obtained. A numerical example for the introduced model is considered. The results of numerical analysis showed that if URLLC traffic users arrive more often, then the probability of service interruption for eMBB users will be greater, and accordingly, the average service rate will be lower.

References

1. Navarro-Ortiz, J., Romero-Diaz, P., Sendra, S., Ameigeiras, P., Ramos-Munoz, J.J., Lopez-Soler, J.M.: A survey on 5G usage scenarios and traffic models. IEEE Commun. Surv. Tutor. **22**(2), 905–929 (2020). https://doi.org/10.1109/COMST. 2020.2971781
2. ITU-R M.2412. Guidelines for evaluation of radio interface technologies for IMT-2020 (2021)
3. Kochetkov, D., Almaganbetov, M.: Using patent landscapes for technology benchmarking: a case of 5G networks. Adv. Syst. Sci. Appl. **21**(2), 20–28 (2021). https://doi.org/10.25728/assa.2021.21.2.988

4. Kochetkov, D., Vuković, D., Sadekov, N., Levkiv, H.: Smart cities and 5G networks: an emerging technological area? J. Geogr. Inst. Jovan Cvijic SASA **69**(3), 289–295 (2019). https://doi.org/10.2298/IJGI1903289K

5. Guo, S., Lu, B., Wen, M., Dang, S., Saeed, N.: Customized 5G and beyond private networks with integrated URLLC, eMBB, mMTC, and positioning for industrial verticals. IEEE Commun. Stand. Mag. **6**(1), 52–57 (2022). https://doi.org/10.1109/MCOMSTD.0001.2100041

6. Varsier, N., Dufrene, L.-A., Dumay, M., Lampin, Q., Schwoerer, J.: A 5G new radio for balanced and mixed IoT use cases: challenges and key enablers in FR1 band. IEEE Commun. Mag. **59**(4), 82–87 (2021). https://doi.org/10.1109/MCOM.001.2000660

7. Anand, A., De Veciana, G., Shakkottai, S.: Joint scheduling of URLLC and eMBB traffic in 5G wireless networks. IEEE/ACM Trans. Netw. **28**(2), 477–490 (2020). https://doi.org/10.1109/TNET.2020.2968373

8. Yarkina, N., Correia, L.M., Moltchanov, D., Gaidamaka, Y., Samouylov, K.: Multitenant resource sharing with equitable-priority-based performance isolation of slices for 5G cellular systems. Comput. Commun. **188**, 39–51 (2022). https://doi.org/10.1016/j.comcom.2022.02.019

9. Ivanova, D., Markova, E., Moltchanov, D., Pirmagomedov, R., Koucheryavy, Y., Samouylov, K.: Performance of priority-based traffic coexistence strategies in 5G mmWave industrial deployments. IEEE Access **10**, 9241–9256 (2022). https://doi.org/10.1109/ACCESS.2022.3143583

10. Sopin, E., Begishev, V., Moltchanov, D., Samuylov, A.: Resource queuing system with preemptive priority for performance analysis of 5G NR systems. In: Vishnevskiy, V.M., Samouylov, K.E., Kozyrev, D.V. (eds.) DCCN 2020. LNCS, vol. 12563, pp. 87–99. Springer, Cham (2020). https://doi.org/10.1007/978-3-030-66471-8_8

11. Naumov, V., Samouylov, K.: Resource system with losses in a random environment. Mathematics **9**(21), 2685 (2021). https://doi.org/10.3390/math9212685

12. Kushchazli, A., Ageeva, A., Kochetkova, I., Kharin, P., Chursin, A., Shorgin, S.: Model of radio admission control for URLLC and adaptive bit rate eMBB in 5G network. In: CEUR Workshop Proceedings, vol. 2946, pp. 74–84 (2021)

13. Makeeva, E., Polyakov, N., Kharin, P., Gudkova, I.: Probability model for performance analysis of joint URLLC and eMBB transmission in 5G networks. In: Galinina, O., Andreev, S., Balandin, S., Koucheryavy, Y. (eds.) NEW2AN/ruSMART -2019. LNCS, vol. 11660, pp. 635–648. Springer, Cham (2019). https://doi.org/10.1007/978-3-030-30859-9_55

14. 3GPP TR 38.901 Study on channel model for frequencies from 0.5 to 100 GHz (2022)

Firewall Simulator Development for Performance Evaluation of Ranging a Filtration Rules Set

A. Yu. Botvinko$^{(\boxtimes)}$ and K. E. Samouylov ⓘ

Peoples' Friendship University of Russia, Miklukho-Maklaya.6, Moscow, Russia
{botviay,ksam}@sci.pfu.edu.ru

Abstract. This paper is written as a continuation of works devoted to solving the task of increasing the firewall performance in conditions of high heterogeneity and variability of the parameters of the filtered network traffic. The paper shows a simulation model that is intended for the evaluation of the major performance indicators of a firewall when ranging a filtration rule set. We've evaluated the effectiveness of the method for ranging a filtration rule set (it was developed earlier by the authors) for various parameters of the simulation model and different scenarios of network traffic behavior.

Keywords: Firewall · Ranging the filtration rules · Network traffic · Simulation model · Queuing system · Phase service · Local approximation method

1 Introduction

A firewall is a local or functional distributing tool that provides control over the incoming and/or outgoing information in the automated system (AS), and ensures the protection of the AS by filtering the information, i.e., providing analysis of the information by the criteria set and making a decision on its distribution.

Firewall is one of the major components of the network architecture. It ensures network security, including for special-purpose AS, the uninterrupted operation of which is critical for ensuring the security and defensive capability of any state. By using firewalls, you can solve such problems as preventing unauthorized access, and deleting, modifying, blocking, copying, providing and distributing information under protection. It's very important to ensure high performance of the firewall since there is a huge need to provide the stable operation of a critically important information infrastructures in the conditions of an avalanche-like growth in the volume of public network information flows, high heterogeneity and variability of network traffic parameters, widespread use of multimedia protocols that are sensitive to the data transmission delays, and significant increase in the number of computer attacks.

Information filtering is executed by a certain rule set, determined in accordance with the security policy of the AS under protection. The filtration rules are

V. M. Vishnevskiy et al. (Eds.): DCCN 2022, LNCS 13766, pp. 190–201, 2022.
https://doi.org/10.1007/978-3-031-23207-7_15

a list of conditions under which further transmission of information is allowed or prohibited and a number of actions is performed by the firewall for registration and/or implementation of additional protective functions.

One of the major factors affecting the search time for rules, and therefore the firewall performance, is the order in which the filtration rules are arranged in sets that are linear lists of large dimensions. This is due to the fact that the search time for any rule corresponding to the filtered data, is in proportion to the number of checked rules. And the information flow filtering time that meets the conditions contained at the end of a large dimension set, will be much longer than the time required to filter data that meet the conditions contained at the beginning of a rule set. Therefore, the use of optimization methods for a filtration rule set is in demand for AS with a complex network architecture and large volumes of heterogeneous network traffic [1,2].

The papers [1,3] published earlier by the authors, describe the developed method for optimizing a filtration rule set (method for ranging the rules). This method takes changes in the parameters of information flows into account. An increase in the efficiency of traffic filtration can be provided by periodical ranging the filtration rules in descending order of their weights, obtained in accordance with the estimates of the parameters of the filtered information flows. A particularity of the developed approach is the use of the non-parametric method of local approximation (MLA) [4] to evaluate the parameters of filtered information flows. In the ranging process for a rule set, the current characteristics and dynamics of changes in the parameters of information flows are considered. At the same time, there is no need to select a parametric model that is acceptable for all evaluated parameters of information flows.

The implementation of MLA has provided the adaptability of the method, as well as a high response speed for changes in the parameters of filtered information flows thanks to the such specifics of MLA estimates as the use of:

- a local parametric model with a sliding region of parameter constancy and a controlled locality parameter that determines the dimension of the locality region;
- a special locality function to set the estimates of previous values when calculating the estimates of the parameters of filtered information flows.

In practice, the increased computational complexity of the developed approach when using modern microprocessors, doesn't have a significant affect on ensuring the stable firewall operation.

Therefore, a relevant task is to create a simulation model intended for the evaluation of the firewall performance with ranging the filtration rules. In contrast to earlier works [1–3], the developed model allows us to obtain estimates for the firewall performance for different MLA parameters, as well types and scenarios of network traffic behavior.

2 Previous Works. Firewall in 5G Networks

In first papers devoted to solving the problem of improving the firewall performances, static methods to optimize the filtration rule set were used. These methods are deterministic and based on hardware solutions, heuristic algorithms and specialized data storage structures that don't depend on the parameters of filtered information flows [5]. Static optimization methods also include methods for finding and correcting inconsistency and redundancy of rules, which are defined as filtration rule configuration errors by a number of authors [6]. Other well-known optimization methods include early packet rejection optimization methods that create additional small dimension filters designed to early reject unwanted traffic before packets are verified by the main rule set. The use of such filters increases the firewall performance, since the rejection of unwanted traffic by the main rule set is executed after verifying all the rules of the set and takes the maximum possible filtering time [6,7].

Currently, a relevant task is to solve the problem of increasing the firewall performance in fifth-generation mobile communication networks (5G/IMT-2020) [8–10]. A particularity of the use of firewall in 5G networks is high requirements for the duration of the data transmission delay to ensure ultra-reliable communication with low latency ULLRC (Ultra Low Latency Reliable Communication), as well as a very large number of filtration rules that are necessary to provide the secure functioning of massive computer-to-computer communications—Massive IoT/IIoT [11].

In paper [9], a new firewall is being developed for filtering traffic in 5G networks. It's intended for the analysis and filtration of network traffic in a specific segment (edge-to-core network segment of a 5G infrastructure). Field Programmable Gate Arrays (FPGA) and a P4 language (programming language) for programming packet routing rules are used to create the firewall.

The paper [11] is devoted to the development of a software firewall for 5G networks based on a software multi-layer switch with an open source text (OVS) designed to work in hypervisors and on computers with virtual devices. Developed firewall is intended to provide network traffic filtration for 5G Internet of Things devices. The firewall performance makes it possible to set and use up to 1 million filtration rules for NB-IoT devices with 4 Gbps traffic.

The paper [10] is devoted to the efficient distribution of filtration rules between the firewalls within Internet of things (IoT). Automatic distribution, as well coordination of virtual firewalls is proposed. This is executed due to network function virtualization (NFV) management and orchestration (MANO) technology to protect NB-IoT mMTC communications. The major idea was to use NFV for efficient distribution of rules between firewalls based on VNFs to achieve scalability in terms of the number of IoT devices under control.

Also, it should be noted that most of the works devoted to the study of methods for optimizing a filtration rule set, don't imply the use of mathematical models to obtain estimates for the performance/efficiency indicators of the firewalls. The development of the firewall model with the possibility of changing the rule set will allow us to evaluate the effectiveness of optimization methods regardless of the specifics of the set rules and the characteristics of the hardware and software platform, as well as to analyze the dependence of performance indicators on the parameters of the rule optimization method.

3   Simulator Development

The use of simulation modeling methods made it possible to eliminate restrictions on the type of the distribution function (DF) for the incoming packet flow and packet service for the earlier developed [1–3] firewall analytical models presented in the form of single-line queuing systems (QS), with a storage drive of limited capacity, heterogeneity of Poisson incoming flow, packet service with a phase-type service duration distribution function that depends on the order of the filtration rules. According to the Basharin-Kendall classification, QS data belong to $M_N/PH/1$ class. Hence, the implementation of the simulation model makes it possible to obtain estimates for the firewall performance when filtering various types and scenarios of network traffic behavior, as well as to evaluate the effectiveness of ranging the filtration rule set.

To create a simulation model that reproduces the filtration process for information flows and maintains the logical structure and sequence in terms of time, discrete-event modeling was chosen. Model traffic is given as a sum of Poisson flows for packets of various types and intensities. Packets incoming to the model are serviced in accordance with the FCFS principle. Only one packet can be served at a time, the other packets await for the service to start in a limited capacity storage drive. The packet service time distribution function (DF) is a phase-type function.

Since the purpose of modeling is to obtain estimates of the firewall performance indicators for various scenarios of the network traffic behavior and parameters of the ranging method for the filtration rule set, we will describe a number of basic parameters that are necessary for modeling (see Table 1).

The 'QS parameters' group corresponds to the parameters of the firewall mathematical model described in [1,3] in the form of QS The 'Filtration rule set' parameter group allows us to experiment with a randomly generated filtration rule set or, when it matters, with a predefined rule set that corresponds to a real one.

The 'Parameters of the ranging method for the filtration rule set' group is intended to set all the necessary parameters for the functioning of the ranging method for the filtration rule set. At the same time, for the convenience of obtaining comparative estimates of the firewall performance, some of the parameters are predetermined by the 'calculation of filtration rule weights'.

'Modeling time parameters' set the time intervals, after which the system states change. Such values as the number of packet groups and data segments are necessary to implement the accumulation of experimental data and calculate the MLA estimates.

In most cases, writing an algorithm intended for modelling a complex process, in the form of a program, has significant difficulties and can lead to various errors. A large number of operators related to computational procedures and various support functions, make such a record obscure, and it becomes difficult to navigate in the modeling algorithm structure. That's why the developed modeling algorithm for the firewall model with the ranging the filtration rule set reflects only the features of its structure, without unnecessary secondary details.

Table 1. Simulation model parameters

Parameter groups	Description of the parameters	Possible values
QS parameters	μ_0—intensity of packet service during initial processing	Double
	μ—intensity of service when checking whether a packet matches a specific (single) rule of the filtration rule set	Double
	C—storage capacity of the system drive	uint8. Storage drive capacity is not less than 10% of the number of filtration rules in the set
Parameters of the filtration rule set	N—number of filtration rules in the set	uint8
	$\mathbf{r}_1$—initial set of the filtration rules	$1 \times N$ uint8
Model traffic parameters	m—number of priority packets. Priority packets are packets with a high incoming intensity	uint8. The number of priority packets is less than the number of non-priority packets, $m < N - m$
	λ_1^{high}— initial intensity for the incoming priority packets	Double
	λ_1^{low}—initial intensity for the incoming non-priority packets	Double
	$f : (\lambda_{k,1}, ..., \lambda_{k,1}) \rightarrow (\lambda_{k+1,1}, ..., \lambda_{k+1,1})$	A function that sets the change in the intensity of the incoming packet flow after servicing the k-th packet group
Parameters of the ranging method	ω—calculation of filtration rule weights in accordance with the selected ranging method	uint8
Modelling time parameters	t—modelling time corresponding to the time of arrival of packets that make up one group	Double
	t_0—initial modelling time	Double
	N_1—number of packet groups in the data segment	uint8. $N_1 \geq 2$
	N_2—number of data segments	uint8. $N_2 \geq 2$

The algorithm is presented in the form of a pseudo-code similar, in terms of syntax, to the operators of the MATLAB system. Structurally, this modeling algorithm is divided into two algorithms—'Firewall simulation model' and 'Firewall functioning'.

Algorithm 3.1. 'Firewall simulation model' displays the filtration process for packet groups and data segments of the firewall, as well as the process of ranging the rule set and calculating the firewall performance indicators.

Algorithm 3.1. 'Core algorithm'

Input parameters: N, $\mathbf{r}_1$, m, , , N_1, N_2, t, $f : (\lambda_{k,1}, ..., \lambda_{k,1}) \rightarrow (\lambda_{k+1,1}, ..., \lambda_{k+1,1})$
Output parameters: V_k, W_k, U_k, Q_k, ρ_k

1:    Initialization: $k = 1$, $\Lambda = 0$
2:    Calculating the probability of packet arrival: $\Lambda_. = m \cdot \lambda_1^{high} + (N - m)\lambda_1^{low}$
3:    $\mathbf{P}(1 : m) = \lambda_1^{high}/\Lambda_.$
4:    $\mathbf{P}(m + 1 : N) = \lambda_1^{low}/\Lambda_.$
5:    for $j_2 = 1 : 1 : N_2$
6:        for $j_1 = 1 : 1 : N_1$
7:            k-th launch of the Simulink model for the t time duration (Algorithm 3.2. 'Firewall algorithm')
8:            $S_k \leftarrow$ get the values of the vector containing the numbers of the rules that matched the served packet from the Simulink model for the N_1-th group of the N_2 data segment
9:            Calculate x_i^k - the number of packets that match the r_i^k rule for i=1:1:N
10:           $x_i^k =$ sum $(S_k(:) = i)$
11:           end
12:           Calculate δ
13:           if $j_2=$ the first data segment
14:               $\widehat{x}_k = 0, i = 1, ..., N, \mathbf{p}_k = [p_1^k, ..., p_N^k] = \mathbf{0}$
15:           else
16:               calculate $\widehat{x}_{k+1}, i = 1, ..., N$
17:               Set rule weights: $\mathbf{p}_k = [p_1^k, ..., p_N^k] = [\widehat{x}_{k+1}, ..., \widehat{x}_{k+1}]$.
18:           end
19:           Calculate: $\gamma_j(k) \in \{1, ..., N\}, j = 1, 2, ..., N$
20:           Get a ranged set using $\mathbf{r}_k$: $\mathbf{r}_{k+1} = [r_{\gamma_1(k)}^k, ..., r_{\gamma_N(k)}^k]$
21:           $V_k, W_k, Q_k \leftarrow$ get values of the performance indicators from the Simulink model
22:           Calculate performance indicators: U_k, ρ_k
23:           Set the intensity of incoming packets for the following packet group: $\lambda_{1,k+1}, ..., \lambda_{N,k+1} = f(\lambda_{1,k}, ..., \lambda_{N,k})$
24:           $k = k + 1$
25:       End
26:   End

Algorithm 3.2. 'Firewall algorithm' displays only the main procedures that ensure the functioning of the earlier considered QS with phase-type packet service. At the same time, the sequence of the processes of packet arrival, waiting, and servicing, and the transitions between blocks implementing the QS, are not shown in the algorithm.

Algorithm 3.2. 'Firewall functioning'
Input parameters: N, $\mathbf{P} = [p_{1,i}, ..., p_{1,N}]$, C, μ_0, μ, Λ.
Output parameters: V_k, W, Q_k, S_k

1:    Generate a number with a equal distribution on the interval $[0,1]$ $a \sim U(0,1)$
2:    Calculate the arrival time for the packets; dt $= \Lambda^{-1} \ln(1-a)$
3:    Generate a number with an equal distribution on the interval $[0,1]$ $b \sim U(0,1)$
4:    Get the type of the packet that incomes the QS:

$$e = \mathbf{sum}(b \geq [0, \sum_{i=1}^{1} p_{1,i}, ..., \sum_{i=1}^{N} p_{1,i}])$$

5:    Generate a packet of $e$type using the 'Entity Generator' block
      of the Simulink model
6:    Generate a final packet queue with length using the 'Entity Queue' block
      of the Simulink model
7:    With a free 'Entity Server' block
      — determine the i-th number of the rule corresponding to the
      incoming packet of e type;
      — calculate the service time (v_e) using the packet service function
      $g : (e, \mu_0, \mu, \mathbf{r}_k, N) \rightarrow [v_e, i]$.
8:    Over the v_e time, provide the packet service in the 'Entity Server' block
      of the Simulink environment
9:    Complete servicing the packets in the 'Entity Terminator' block
10:   Proceed to servicing and generating the next packets
11:   After completion of t interval, using the statistics of the 'Entity Queue'
      and 'Entity Server' blocks and the Simulink model, transfer the following
      values to the Algorithm 3.1 — 'Core algorithm': V_k, W_k, Q_k, S_k .

For the software implementation of the modeling algorithm, we used the Simulink environment for dynamic interdisciplinary modeling of complex technical systems with the SimEvents library of discrete states that uses the apparatus of the theory of queues and queuing systems.

A particularity of the Simulink environment is a high degree of integration into the MATLAB matrix calculation system. This makes it possible to launch a model from a MATLAB script file, get the parameters from it, as well as send the modelling results back to the working environment. It allowed us to use built-in mathematical algorithms and tools when implementing the modeling algorithm and processing the experimental results. With the MATLAB system, the scalability of the model was also provided. It became possible to add and expand the functionality of the simulation model without significant changes in the existing project architecture.

4   Case Study and Performance Analysis

The purpose of the experiment was to evaluate the firewall performance when ranging the filtration rules by the method proposed in papers [1,3] for various traffic types and MLA parameters.

When modeling, two types of system operation conditions were under consideration. They correspond to the normal functioning of the system and the firewall functioning the under overload conditions.

The normal conditions mean such a ratio of intensities of network packet arrival, intensities of packet service and storage drive capacity that doesn't lead to system operation with a load being $\rho > 0.9$.

The functioning of the system under overload conditions was considered for incoming traffic with a harmonically changing intensity of the incoming flows depending on the number of the packet group being serviced, for which the system operates with overload or is close to the packet loss limit. There are studies [12,13] of similar QS models with harmonic fluctuations of the incoming flow intensity, where the analysis of the stability of the packet queue characteristics was performed. Approximately, these conditions can be considered as the modeling process for the firewall operation with overloads, for example, when implementing DDoS attacks.

Depending on the values of the MLA parameters and the selected weights, the following methods for ranging the filtration rule set were considered:

1. Ranging with adaptive δ was performed in accordance with the 1st order MLA estimates with an adaptive locality parameter δ.
2. Ranging by least-square method (LSM) was executed in accordance with the 1st order MLA estimates at $\delta = \infty$. The estimates obtained are LSM estimates.
3. Ranging without levelling out was executed in accordance with the 1st order MLA estimates at $\delta \to 0$.
4. No ranging . The weights of the rules are $p_i^k = 0, i = 1, ..., N$.

To evaluate the effectiveness of the ranging method, the average values—$M(\Delta Z^{\omega})$, $M(\Delta \delta_Z^{\omega})$—and maximum values —$\max(\Delta Z^{\omega})$, $\max(\Delta \delta_Z^{\omega})$—of absolute and relative errors of performance/efficiency indicators calculated for packet groups without applying the $Z^0 = \{V_i^0, W_i^0, U_i^0, Q_i^0, \rho_i^0\}$ rule ranging, and $Z_i^{\omega} = \{V_i^{\omega}, W_i^{\omega}, U_i^{\omega}, Q_i^{\omega}, \rho_i^{\omega}\}$ performance/efficiency indicators calculated for packet groups after applying the ranging method are used with being the ranging method and n being the number of serviced packet groups since the first ranging.

In the experiment with the normal functioning of the system, the following initial data were chosen to calculate the performance indicators: the number of filtration rules in the set—$N = 1000$, the storage drive capacity in the system—$C = 1000$, which is 10% of the number of rules, the time for the packet initial processing—$\mu_0^{-1} = 2.7 \cdot 10^{-3}$ [ms], the time for checking one rule—$\mu^{-1} = 5 \cdot 10^{-5}$ [ms]. The intensity of incoming packets (see Fig. 4.1) changes after servicing each packet group so that $\Lambda_{k+1} = \Lambda_k + X_{k+1} \cdot \Lambda_k$—with X_k being a random variable equally distributed on the interval of $[-0.1, 0.2]$. The initial the intensity value is $\Lambda_1 = 20$ [ms^{-1}], the number of packet groups is $k = 60$, the time interval of the system operation is $T = 60$ [s] The packet flow that incomes to the system is the sum of the Poisson packet flows. The number of flows with a high intensity is 15% ($m = 150$), their corresponding intensities are $\lambda_{i,k} = \lambda_k$, $i = 1, ..., m$. The intensities of the remaining flows are $\lambda_{i,k} = \lambda_k$, $i = m+1, ..., N$. The maximum value of the total intensity is 31.21 [ms^{-1}], the average value is 19.53 [ms^{-1}]. The maximum intensity of the priority packet flow is 0.0975 [ms^{-1}], for non-priority packets it is 0.0195 [ms^{-1}].

In the experiment with a harmonically changing intensity, depending on the number of packet group being serviced, the initial data were equal to those used when modeling the traffic filtration process with normal system functioning conditions, except setting the intensity of packet arrival—the intensity changes after servicing each packet group, so $\Lambda_k = \frac{a_0}{2} + \sum_{l=1}^{2} \left\{ a_l \cos\left(\frac{2\pi}{T} lk\right) + b_l \sin\left(\frac{2\pi}{T} lk\right) \right\}$ with $T = 30$, $a_0 = 110$, $a_1 = 1$, $a_2 = 9$, $b_1 = -4$, $b_2 = -4$, $k = 1, ..., 60$. The maximum value of the total intensity is 31.21 [ms^{-1}], the average value is 19.53 [ms^{-1}]. The maximum intensity of the priority packet flow is 0.0975 [ms^{-1}], the maximum intensity for the non-priority packet flow is 0.0195 [ms^{-1}].

5   Analysis of Firewall Performance Indicators

The estimates for the firewall performance indicators (with ranging the rules), obtained during modelling under system overload conditions, are given in Figs. 1, 2, 3 and 4. The maximum values of absolute and relative errors of performance indicators for various MLA parameters are given in Table 2.

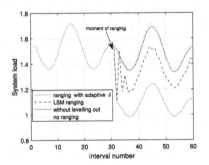
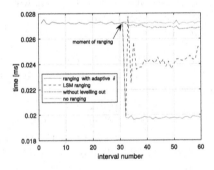

Fig. 1. System load   **Fig. 2.** Average service time

Optimization of the filtration rule set led to a decrease in the load within the system (from the moment the first ranging was launched)—on average from 1.539 to 1.103. Figure 1 shows the load graphs in the system for various MLA parameters. The maximum load in the system without ranging was 1.72. The change in the load (showed in Fig. 1) corresponds to a change in the total intensity with a significant decrease in the load when ranging with adaptive δ and LSM ranging.

When the filtration rule set is optimized, there is a decrease in service time from the moment of the first ranging. Figure 2 shows the average service time

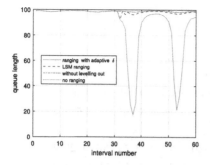

Fig. 3. Average queue length

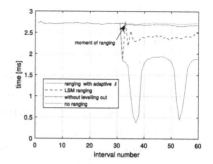

Fig. 4. Average residence time in the system

for different MLA parameters. Qualitatively, the obtained results do not differ from those obtained for the system operation without overloads—a change in load doesn't change the service time, while a significant decrease in service time is obtained when ranging with adaptive δ and LSM ranging.

As shown in Fig. 3, ranging the filtration rule set significantly decreases the residence time from the moment of the first ranging. Similar to the case without overloads, the system eventually turns into an established mode when servicing one packet group. It should be noted that despite the fact that the load in the system is more than 1, the average packet residence time in the network doesn't become infinitely large because of the limited storage drive capacity.

Ranging with adaptive δ and LSM ranging significantly reduce the queue length, as demonstrated in Fig. 4. The best results were obtained for ranging with adaptive δ (see Table 2).

Modelling the process of traffic filtration with a harmonically changing intensity of the incoming flow showed the results as follows:

1. The use of ranging allowed us to reduce the average service time, which provided significant decreases in the system load and in the values of all performance indicators.
2. On average, ranging with adaptive δ demonstrated higher performance/efficiency compared to other methods.
3. For ranging with adaptive δ:
   (a) Decrease in the system load is 0.39 (26.5%).
   (b) Decrease in the average packet service time is 0.007 ms (26.5%).
   (c) Decrease in the average packet residence time is 1.16 ms (42.6%).
   (d) Decrease in the average waiting time for service is 1.15 ms (42.8%).
   (e) Decrease in the average queue length is 22.23 (22.6%).

Compared to the results obtained for modelling the traffic filtration process under normal operating conditions, the absolute errors of performance indicators have higher values, except for the average service time, since it doesn't depend on the system load.

Table 2. Performance indicators

System load				
Ranging type, ω	$M(\varDelta\rho^{\omega})$	$\max(\varDelta\rho^{\omega})$	$M(\delta_{\rho}^{\omega})$	$\max(\delta_{\rho}^{\omega})$
With adaptive δ	0.397436	0.952226	26.54746	56.26077
Least-square method (LSM)	0.161254	0.834968	10.76459	54.9313
Without levelling out	0.017484	0.049596	1.161192	3.528732
Average service time				
With adaptive δ	0.007232	0.01541	26.54725	56.26145
Least-square method (LSM)	0.002932	0.015016	10.76413	54.93123
Without levelling out	0.000316	0.00096	1.161012	3.523324
Average residence time in the system				
With adaptive δ	1.157548	4.634046	42.65268	172.0129
Least-square method (LSM)	0.316624	1.755434	11.64846	64.26404
Without levelling out	0.032942	0.103576	1.211598	3.825138
Average queue length				
With adaptive δ	22.23406	161.2538	22.55491	163.9302
Least-square method (LSM)	1.052322	12.94445	1.065178	13.06728
Without levelling out	0.053542	0.32374	0.054194	0.32865

6   Conclusion

In this paper, a simulation model has been created to evaluate the main firewall performance indicators when ranging a filtration rule set. An estimate of the effectiveness of the ranging method for the filtration rule set is obtained for various parameters of the simulation model, as well as for various scenarios of network traffic behavior and MLA parameters. On average, for normal conditions and system overload conditions, ranging with adaptive δ demonstrated higher efficiency compared to other methods. This can be explained by a smaller approximation error of MLA with adaptive δ compared to the LSM evaluation [4]. Thus, the results of this work confirm the assumption about the increase in the firewall performance due to the use of the method of ranging a rule set that is adaptive to changing parameters of information flows, including the case of DDOS attacks.

References

1. Botvinko, A., Samouylov, K.: Evaluation of firewall performance when ranging a filtration rule set. Discrete Continuous Models Appl. Comput. Sci. **29**(3), 230–241 (2021)
2. Botvinko, A., Samouylov, K.: Evaluation of the firewall influence on the session initiation by the sip multimedia protocol. Discrete Continuous Models Appl. Comput. Sci. **29**(3), 221–229 (2021)

3. Botvinko, A., Samouylov, K.: Firewall simulation model with filtering rules ranking. In: Vishnevskiy, V.M., Samouylov, K.E., Kozyrev, D.V. (eds.) DCCN 2020. CCIS, vol. 1337, pp. 533–545. Springer, Cham (2020). https://doi.org/10.1007/978-3-030-66242-4_42

4. Hardle, W.: Applied Nonparametric Regression. Cambridge University Press, Cambridge (1990)

5. Zhu, Y. C.: Optimization design and implementation of gateway based on firewall for access control. In: Proceedings 6th International Conference on Information Science and Technology, ICIST 2016, Dalian, pp. 100–104. IEEE (2016)

6. Al-Shaer, E.: Automated Firewall Analytics: Design, Configuration and Optimization. Springer, Cham (2014). https://doi.org/10.1007/978-3-319-10371-6

7. Bagheri, S., Shameli, S.A.: Dynamic firewall decomposition and composition in the cloud. IEEE Trans. Inf. Forensics Secur. **15**, 3526–3539 (2020)

8. Chomsiri, T., He, X., Nanda, P.: Hybrid tree rule firewall for high speed data transmission. IEEE Trans. Cloud Comput. **8**(4), 1237–1249 (2016)

9. Ricart-Sanchez, R.: Hardware-accelerated firewall for 5G mobile networks. In: IEEE 26th International Conference on Network Protocols (ICNP), pp. 446–447 (2018)

10. Ricart-Sanchez R.: NetFPGA-based firewall solution for 5G multi-tenant architectures. In: IEEE International Conference on Edge Computing (EDGE), pp. 132–136 (2019)

11. Salva-Garcia, P.: Towards automatic deployment of virtual firewalls to support secure mMTC in 5G networks. IEEE INFOCOM 2019-IEEE Conference on Computer Communications Workshops (INFOCOM WKSHPS), pp. 385–390 (2019)

12. Escolar, A.: Highly-scalable software firewall supporting one million rules for 5G NB-IoT networks. In: ICC 2020-2020 IEEE International Conference on Communications (ICC), pp. 1–6 (2020)

13. Markova, E.: Queuing system with unreliable servers and inhomogeneous intensities for analyzing the impact of non stationarity to performance measures of wireless network under licensed shared access. Mathematics **8**(5), 800 (2020)

14. Gudkova, I., Korotysheva, A., Zeifman, A.: Modeling and analyzing licensed shared access operation for 5G network as an inhomogeneous queue with catastrophes. In: Proceedings 8th International Congress on Ultra Modern Telecommunications and Control Systems and Workshops (ICUMT), 18–20 October 2016, Lisbon, vol. 2016, pp. 282–287. IEEE (2016)

Analytical Modeling of Distributed Systems

Performance Analysis for Tethered HAP Systems: An Analytical Approach

Dharmaraja Selvamuthu[1(✉)], Vidyottama Jain[2], and Raina Raj[1]

[1] Indian Institute of Technology Delhi, New Delhi, India
dharmar@maths.iitd.ac.in, rainaraj.curaj@gmail.com
[2] Central University of Rajasthan, Ajmer, India
vidyottama.jain@curaj.ac.in

Abstract. An analytical model for tethered high altitude platform (HAP) systems, that handle voice, video, and data traffic, is presented in this work. The Markovian arrival process (MAP) is used to depict the arrival of traffic due to the burstiness and self-similar properties of the incoming traffic. Guard channel policy has been employed for the prioritization of the higher priority traffic. If the incoming traffic is blocked due to the lack of idle channel, it will retry for the service after some random amount of time. Phase-type (PH) distributions with distinct parameters have been employed for the representation of the service and retrial processes. The suggested model's behaviour is represented by the level-dependent quasi-birth-death ($LDQBD$) process. Expressions for some significant performance measures are derived. Additionally, queueing examples are offered, which can be acquired as the special cases of the suggested analytical model.

Keywords: Blocking probabilities · High altitude platforms · Level dependent quasi-birth-death process · Markovian arrival process · Phase type distributions

1 Introduction

The confluence of customer demand for personal communication systems has contributed to the global adoption related to wireless communications. The deployment of wireless services also come with the emerging challenges of proper resource allocation, efficient use of spectrum, quality of service for users, etc. The challenge of wireless communication may overcome by aerial platforms carrying communications relay payloads and moving in a largely stationary posture. In contrast to the majority of satellites, a payload might be as simple as a transparent transponder or as complex as a full-fledged base station [14]. A single aerial platform can replace a large number of terrestrial masts, as well as the costs, environmental effect, and backhaul limits that come with them. These aerial

The publication has been prepared with the support of DST-RSF research project no. 22-49-02023 (RSF) and research project no. 64800 (DST).

platforms can be manned or unmanned, with autonomous operation combined with ground control. High-altitude platforms (HAPs) are unmanned aerial base stations which are designed to operate in the stratosphere at a height of between 17 and 22 km. The HAP system is further classified as tethered and untethered. The HAP systems are intended to carry voice, multimedia, file transfer, and web access [11]. To provide QoS for a diverse variety of traffic, service providers must make optimum use of available radio resources. Call Admission Control (CAC) schemes manage radio resources by admitting an arriving traffic based on a pre-defined criterion. References [1,13] present a comprehensive survey for a thorough examination of CAC issues in wireless communication.

The statistical characteristics of self-similarity and burstiness that characterise contemporary voice, video, and data traffic lead to a correlation in the times between arriving packet. As a result, the exponentially distributed call holding times (CHT) and inter-arrival times assumptions are rendered inaccurate. Furthermore, with new applications, the length of time a mobile user remains or the time spent in a cell, known as cell residence time (CRT), is determined by the user's mobility and the geographic location. To estimate realistic performance measures, traffic arrivals can be modelled using the Markovian arrival process (MAP), which has the advantage of smoothing total traffic through statistical aggregations [7]. The CHT and CRT should be represented by phase-type (PH) distribution, which is the convolution of different exponential distributions [7]. When the incoming traffic finds no idle channel available in the system, it will join a virtual space termed as orbit to retry for the services after some random amount of time. The phenomenon is referred as a retrial phenomenon and is very ubiquitous in wireless communication scenarios. Though, the retrial phenomenon has been explored in a number of queueing models with the consideration of exponential distribution, the literature considering a general distributed or PH distributed retrial times is still moderate (see, [4]).

In this study, we have constructed an analytical retrial queueing model which represents the HAP systems with four types of traffic. Also, to consider a more realistic scenario, the inter-arrival times is represented by MAP, and the service and retrial times are denoted by PH distributions. The proposed model's behaviour is represented by level dependent quasi-birth-death ($LDQBD$) process. Further, some special cases of the proposed model has been described and the explicit expressions of the generator matrix have been obtained.

The remainder of this article is structured as follows. The suggested multi-server retrial queueing mechanism is briefly described and mathematically developed in Sect. 2. In Sect. 3, explicit expressions of the infinitesimal generator matrix and significant performance measures required to examine the network efficiency are provided by modelling the stochastic behaviour into a $LDQBD$ process. The crucial performance measures of the proposed model is covered in Sect. 4. Section 5 presents in-depth presentations of specific cases of the proposed model. Section 6 provides the conclusions and suggestions for further research.

2   Model Description

The HAP will respond to any incoming traffic, including video (vo), voice (ve), and data queries (d). The quality of service (QoS) requirements for video traffic is considered as both soft and hard, depending on the particular application. Video traffic has soft QoS for a few specific applications (like, video on demand), whereas it can have hard QoS for some other ones (like, live video). The video traffic is therefore divided into video 1 (vo1) and video 2 (vo2) traffic. Each inbound traffic is classified as new (n_i, i = ve, vo1, vo2, d) and handoff (h_i, i = ve, vo1, vo2, d). A continuous time MAP is used to represent each traffic's arrival since it enables correlation between the traffic's individual arrival times. Numerous prioritization policies have been put out in the literature for traffic prioritization. One of the most used techniques, fixed guard channel policy (GC), is a decent balance between performance and complexity. In this study, the GC policy is used to maintain connectivity and provide various types of traffic preference. According to this policy, a specific number of channels, let's say g, (g changes for each type of traffic), are set aside for each type of traffic. Some other detailed assumptions are as follows:

- **Arrival Process:** The arrival of each traffic is modeled by using MAP and represented by MAP_1 (ve), MAP_2 (vo1), MAP_3 (vo2) and MAP_4 (d). The arrival process for ve traffic is stated briefly. Let C^{ve} be the infinitesimal generator matrix with dimension K. The arrival rates for new and handoff ve traffic are given as $\lambda_{nve} = \pi C_n^{ve} e$ and $\lambda_{hve} = \pi C_h^{ve} e$, respectively. Here, π is the stationary probability vector of the generator C^{ve} and e is column vector with ones entries. On the similar track, the arrival processes for vo1, vo2, and d traffic can be defined.
- **Cell Residence Time:** The CRT (R_s) is PH distributed with representation (α, T) and dimension r.
- **Call Holding Time:** The CHT (H_i, i = ve, vo, d) of type i traffic has PH distribution with representation (β_i, S_i) and dimension h_i.
- **Service Process:** According to Neuts [12], the residual CRT and residual CHT are also PH distributed. Let $S^{i,N}$ be the service times of new traffic of type i, defined as $S^{i,N} = min(H_i, \overline{R}_s)$. Since, the minimum of two PH distribution is also a PH distribution. Therefore, $S^{i,N}$ is also PH distributed with representation (δ_{N_i}, L_{N_i}) and dimension rh_i. Similarly, $S^{i,H}$ is PH distributed with representation (δ_{H_i}, L_{H_i}) and dimension rh_i.
- **Retrial Process:** The retrial process is PH distributed with representation (δ, L) and dimension n. The inter-retrial times or departure times denoted by $R = min(R_t, \overline{R}_s)$, which ever happens first. The retrial process has two absorbing states either due to successful retrial or departure from the system
- **Call Admission Control Scheme:** As per the proposed CAC scheme, the incoming traffic is classified in the decreasing order of priority as h_{ve}, n_{ve}, h_{vo1}, n_{vo1}, h_{vo2}, n_{vo2}, h_d, n_d. It has been assumed that ve and d traffic are assigned only one channel whereas vo are assigned four channels. Therefore, h_1, h_2, h_3 and h_4 are reserved for h_{ve}, h_{vo1}, h_{vo2}, h_d, whereas n_1, n_2 and

n_3 channels are reserved for n_{ve}, n_{vo1}, n_{vo2}, respectively. Since, n_d is of least priority and can tolerate the delay, no channel has been reserved for it. Let M be the total number of available channels. We define the following notations in order to carry out the analysis:

$$M_1 = M - \sum_{i=1}^{3}(h_i + n_i) - h_4; \qquad M_2 = M - \sum_{i=1}^{3}(h_i + n_i)$$

$$M_3 = M - \sum_{i=1}^{2}(h_i + n_i) - h_3; \qquad M_4 = M - \sum_{i=1}^{2}(h_i + n_i)$$

$$M_5 = M - h_1 - n_1 - h_2; \qquad M_6 = M - h_1 - n_1; \qquad M_7 = M - h_1.$$

3   Analytical Model

Consider a HAP system with M channels and J orbit size. Let $\{X(t), t \geq 0\}$ be a stochastic process for a given system with the state space

$$\Omega = \{i, j, k, l, u_{ve}, u_{vo1}, u_{vo2}, u_d, \mathbf{S}^{ve}_{i-j-k}, \mathbf{S}^{vo}_j, \mathbf{S}^d_k, \mathbf{r}_l; \ 0 \leq i \leq M, \ 0 \leq j \leq j_m,$$
$$0 \leq k \leq k_m, \ 0 \leq l \leq J, \ 1 \leq u_m \leq K_m, \ 1 \leq u_d \leq 3\},$$

where,

- i is the number of traffic receiving service,
- j is the number of vo traffic,
- k is the number of d traffic,
- l is the total number of new ve, handoff vo2, new vo2, handoff d and new d traffic in the orbit,
- $u_{ve}, u_{vo1}, u_{vo2}, u_d$ are the phases of $MAPs$ for ve, vo1, vo2 and d traffic arrival process, respectively,
- $\mathbf{S}^{ve}_{i-j-k}, \mathbf{S}^{vo}_j, \mathbf{S}^d_k$ are the set of phase of service for ve, vo (1 and 2) and d traffic, respectively,
- $\mathbf{r}_l$ is the set of retrial phase of l traffic in the orbit,

and, $m = $ ve, vo1, vo2, $j_m = min\{i, \lceil \frac{M_6}{4} \rceil\}$ and $k_m = min\{i - j, M_2\}$. Here, $[x]$ means the greatest integer function of x. The infinitesimal generator matrix Q for this $LDQBD$ process and the blocks of the matrix Q are provided as follows.

$$Q = \begin{bmatrix} Q_{01} & Q_{00} & & & & \\ \ddots & \ddots & \ddots & & & \\ & Q_{i2} & Q_{i1} & Q_{i0} & & \\ & & \ddots & \ddots & \ddots & \\ & & & & Q_{M2} & Q_{M1} \end{bmatrix}$$

$$Q_{i0} = \begin{bmatrix} A_{01}^i & A_{00}^i & & & \\ \ddots & \ddots & \ddots & & \\ 0 & A_{j1}^i & A_{j0}^i & & \\ & \ddots & \ddots & \ddots & \\ & & 0 & A_{j_m1}^i \end{bmatrix} \quad Q_{i2} = \begin{bmatrix} C_{01}^i & 0 & & & \\ \ddots & \ddots & \ddots & & \\ C_{j2}^i & C_{j1}^i & 0 & & \\ & \ddots & \ddots & \ddots & \\ & & C_{j_m2}^i & C_{j_m1}^i \end{bmatrix}$$

and $Q_{i1} = diag(B_0^i, \dots, B_j^i, \dots, B_{j_m}^i)$.

Note that the notations $I_w, W_r(i), V_r(i,1), I(w,s)$ and $\widehat{V_r}(i)$ as defined in [3] have been used to carry out the analysis. A detailed explanation for the blocks of the generator matrix is given as follows.

- Q_{i0} **block:** The blocks A_{j1}^i and A_{j0}^i represents the case when due to successful retrial or arrival of fresh traffic, the number of ve or d traffic and vo traffic goes up by one. The main diagonal elements $A_{k1}^{i,j,1}$ represents the case, when the number of ve traffic goes up by one. Similarly, the upper diagonal elements $A_{k0}^{i,j,1}$ shows increment in the d traffic. Each of $A_{k1}^{i,j,1}$ and $A_{k0}^{i,j,1}$ is a lower diagonal block matrices with main diagonal elements $A_{l,1}^{i,j,1,k,1}$ and $A_{l,1}^{i,j,1,k,0}$, respectively and lower diagonal elements as $A_{l,2}^{i,j,1,k,1}$ and $A_{l,2}^{i,j,1,k,0}$.
  - $A_{l,1}^{i,j,1,k,1}$ represents the increment due to the arrival of fresh ve traffic. Similarly, $A_{l,1}^{i,j,1,k,0}$ denotes the increment in the fresh d traffic.
  - $A_{l,2}^{i,j,1,k,1}$ and $A_{l,2}^{i,j,1,k,0}$ represents the increment due to the successful retrial of new ve and d traffic from the orbit.

  The upper diagonal block A_{j0}^i is a diagonal matrix with $A_{k1}^{i,j,0}$, $0 \leq k \leq k_m$ as diagonal entries. Each of $A_{k1}^{i,j,0}$ is a lower diagonal matrix with main diagonal elements $A_{l,1}^{i,j,0,k,1}$ and lower diagonal elements $A_{l,2}^{i,j,0,k,1}$.
  - $A_{l,1}^{i,j,0,k,1}$ represents the increase in the number of vo traffic due to the fresh arrivals of vo (1 and 2) traffic.
  - $A_{l,2}^{i,j,0,k,1}$ shows the case when increase in the vo traffic is due to successful retrials by the vo traffic from the orbit.

The expressions of the block entries are given as follows for $0 \leq i \leq M-1$, $0 \leq j \leq j_m$, $0 \leq k \leq k_m$ and $0 \leq l \leq J$:

$$A_{l,1}^{i,j,1,k,1} = \lambda_{VE}^{i,j} \oplus C_0^{vo1} \oplus C_0^{vo2} \oplus C_0^d \oplus W_{ve}(i-j-k) \oplus W_{vo}(j) \oplus W_d(k) \oplus W_r(l),$$

$$A_{l,1}^{i,j,1,k,0} = C_0^{ve} \oplus C_0^{vo1} \oplus C_0^{vo2} \oplus \lambda_D^{i,j} \oplus W_{ve}(i-j-k) \oplus W_{vo}(j) \oplus W_d(k) \oplus W_r(l),$$

$$A_{l,1}^{i,j,0,k,1} = C_0^{ve} \oplus \lambda_{VO}^{i,j} \oplus C_0^d \oplus W_{ve}(i-j-k) \oplus W_{vo}(j) \oplus W_d(k) \oplus W_r(l),$$

$$A_{l,2}^{i,j,1,k,1} = C_0^{ve} \oplus C_0^{vo1} \oplus C_0^{vo2} \oplus C_0^d \oplus W_{ve}(i-j-k) \oplus W_{vo}(j) \oplus W_d(k) \oplus g^{i,j}\widehat{V_r}(l),$$

$$A_{l,2}^{i,j,1,k,0} = C_0^{ve} \oplus C_0^{vo1} \oplus C_0^{vo2} \oplus C_0^d \oplus W_{ve}(i-j-k) \oplus W_{vo}(j) \oplus W_d(k) \oplus h^{i,j}\widehat{V_r}(l),$$

$$A_{l,2}^{i,j,0,k,1} = C_0^{ve} \oplus C_0^{vo1} \oplus C_0^{vo2} \oplus C_0^d \oplus W_{ve}(i-j-k) \oplus W_{vo}(j) \oplus W_d(k) \oplus m^{i,j}\widehat{V_r}(l),$$

where

$$\lambda_{VE}^{i,j} = \begin{cases} C_{hv}^{ve} + C_{nv}^{ve}, & 0 \le i + 3j < M_7 \\ C_{hv}^{ve}, & M_7 \le i + 3j < M \\ 0, & \text{otherwise} \end{cases} ; \quad g^{i,j} = \begin{cases} p_{nve}, & 0 \le i + 3j < M_7 \\ 0, & \text{otherwise} \end{cases} ;$$

$$\lambda_{D}^{i,j} = \begin{cases} C_{hd}^{d} + C_{nd}^{d}, & 0 \le i + 3j < M_1 \\ C_{hd}^{d}, & M_1 \le i + 3j < M_2 \\ 0, & \text{otherwise} \end{cases} ; \quad h^{i,j} = \begin{cases} p_{hd} + p_{nd}, & 0 \le i + 3j < M_1 \\ p_{hd}, & M_1 \le i + 3j < M_2 \\ 0, & \text{otherwise} \end{cases} ;$$

]

$$\lambda_{VO}^{i,j} = \begin{cases} (C_{hvo1}^{vo1} + C_{nvo1}^{vo1}) \oplus (C_{hvo2}^{vo2} + C_{nvo2}^{vo2}), & 0 \le i + 3j + 4 < M_3 \\ (C_{hvo1}^{vo1} + C_{nvo1}^{vo1}) \oplus C_{hvo2}^{vo2}, & M_3 \le i + 3j + 4 < M_4 \\ (C_{hvo1}^{vo1} + C_{nvo1}^{vo1}) \oplus C_{0}^{vo2}, & M_4 \le i + 3j + 4 < M_5 \\ C_{hvo1}^{vo1} \oplus C_{0}^{vo2}, & M_5 \le i + 3j + 4 < M_6 \\ 0, & \text{otherwise} \end{cases} ;$$

$$m^{i,j} = \begin{cases} p_{hvo2} + p_{nvo2}, & 0 \le i + 3j + 4 < M_3 \\ p_{hvo2}, & M_3 \le i + 3j + 4 < M_4 \\ 0, & \text{otherwise.} \end{cases} ;$$

– Q_{i2} **block:** For each $1 \le i \le M$, the blocks C_{j1}^{i} corresponds to the case of ve or d traffic completion whereas C_{j2}^{i} represents the traffic completion by the vo traffic. C_{j1}^{i} is a lower diagonal matrix with $C_{k1}^{i,j,1}$ as main diagonal and $C_{k2}^{i,j,1}$ as lower diagonal elements. Each of $C_{k1}^{i,j,1}$ and $C_{k2}^{i,j,1}$ is a diagonal matrix with diagonal elements $C_{l,1}^{i,j,1,k,1}$ and $C_{l,1}^{i,j,1,k,2}$, respectively.

- $C_{l,1}^{i,j,1,k,1}$ shows reduction in the number of ongoing ve traffic.
- $C_{l,1}^{i,j,1,k,2}$ represents the reduction in the d traffic due to service completion.

For $1 \le i \le M, 0 \le j \le j_m, 0 \le k \le k_m, 0 \le l \le J$,

$$C_{l,1}^{i,j,1,k,1} = C_0^{ve} \oplus C_0^{vo1} \oplus C_0^{vo2} \oplus C_0^{d} \oplus V_{ve}(i-j-k) \oplus W_{vo}(j) \oplus W_d(k) \oplus W_r(l)$$

$$C_{l,1}^{i,j,1,k,2} = C_0^{ve} \oplus C_0^{vo1} \oplus C_0^{vo2} \oplus C_0^{d} \oplus W_{ve}(i-j-k) \oplus W_{vo}(j) \oplus V_d(k) \oplus W_r(l).$$

C_{j2}^{i} is a diagonal matrix with diagonal entries as $C_{k,1}^{i,j,2}$. Each of $C_{k,1}^{i,j,2}$ is a diagonal matrix with its diagonal elements $C_{l,1}^{i,j,2,k,1}$. $C_{l,1}^{i,j,2,k,1}$ represents the decrement in the number of vo traffic.

For $1 \le i \le M, 1 \le j \le j_m, 0 \le k \le k_m$ and $0 \le l \le J$,

$$C_{l,1}^{i,j,2,k,1} = C_0^{ve} \oplus C_0^{vo1} \oplus C_0^{vo2} \oplus C_0^{d} \oplus W_{ve}(i-j-k) \oplus V_{vo}(j) \oplus W_d(k) \oplus W_r(l).$$

– Q_{i1} **block:** B_j^i represents no change in the number of traffic receiving service where, $B_j^i = diag(B_0^{i,j}, \ldots, B_k^{i,j}, \ldots, B_{k_m}^{i,j})$. Each of $B_k^{i,j}$ is a tri-diagonal matrix.

- The main diagonal element $B_{l,1}^{i,j,k}$ exhibits no change in the traffic in service as well as no change in the number of traffic in the orbit.
  For $0 \leq i \leq M$, $0 \leq j \leq j_m$, $0 \leq k \leq k_m$, and $0 \leq l \leq J$,

$$B_{l,1}^{i,j,k} = C_0^{ve} \oplus C_0^{vo1} \oplus C_0^{vo2} \oplus C_0^d \oplus W_{ve}(i-j-k) \oplus W_{vo}(j) \oplus W_d(k) \oplus W_r(l).$$

- The lower diagonal elements $B_{l,2}^{i,j,k}$ shows departure of a traffic from the orbit without service.
  For $0 \leq i \leq M$, $0 \leq j \leq j_m$, $0 \leq k \leq k_m$, and $1 \leq l \leq J$,

$$B_{l,2}^{i,j,k} = C_0^{ve} \oplus C_0^{vo1} \oplus C_0^{vo2} \oplus C_0^d \oplus W_{ve}(i-j-k) \oplus W_{vo}(j) \oplus W_d(k) \oplus V_r(l,1).$$

- The upper diagonal elements $B_{l,0}^{i,j,k}$ represents increase in the number of traffic in the orbit.
  For $0 \leq i \leq M$, $0 \leq j \leq j_m$, $0 \leq k \leq k_m$, and $0 \leq l \leq J-1$,

$$B_{l,0}^{i,j,k} = C_0^{vo1} \oplus n^{i,j,k} \oplus W_{ve}(i-j-k) \oplus W_{vo}(j) \oplus W_d(k) \oplus W_r(l),$$

where

$$n^{i,j,k} = \begin{cases} C_0^{nd} \oplus C_0^{ve} \oplus C_0^{vo2}, & M_1 \leq i+3j < M_2 \\ (C_{hd}^d + C_{nd}^d) \oplus C_0^{ve} \oplus C_0^{vo2}, & M_2 \leq i+3j < M_3 \\ C_{nvo2}^{vo2} \oplus (C_{hd}^d + C_{nd}^d) \oplus C_0^{ve}, & M_3 \leq i+3j < M_4 \\ (C_{nvo2}^{vo2} + C_{hvo2}^{vo2}) \oplus (C_{hd}^d + C_{nd}^d) \oplus C_0^{ve}, & M_4 \leq i+3j < M_7 \\ C_{nv}^{ve} \oplus (C_{nvo2}^{vo2} + C_{hvo2}^{vo2}) \oplus (C_{hd}^d + C_{nd}^d) \oplus C_0^{ve}, & M_7 \leq i+3j < M \\ 0, & \text{otherwise.} \end{cases}$$

4   Performance Measures

Let $\mathbf{x}$ be the steady state probability vector of the system. Some important performance measures are given as follows:

1. The probability that there are l traffic in the orbit is given by

$$P_r(l) = \sum_{i=0}^{M} \sum_{j=0}^{j_m} \sum_{k=0}^{k_m} \mathbf{x}(i,j,k,l)\mathbf{e}, \quad 0 \leq l \leq J.$$

2. The dropping probabilities of handoff ve and vo1, respectively are given as

$$P_{dv} = \sum_{i} \sum_{j} \sum_{k=0}^{k_m} \sum_{l=0}^{J} \mathbf{x}(i,j,k,l)\mathbf{e}, 0 \leq i+3j \leq M, 0 \leq i \leq M,\ 0 \leq j \leq j_m,$$

$$P_{dvo1} = \sum_{i} \sum_{j} \sum_{k=0}^{k_m} \sum_{l=0}^{J} \mathbf{x}(i,j,k,l)\mathbf{e}, M_6 \leq i+3j \leq M, 0 \leq i \leq M,\ 0 \leq j \leq j_m.$$

3. The blocking probabilities of new vo1 traffic is given by

$$P_{bvo1} = \sum_{i} \sum_{j} \sum_{k=0}^{k_m} \sum_{l=0}^{J} \mathbf{x}(i,j,k,l)\mathbf{e}, M_6 \leq i+3j \leq M, 0 \leq i \leq M,\ 0 \leq j \leq j_m.$$

4. The probability that new ve, handoff vo2, handoff d and new d traffic, respectively has to join the orbit is given as follows

$$P_{nv} = (1 - P_r(J)) \sum_i \sum_j \sum_{k=0}^{k_m} \sum_{l=0}^{J} \mathbf{x}(i,j,k,l)\mathbf{e}, \quad M_7 \le i + 3j \le M,$$

$$P_{hvo2} = (1 - P_r(J)) \sum_i \sum_j \sum_{k=0}^{k_m} \sum_{l=0}^{J} \mathbf{x}(i,j,k,l)\mathbf{e}, \quad M_4 \le i + 3j \le M,$$

$$P_{nvo2} = (1 - P_r(J)) \sum_i \sum_j \sum_{k=0}^{k_m} \sum_{l=0}^{J} \mathbf{x}(i,j,k,l)\mathbf{e}, \quad M_3 \le i + 3j \le M,$$

$$P_{hd} = (1 - P_r(J)) \sum_i \sum_j \sum_{k=0}^{k_m} \sum_{l=0}^{J} \mathbf{x}(i,j,k,l)\mathbf{e}, \quad M_2 \le i + 3j \le M,$$

$$P_{nd} = (1 - P_r(J)) \sum_i \sum_j \sum_{k=0}^{k_m} \sum_{l=0}^{J} \mathbf{x}(i,j,k,l)\mathbf{e}, \quad M_1 \le i + 3j \le M.$$

Note that $(1 - P_r(J))$ is the probability that the orbit is not full.

5. The average value of waiting time of traffic in the retrial queue is given by

$$E(T) = L/[\lambda_{nv}(1 - P_{nv}) + \lambda_{hvo2}(1 - P_{hvo2}) + \lambda_{nvo2}(1 - P_{nvo2})$$
$$+ \lambda_{hd}(1 - P_{hd}) + \lambda_{nd}(1 - P_{nd})].$$

5   Particular Cases

CASE I: Poisson Arrivals, Exponential Service and Retrial Times, Four Type of Traffic [6]

Let the state space for the stochastic process $\{X(t), t \ge 0\}$ is defined as : $\Omega = \{(i,j,k,l); \ 0 \le i \le M; \ 0 \le j \le j_m; \ 0 \le k \le k_m; \ 0 \le l \le J\}$. The expression for the block matrices of the generator matrix Q for this particular case is given as follows.

– First, we give the expression of $A_{l,1}^{i,j,1,k,1}$ and $A_{l,2}^{i,j,1,k,1}$. For $0 \le i \le M-1$, $0 \le j \le j_m$, $0 \le k \le k_m$, $A_{l,1}^{i,j,1,k,1} = \lambda_{VE}^{i,j}$, $l = 0, 1, \ldots, J$; $A_{l,2}^{i,j,1,k,1} = g^{i,j} * l * \mu_{rc}$, $l = 1, \ldots, J$, where

$$\lambda_{VE}^{i,j} = \begin{cases} \lambda_{hv} + \lambda_{nv}, & 0 \le i + 3j < M_7 \\ \lambda_{hv}, & M_7 \le i + 3j < M; \\ 0, & \text{otherwise} \end{cases} \quad g^{i,j} = \begin{cases} p_{nv}, & 0 \le i + 3j < M_7 \\ 0, & \text{otherwise}. \end{cases}$$

– For $0 \le i \le M-1$, $0 \le j \le j_m$, $0 \le k \le k_m - 1$, $A_{l,1}^{i,j,1,k,0} = \lambda_D^{i,j}$, $l = 0, 1, \ldots, J$; $A_{l,2}^{i,j,1,k,0} = h^{i,j} * l * \mu_{rc}$, $l = 1, 2, \ldots, J$ where,

$$\lambda_D^{i,j} = \begin{cases} \lambda_{hd} + \lambda_{nd}, & 0 \le i + 3j < M_1 \\ \lambda_{hd}, & M_1 \le i + 3j < M_2; \\ 0, & \text{otherwise}, \end{cases} \quad h^{i,j} = \begin{cases} p_{hd} + p_{nd}, & 0 \le i + 3j < M_1 \\ p_{hd}, & M_1 \le i + 3j < M_2 \\ 0, & \text{otherwise}. \end{cases}$$

- For $0 \leq i \leq M-1$, $0 \leq j \leq j_m-1$, $0 \leq k \leq k_m$, the expressions for $A_{l,1}^{i,j,0,k,1}$ and $A_{l,2}^{i,j,0,k,1}$: $A_{l,1}^{i,j,0,k,1} = \lambda_{VO}^{i,j}$, $l = 0,1,\ldots,J$; $A_{l,2}^{i,j,0,k,1} = m^{i,j} * l * \mu_{rc}$, $l = 1,\ldots,J$, where

$$\lambda_{VO}^{i,j} = \begin{cases} \lambda_{hvo1} + \lambda_{nvo1} + \lambda_{hvo2} + \lambda_{nvo2}, & 0 \leq i+3j+4 < M_3 \\ \lambda_{hvo1} + \lambda_{nvo1} + \lambda_{hvo2}, & M_3 \leq i+3j+4 < M_4 \\ \lambda_{hvo1} + \lambda_{nvo1}, & M_4 \leq i+3j+4 < M_5 \\ \lambda_{hvo1}, & M_5 \leq i+3j+4 < M_6 \\ 0, & \text{otherwise} \end{cases}$$

$$m^{i,j} = \begin{cases} p_{hvo2} + p_{nvo2}, & 0 \leq i+3j+4 < M_3 \\ p_{hvo2}, & M_3 \leq i+3j+4 < M_4 \\ 0, & \text{otherwise.} \end{cases}$$

- For $1 \leq i \leq M$, $0 \leq j \leq j_m$, $0 \leq k \leq k_m$ and $0 \leq l \leq J$, $C_{l,1}^{i,j,1,k,1} = (i-j-k) * \mu_{ve}$; $C_{l,1}^{i,j,2,k,1} = j * \mu_{vo}$; $C_{l,1}^{i,j,1,k,2} = k * \mu_d$. Finally, we consider the blocks of Q_{i1}. For $0 \leq i \leq M$, $0 \leq j \leq j_m$, $0 \leq k \leq k_m$ and $0 \leq l \leq J$,

$$B_{l,1}^{i,j,k} = -(\lambda_{VE}^{i,j} + \lambda_{VO}^{i,j} + \lambda_D^{i,j}) - (i-j-k) * \mu_{ve} - j * \mu_{ve} - k * \mu_d - l * \mu_{rq} - l * \mu_{rc}$$
$$B_{l,0}^{i,j,k} = n^{i,j,k}; \quad B_{l,2}^{i,j,k} = l * \mu_{rq},$$

where

$$n^{i,j,k} = \begin{cases} \lambda_{nd}, & M_1 \leq i+3j < M_2 \\ \lambda_{hd} + \lambda_{nd}, & M_2 \leq i+3j < M_3 \\ \lambda_{nvo2} + \lambda_{hd} + \lambda_{nd}, & M_3 \leq i+3j < M_4 \\ \lambda_{hvo2} + \lambda_{nvo2} + \lambda_{hd} + \lambda_{nd}, & M_4 \leq i+3j < M_7 \\ \lambda_{nv} + \lambda_{hvo2} + \lambda_{nvo2} + \lambda_{hd} + \lambda_{nd}, & M_7 \leq i+3j < M \\ 0, & \text{otherwise.} \end{cases}$$

CASE II: MAP Arrivals, Exponential Service and Retrial Times, One Type of Traffic [5]

Let the state space of the underlying stochastic process is given as:

$$\{(i,l,u),\ 0 \leq i \leq M,\ 0 \leq l \leq J,\ 1 \leq u \leq K\}.$$

The block matrix entries are given as follows.

- Q_{i0} is a lower diagonal matrix with A_{l1}^i as main diagonal and A_{l2}^i as lower diagonal entries. The matrices A_{l1}^i and A_{l2}^i corresponds to the increment in the number of traffic in service due to fresh arrivals of traffic and due to successful retrials of traffic, respectively. The expressions are given as:

$$A_{l1}^i = C_N + C_H, \quad 0 \leq l \leq J; \quad A_{l2}^i = l\mu_{rc}I, \quad 1 \leq l \leq J.$$

- Consider the block Q_{i2}. It is a diagonal matrix with diagonal elements C_{l1}^i. The expression for C_{l1}^i is given as $C_{l1}^i = i\mu I$.

- Finally, consider the blocks Q_{i1}. It is a tri-diagonal matrix with main diagonal elements B_{l1}^i, as

$$B_{l1}^i = \begin{cases} C_0 - (i\mu + l\mu_{rq} + l\mu_{rc})I, i = 0, 1, \ldots, M-1, \ l = 0, 1, \ldots, J, \\ C_0 + C_H - (i\mu + l\mu_{rq})I, \ i = M, \ l = 0, 1, \ldots, J-1, \\ -(i\mu + l\mu_{rq})I, \qquad\qquad i = M, \ l = J. \end{cases}$$

Similarly, B_{l0}^i represents the upper diagonal given as

$$B_{l0}^i = \begin{cases} \mathbf{0} \quad 0 \le i \le M-1, \ 0 \le l \le J-1, \\ C_N, i = M, \ 0 \le l \le J-1. \end{cases}$$

Lastly, B_{l2}^i is given by : $B_{l2}^i = l\mu_{rq}; \ i = 0, 1, \ldots, M, \ l = 1, 2, \ldots, J.$

CASE III: Poisson Arrivals, Phase Type Service and Exponential Retrial Times, One Type of Traffic [2]

Let the state space for the stochastic process $\{X(t), t \ge 0\}$ is given with the state space $\{(i, l, \mathbf{S}_i), \ 0 \le i \le M, \ 0 \le l \le J\}$.

- The expressions of A_{l1}^i and A_{l2}^i, for $0 \le i \le M-1$, are $A_{l1}^i = \lambda I, \ 0 \le l \le J, \ A_{l2}^i = l\mu_{rc}I, \ 1 \le l \le J.$
- C_{l1}^i represents decrement in the ongoing traffic due to service completion by one of the i, $1 \le i \le M$ traffic and is given by $V(i)$, for $l = 0, 1, \ldots, J$.
- B_{l1}^i represents no service completion whereas B_{l0}^i represents the increment in the number of traffic in orbit. The decrement in the number of traffic in orbit is possible when the traffic leaves the orbit and is given by B_{l2}^i. The expressions are given below.

$$B_{l1}^i = \begin{cases} -(\lambda + l\mu_{rq} + l\mu_{rc})I, \qquad i = 0, \ 0 \le l \le J \\ W(i) - (\lambda + l\mu_{rq} + l\mu_{rc})I, \ 1 \le i \le M-1, \ 0 \le l \le J \\ W(i) - (\lambda - l\mu_{rq})I, \qquad i = M, \ 0 \le l \le J-1. \\ W(i) - l\mu_{rq}I, \qquad\qquad i = M, \ l = J \end{cases}$$

$$B_{l0}^i = \begin{cases} 0, \ 0 \le i \le M-1 \\ \lambda I, i = M \end{cases}; \quad B_{l2}^i = l\mu_{rq}, 0 \le i \le M, \ 1 \le l \le J.$$

CASE IV: MAP Arrivals, Exponential Service, No Retrials, One Type of Traffic [8]

Let the state space for the stochastic process $\{X(t), t \ge 0\}$ is given with the state space $\{(i, u), \ 0 \le i \le M, \ 1 \le u \le K\}$.

- The lower diagonal elements Q_{i2} represent the decrement in the number of ongoing traffic. For $i = 1, 2, \ldots, M$, $Q_{i2} = i\mu I$.
- The upper diagonal element Q_{i0} represent increment in the number of ongoing traffic due to handoff or new ve traffic arrival as

$$Q_{i0} = \begin{cases} C_H + C_N, 0 \le i < M-g \\ C_H, \qquad M-g \le i < M \\ 0, \qquad\quad \text{otherwise.} \end{cases}$$

- The main diagonal elements Q_{i1} corresponds to no change in the number of ongoing traffic given as follows.

$$Q_{i1} = \begin{cases} C_0 - i\mu I, & 0 \leq i < M - g \\ C_0 + C_N - i\mu I, & M - g \leq i < M \\ -i\mu I, & i = M. \end{cases}$$

CASE V: Poisson Arrivals, Exponential Service, No Retrials, One Type of Traffic [9]

Let $\{X(t), t \geq 0\}$ be the stochastic process, where $X(t)$ represents the number of ongoing traffic at time t.

- The lower diagonal elements Q_{i2} represent the decrement in the number of ongoing traffic. For $i = 1, 2, \ldots, M$, $Q_{i2} = i\mu$.
- The upper diagonal element Q_{i0} represent increment in the number of ongoing traffic due to handoff or new ve traffic arrival as

$$Q_{i0} = \begin{cases} \lambda_N + \lambda_H, & 0 \leq i < M - g \\ \lambda_H, & M - g \leq i < M \\ 0, & \text{otherwise.} \end{cases}$$

- The main diagonal elements Q_{i1} corresponds to no change in the number of ongoing traffic given as follows.

$$Q_{i1} = \begin{cases} -\lambda_N - \lambda_H - i\mu, & 0 \leq i < M - g \\ -\lambda_H - i\mu, & M - g \leq i < M \\ -i\mu, & i = M. \end{cases}$$

CASE VI: Poisson Arrivals, Exponential Service, Exponential Retrials, One Type of Traffic [10]

Let $\{X(t), t \geq 0\}$ be the stochastic process with a state space $\{(j,k) : 0 \leq j \leq M; 0 \leq k \leq J\}$.

- The elements of the upper diagonal Q_{i0} are represented by

$A_{i2} = jp\mu_r, 0 \leq j \leq J; A_{i1} = \begin{cases} \lambda = \lambda_H + \lambda_N, & 0 \leq l \leq M - g - 1, \\ \lambda_H, & M - g \leq l \leq M - 1. \end{cases}$

- The elements of the lower diagonal are given by $Q_{i2} = i\mu$.
- The elements of the main diagonal, denoted as Q_{i1} represented as

$B_{i2} = k(1-p)\mu_r, 0 \leq k \leq J; B_{i1} = -(\lambda + i\mu + k\mu_r), 0 \leq k \leq J;$

$B_{i0} = \begin{cases} 0, & 0 \leq l \leq M - g - 1, \\ \lambda_N, & M - g \leq l \leq M. \end{cases}$

After performing the numerical illustration for $g = 1, 2, 3$ and $J = 0$, Fig. 1(a) and Fig. 1(b) plot loss probabilities (the blocking probability P_b and the dropping probability P_d) with respect to M. We observe by these figures that for no retrial, i.e., for $J = 0$, P_b and P_d decreases with M for a fixed value of g, P_b increases with g for a fixed value of M, and P_d decreases with g for a fixed value of M.

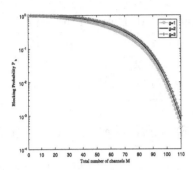 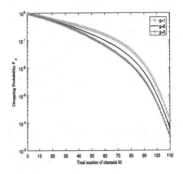

(a) Dependence of P_b over M for $J = 0$.   (b) Dependence of P_d over M for $J = 0$.

Fig. 1. Dependence of loss probabilities over M for $J = 0$.

6   Conclusions and Future Directions

In this paper, we consider a ve, vo and d traffic supported tethered HAP systems with MAP and PH distributions for arrival, service and retrial processes. A very well known GC scheme has been implemented for the prioritization of the higher priority traffic. The proposed study has been modeled by a $LDQBD$ process. An explicit expression for the generator matrix Q has been obtained. Further, formulae of some important performance measures have been derived. The suggested model is a generalization of various queueing models, a few of these are included here as special cases. For the sake of numerical illustration, a few results have been shown for the special case queueing model.

In the future, a CAC scheme with dynamic resource reservations for ve, vo and d traffic will be proposed and an extensive numerical illustration will be presented with efficient algorithms for the computation. For the reported model, resource allocation and energy optimization problems may be constructed.

References

1. Ahmed, M.H.: Call admission control in wireless networks: a comprehensive survey. IEEE Commun. Surv. Tutorials **7**(1–4), 50–69 (2005)
2. Alfa, A.S., Isotupa, K.S.: An M/PH/k retrial queue with finite number of sources. Comput. Oper. Res. **31**(9), 1455–1464 (2004)
3. Alfa, A.S., Li, W.: A homogeneous pcs network with Markov call arrival process and phase type cell residence time. Wireless Netw. **8**(6), 597–605 (2002). https://doi.org/10.1023/A:1020329719692
4. Alfa, A.S., Li, W.: PCS networks with correlated arrival process and retrial phenomenon. IEEE Trans. Wireless Commun. **1**(4), 630–637 (2002)
5. Artalejo, J.R., Chakravarthy, S.R.: Computational analysis of the maximal queue length in the MAP/M/c retrial queue. Appl. Math. Comput. **183**(2), 1399–1409 (2006)

6. Dharmaraja, S., Jindal, V., Alfa, A.S.: Phase-type models for cellular networks supporting voice, video and data traffic. Math. Comput. Model. **47**(11–12), 1167–1180 (2008)
7. Dudin, A.N., Klimenok, V.I., Vishnevsky, V.M.: The Theory of Queuing Systems with Correlated Flows. Springer, Cham (2020). https://doi.org/10.1007/978-3-030-32072-0.pdf
8. Gaver, D., Jacobs, P., Latouche, G.: Finite birth-and-death models in randomly changing environments. Adv. Appl. Probab. **16**(4), 715–731 (1984)
9. Harine, G., Marie, R., Puigjaner, R., Trivedi, K.: Loss formulas and their application to optimization for cellular networks. IEEE Trans. Veh. Technol. **50**(3), 664–673 (2001)
10. Jain, V., Raj, R., Dharmaraja, S.: Numerical optimization of loss system with retrial phenomenon in cellular networks. Int. J. Oper. Res. (2022). In Press
11. Kurt, G.K., et al.: A vision and framework for the high altitude platform station (haps) networks of the future. IEEE Commun. Surv. Tutorials **23**(2), 729–779 (2021)
12. Neuts, M.F.: Matrix-Geometric Solutions in Stochastic Models: An Algorithmic Approach. Courier Corporation, North Chelmsford (1994)
13. Slalmi, A., Chaibi, H., Saadane, R., Chehri, A., Jeon, G.: 5G NB-IoT: efficient network call admission control in cellular networks. Concurr. Comput. Pract. Exp. **33**(22), e6047 (2021)
14. Tozer, T., Grace, D.: High-altitude platforms for wireless communications. Electron. Commun. Eng. J. **13**(3), 127–137 (2001)

Analysis of Power Management in a Tethered High Altitude Platform Using $MAP/PH[3]/1$ Retrial Queueing Model

Vidyottama Jain[1]([✉]), V. M. Vishnevsky[2], Dharmaraja Selvamuthu[3], and Raina Raj[3]

[1] Central University of Rajasthan, Ajmer, India
vidyottama.jain@curaj.ac.in
[2] Institute of Control Sciences of Russian Academy of Sciences, Moscow, Russia
vishn@inbox.ru
[3] Indian Institute of Technology Delhi, New Delhi, India
dharmar@maths.iitd.ac.in, rainaraj.curaj@gmail.com

Abstract. Due to the increasing demand of the tethered high altitude platform (HAP) systems, it is imperative to assess their power consumption along with their deployment. This study considers the power management of lithium-ion batteries based tethered HAP systems for wireless communications service provisioning. This article discusses a novel model based on a multi-dimensional Markov process applied for the evaluation of the power consumption characteristics of a tethered HAP system. The proposed model takes into account the increment in the load on the functioning of system after the consumption of power in batteries. The underlying study introduces functioning of the system in three modes along with the retrial phenomenon. The arrival of packets follows Markovian arrival process (MAP) and the service time is phase-type (PH) distributed with distinct parameters in three different modes. The stationary distributions and stability of the system have been derived using the matrix-geometric method. Further, by deriving key system performance measures, numerical examples have been illustrated.

Keywords: High altitude platform systems · Degraded service rate · Markovian arrival process · Phase type distribution · Power management

1 Introduction

For more than two decades, high expectations have been associated with the deployment of satellites or high altitude platform (HAP) for ubiquitous, global

The publication has been prepared with the support of DST-RSF research project no. 22-49-02023 (RSF) and research project no. 64800 (DST).

broadband communications availability [14]. These HAP systems can be classified as tethered HAP or untethered HAP systems [11]. The untethered HAP systems are established at the height of 20 to 50 km whereas the tethered HAP systems are positioned 200–400 metres above land. One of the key obstacles of using HAP to enhance the existing wireless networks is the restricted on-board energy and flight time [2,7,12]. According to the Alam et al. [1], batteries in an untethered HAP can last only for less than an hour. The hovering period is further lowered when the payload, communication, and signal processing use energy. As a result, a tethered HAP system is often used, with power supplied by lithium-ion batteries, solar panels, ground-based energy sources with energy transmission via a thin cable-rope, etc. Over the time the energy density of a new light weight lithium-ion battery is predicted to rise by 3% every year [1].

There exist a number of studies describing and illustrating the reliability of the system through queueing system but the literature over the power and energy management considering the queueing system appears to be rather moderate. In the literature, a few studies have constructed queueing models for the tethered HAP system (refer, [6,9,13]). However, the applicability of these models has been diminished in the present scenario, due to their consideration of Poisson process for arrival and exponential distribution for service times. In the cutting edge wireless technologies, the input flow of calls possesses burstiness and correlation properties rather than the memory-less property of stationary Poisson flow [5]. The flow of incoming traffic can be correlated, homogeneous, heterogeneous, bursty, etc. In the case of homogeneous traffic (when traffic is of the same type), an apt mathematical model for the arrival process is known as Markovian arrival process (MAP) which is a generalization of the stationary Poisson process.

Phase-type (PH) distribution forms a versatile family of probability distributions where a process is completed in multiple phases. The exponential, Erlang and hyper-exponential distributions belong to this family. In the HAP system, the service of an arriving packet can be blocked due to the unavailability of the server or insufficient space in the waiting queue. Hence, the blocked packet waits for some time before receiving the service and retry for its service. This phenomenon is referred as retrial phenomenon and is ubiquitous in many real life applications such as traffic centers, communication systems, optical networks, inventory systems, and so on. Therefore, the consideration of retrial phenomenon is essential while evaluating the performance of the system.

The proposed study considers a retrial queueing model for the power management of the tethered HAP systems. It has been assumed that the proposed system works in three different modes as per the availability of the power status of batteries: normal mode, power saving mode, and ultra power saving mode. To this end, two threshold parameters have been defined, say K_1 and K_2. There can be three possible cases:

- If the power level of the batteries is above K_1, the system will provide service in the normal mode.
- If the power level of the batteries is between K_1 and K_2, the system will provide service in the power saving mode with a degraded rate.

 – If the power level of batteries is below K_2, the system will provide service in the ultra power saving mode with a degraded rate.

The main purpose behind considering these different modes is that the system will keep on providing service to the incoming packets at the same time the power will be saved as the system will provide service with a degraded rate. This factor can be extremely useful from both customer's point of view as well as from the service provider's point of view. In this study, the concept of retrial phenomenon has also been added for those packets which find the queue full. These arriving packets will join the orbit of infinite capacity to retry for their service after some random amount of time or it might exit the system without obtaining the service.

This paper has the following structure. In Sect. 2, the queueing model for the power management in a tethered HAP system is thoroughly described. Section 3 demonstrates the construction of an infinitesimal generator matrix for the underlying process and provides the steady-state analysis of the system. Expressions for a few essential performance measures to assess the system's efficiency are formulated in Sect. 4. Section 5 presents numerical examples to highlight the qualitative behaviour of the proposed queueing model. Finally, a few concluding remarks and insights for further research are presented in Sect. 6.

2   Model Description

The proposed study investigates a queueing model for a tethered HAP system. The HAP will provide service to all the arriving requests which will be in the packet form. In this work, the HAP will be addressed as a server in the context of providing the services to the arriving packets. The proposed system works in three different modes on the basis of power level of batteries. To this end, two threshold levels have been introduced. If the power status of the batteries in the system is above a pre-defined threshold value, say K_1, the system will work in the normal mode. When the power status is in between K_1 and K_2 ($K_1 > K_2$), the system goes in power saving mode. If the power status is below the threshold value K_2, it is considered crucial scenario, therefore, the system operates in ultra power saving mode. In the normal mode, power saving mode, and ultra power saving mode, the server will provide the service with rates, say μ_1, μ_2, and μ_3, respectively. Here, it has been considered that as the system changes its mode of providing the service, the service rate also degrades, i.e., $\mu_1 > \mu_2 > \mu_3$.

The arriving packets, which find the queue full, will join a virtual space named orbit of infinite capacity. From the orbit, the packet will retry for its service following the exponential distribution with rate θ and probability p or it might exit the system also with complementary probability $1 - p$. The batteries will be discharged with a rate, say D_b per sec, following the exponential distribution and the discharged batteries will be recharged with a rate R_b per sec, following the exponential distribution. Figure 1 represents the pictorial representation of the proposed system.

- Arrival Process:
  Arrival of packets follows a continuous time MAP with dimension M. Let D be the irreducible infinitesimal generator of this MAP where $D = D_0 + D_1$. Let π be the unique solution of $\pi D = O$ and $\pi e = 1$, where O is a zero row vector of appropriate size and e is a unit column vector. The arrival intensity of the MAP is defined as $\lambda = \pi D_1 e$.
- Service Process:
  The service times in these three different modes of the system follow PH distribution with different parameters. In the normal mode, the service time follows PH distribution with parameters (δ_1, L_1) and dimension M_1, i.e., $L_1 e + L_1^0 = 0$. In the power saving mode, the service time follows PH distribution with parameters (δ_2, L_2) and dimension M_2, i.e., $L_2 e + L_2^0 = 0$. In the ultra power saving mode, the service time follows PH distribution with parameters (δ_3, L_3) and dimension M_3, i.e., $L_3 e + L_3^0 = 0$. Note that notations $\oplus$ and $\otimes$ are used for the Kronecker sum and the Kronecker product of two matrices, respectively. For more description over Kronecker sum and Kronecker product, authors suggest readers to refer [3].

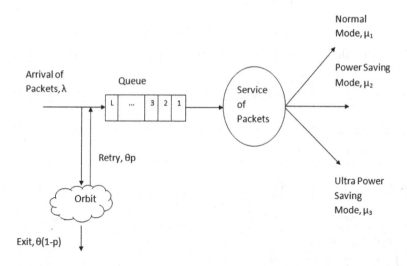

Fig. 1. A schematic diagram of the proposed model.

3   Stochastic Modeling

The state space of the underlying process $\{X(t), t \geq 0\}$ is defined as follows.

$$\Omega = \{(i, j, k, s_1, s_2, s_3, v); i \geq 0, \ 0 \leq j \leq L, \ k = 0, 1, 2, \ 1 \leq s_1 \leq M_1,$$
$$1 \leq s_2 \leq M_2, \ 1 \leq s_3 \leq M_3, \ 1 \leq v \leq M\},$$

where,

- i denotes number of packets in the orbit,
- j denotes number of packets in the queue,
- k denotes the mode of the system,
    - $k = 0$ denotes the normal mode of the system,
    - $k = 1$ denotes the power saving mode of the system,
    - $k = 2$ denotes the ultra power saving mode of the system,
- s_1 is the number of phases in normal mode for PH distribution,
- s_2 is the number of phases in power saving mode for PH distribution,
- s_3 is the number of phases in ultra power saving mode for PH distribution,
- v is the number of phases for MAP.

The generator matrix of the stochastic process $\{X(t), t \geq 0\}$ is given as follows:

$$
Q = \begin{pmatrix}
Q_{0,0} & Q_{0,1} & 0 & 0 & 0 & \cdots \\
Q_{1,0} & Q_{1,1} & Q_{1,2} & 0 & 0 & \cdots \\
0 & Q_{2,1} & Q_{2,2} & Q_{2,3} & 0 & \cdots \\
0 & 0 & Q_{3,2} & Q_{3,3} & Q_{3,4} & \cdots \\
\vdots & \vdots & \vdots & \vdots & \ddots & \ddots & \ddots \\
\vdots & \vdots & \vdots & \vdots & & \ddots & \ddots & \ddots
\end{pmatrix}.
$$

We define the following notations in order to carry out the mathematical analysis.

- I_u is an identity matrix of dimension u.
- $\mathbf{0}$ is a zero matrix of appropriate dimension.

Upper Diagonal:

$$
Q_{i,i+1} = \begin{pmatrix}
C_0 & 0 & 0 & \cdots & 0 \\
0 & C_1 & 0 & \cdots & 0 \\
0 & 0 & C_2 & \cdots & 0 \\
\vdots & \vdots & \vdots & \ddots & \vdots \\
0 & 0 & 0 & \cdots & C_L
\end{pmatrix}; i \geq 0,
$$

$$
C_j = \begin{cases}
\mathbf{0}; 0 \leq j \leq L - 1, \\
\begin{pmatrix}
D_1 \otimes \delta_1 & 0 & 0 \\
0 & D_1 \otimes \delta_2 & 0 \\
0 & 0 & D_1 \otimes \delta_3
\end{pmatrix}; j = L.
\end{cases}
$$

Lower Diagonal:

$$
Q_{i+1,i} = \begin{pmatrix}
A_0 & \hat{A}_0 & 0 & \cdots & 0 & 0 \\
0 & A_1 & \hat{A}_1 & \cdots & 0 & 0 \\
0 & 0 & A_2 & \cdots & 0 & 0 \\
\vdots & \vdots & \vdots & \ddots & \vdots & \vdots \\
0 & 0 & 0 & \cdots & A_{L-1} & \hat{A}_{L-1} \\
0 & 0 & 0 & \cdots & 0 & A_L
\end{pmatrix}; i \geq 0,
$$

$$A_j = \begin{pmatrix} (i+1)\theta(1-p)I & 0 & 0 \\ 0 & (i+1)\theta(1-p)I & 0 \\ 0 & 0 & (i+1)\theta(1-p)I \end{pmatrix} ; 0 \le j \le L,$$

$$\hat{A}_j = \begin{pmatrix} (i+1)\theta pI & 0 & 0 \\ 0 & (i+1)\theta pI & 0 \\ 0 & 0 & (i+1)\theta pI \end{pmatrix} ; 0 \le j \le L.$$

Main Diagonal:

$$Q_{i,i} = \begin{pmatrix} B_0 & \hat{B}_0 & 0 & \cdots & 0 & 0 \\ \bar{B}_1 & B_1 & \hat{B}_1 & \cdots & 0 & 0 \\ 0 & \bar{B}_2 & B_2 & \cdots & 0 & 0 \\ \vdots & \vdots & \vdots & \ddots & \vdots & \vdots \\ 0 & 0 & 0 & \cdots & B_{L-1} & \hat{B}_{L-1} \\ 0 & 0 & 0 & \cdots & \bar{B}_L & B_L \end{pmatrix} ; i \ge 0,$$

$$\bar{B}_j = \begin{pmatrix} I_M \otimes L_1^0 & 0 & 0 \\ 0 & I_M \otimes L_2^0 & 0 \\ 0 & 0 & I_M \otimes L_3^0 \end{pmatrix} ; 1 \le j \le L,$$

$$\hat{B}_j = \begin{pmatrix} D_1 \otimes \delta_1 & 0 & 0 \\ 0 & D_1 \otimes \delta_2 & 0 \\ 0 & 0 & D_1 \otimes \delta_3 \end{pmatrix} ; 0 \le j \le L-1,$$

$$B_j = \begin{cases} \begin{pmatrix} D_0 - (i\theta + D_b)I & D_b I & 0 \\ R_b I & D_0 - (i\theta + D_b + R_b)I & D_b I \\ 0 & R_b I & D_0 - (i\theta + R_b)I \end{pmatrix} ; j = 0, \\ \begin{pmatrix} D_0 - (i\theta + D_b)I \oplus L_1 & D_b I & 0 \\ R_b I & D_0 - (i\theta + D_b + R_b) \oplus L_2 & D_b I \\ 0 & R_b I & D_0 - (i\theta + R_b)I \oplus L_3 \end{pmatrix} ; \\ 1 \le j \le L. \end{cases}$$

3.1 Steady-State Analysis

Let $z_s = \{z_s(0), z_s(1), z_s(2), \ldots\ldots\}$ be the steady-state probability vector of generator matrix Q satisfying

$$z_s Q = 0; z_s e = 1,$$

where,

$$z_s(0) = (z_s(0,0), z_s(0,1), z_s(0,2), z_s(0,3), \ldots, z_s(0,L)),$$
$$z_s(1) = (z_s(1,0), z_s(1,1), z_s(1,2), z_s(1,3), \ldots, z_s(1,L)),$$

$$\vdots$$

$$z_s(i) = (z_s(i,0), z_s(i,1), z_s(i,2), z_s(i,3), \ldots, z_s(i,L)),$$
$$z_s(i,j) = (z_s(i,j,0), z_s(i,j,1), z_s(i,j,2)), i \ge 0, \ 0 \le j \le L.$$

Due to the consideration of PH distributions, it is tedious task to obtain the closed form expression of z_s. Therefore, the methodology provided by Neuts-Rao [10] has been adopted to solve the system. Using the theory of Neuts-Rao truncation method, obtain N, a positive integer such that $Q_{i,i+1} = Q_2$, $Q_{i,i} = Q_1$, and $Q_{i+1,i} = Q_0$. Therefore, the modified generator matrix Q^* will be as follows

$$Q^* = \begin{pmatrix} Q_{0,0} & Q_{0,1} & 0 & 0 & 0 & 0 \\ Q_{1,0} & Q_{1,1} & Q_{1,2} & 0 & 0 & 0 \\ 0 & Q_{2,1} & Q_{2,2} & Q_{2,3} & 0 & 0 \\ 0 & 0 & Q_{3,2} & Q_{3,3} & Q_{3,4} & 0 \\ \vdots & \vdots & \vdots & \ddots & \ddots & \ddots \\ \vdots & \vdots & \vdots & & \ddots & \ddots & \ddots \\ \vdots & \vdots & \vdots & & Q_0 & Q_1 & Q_2 \\ \vdots & \vdots & \vdots & & & Q_0 & Q_1 & Q_2 \\ \vdots & \vdots & \vdots & & & \vdots & \vdots & \vdots \end{pmatrix}.$$

Let z_s' be the steady-state vector of $H = Q_0 + Q_1 + Q_2$. Clearly, the generator matrix H with finite states is irreducible, aperiodic and positive recurrent. In this case, z_s' is the unique solution to the following system of linear equations $z_s' H = 0; z_s' e = 1$, where $z_s' = (z_s'(0), z_s'(1), \ldots, z_s'(L))$ and e is a column vector of ones with the appropriate dimension. The following result provides a necessary and sufficient condition under which the system is stable.

Theorem 1. *The underlying queueing system is stable if and only if*

$$z_s'(L,0)(D_1 \otimes \delta_1)e + z_s'(L,1)(D_1 \otimes \delta_2)e + z_s'(L,2)(D_1 \otimes \delta_3)e$$
$$< [N\theta(1-p) \sum_{j=0}^{L-1} z_s'(j)e + N\theta p \sum_{j=0}^{L} z_s'(j)e].$$

From the [10], the proof of the theorem follows immediately on noting

$$z_s' Q_0 e < z_s' Q_2 e.$$

When the system is considered without retrial phenomenon, the proposed system maps to a classical queueing system and the necessary and sufficient stability condition becomes λ/μ. To compute the steady-state probability vector, the methodology of matrix-geometric method has been applied. For the detailed study over matrix-geometric method, readers are suggested to refer [8,10].

4   Performance Measures

In this section, the important system performance measures along with their expressions for the proposed model are listed as follows.

1. The expected number of packets in queue:

$$E[Q] = \sum_{i=0}^{\infty} \sum_{j=0}^{L} \sum_{k=0}^{2} j z_s(i,j,k)e.$$

2. The expected number of packets in orbit:

$$E[L] = \sum_{i=0}^{\infty} \sum_{j=0}^{L} \sum_{k=0}^{2} i z_s(i,j,k)e.$$

3. The probability that the server is idle:

$$P_{idle} = \sum_{k=0}^{2} z_s(0,0,k)e.$$

4. The probability for the server being in normal mode:

$$P_{normal} = \sum_{i=0}^{\infty} \sum_{j=0}^{L} z_s(i,j,0).$$

5. The probability for the server being in power saving mode:

$$P_{power} = \sum_{i=0}^{\infty} \sum_{j=0}^{L} z_s(i,j,1).$$

6. The probability for the server being in ultra power saving mode:

$$P_{ultra} = \sum_{i=0}^{\infty} \sum_{j=0}^{L} z_s(i,j,2).$$

7. The overall rate of retrial:

$$\theta^* = \sum_{i=0}^{\infty} \sum_{j=0}^{L} \sum_{k=0}^{2} \theta i z_s(i,j,k).$$

5   Numerical Illustration

In this section, illustration of the key performance measures has been discussed that bring out the qualitative aspects of the model under study. For the numerical computation, the representative matrices for the *MAP* are referred from [4] as follows,

$$D_0 = \begin{pmatrix} -0.81 & 0 \\ 0 & -0.02 \end{pmatrix}, \quad D_1 = \begin{pmatrix} 0.8054 & 0.0053 \\ 0.01145 & 0.0116 \end{pmatrix}.$$

The correlation coefficients and variation coefficients are $C_r = 0.2$ and $C_v = 12.34$. Let *PH* distributions parameters for the service rates of normal mode, power saving mode and ultra power saving mode are

$$\delta_1 = \begin{pmatrix} 0.05, 0.95 \end{pmatrix}, \quad L_1 = \begin{pmatrix} -0.03104 & 0 \\ 0 & -2.441 \end{pmatrix},$$

$$\delta_2 = \begin{pmatrix} 0.1, 0.9 \end{pmatrix}, \quad L_2 = \begin{pmatrix} -0.03359 & 0 \\ 0 & -2.5262 \end{pmatrix},$$

$$\delta_3 = \begin{pmatrix} 0.5, 0.5 \end{pmatrix}, \quad L_3 = \begin{pmatrix} -1 & 0 \\ 0 & -1 \end{pmatrix}.$$

To obtain the numerical results, we have set $\theta = 2/s$, $L = 30$, $D_b = 3/s$, $R_b = 3/s$, $p = 0.5$. To demonstrate the feasibility of the developed model, some interesting observations of the proposed system are described through the following numerical experiments. These experiments will present the behaviour of performance measures with respect to arrival, service and retrial rates.

Example 1: Here, the behaviour of the expected number of packets in the system and the idle probability is investigated with respect to arrival rate λ and service rate μ.

In Fig. 2(a), the total number of packets in the system, $E[Q]$ appears to decrease with the increment in service rate μ. It should be noted that $E[Q]$ increases with respect to the increment in the value of arrival rate of packets λ. This finding can be understood as follows. When the arrival rate of packets increases in the system, it will also increase the total number of packets in the system. If the system provides service with an increasing rate, the packets will be served with an increasing rate, and consequently, $E[Q]$ decreases.

Figure 2(b) exhibits the impact of the probability of server being in idle (P_{idle}) situation with respect to λ and μ. From the graph, it can be observed that if the arrival of packets is increased in the system, the possibility of the server being idle decreases. Hence, P_{idle} decreases with respect to the increment in λ. Also, a reverse observation can be seen when the service rate is increased in the system. If the service rate increases in the system for a fixed value of λ, the chances of the server being in idle situation also increases. Therefore, P_{idle} appears to be an increasing function of μ.

Example 2: Here, the behaviour of the expected number of packets in the orbit ($E[L]$) and the overall retrial rate (θ^*) is analyzed with respect to λ, μ, and retrial rate θ.

Figure 3(a) demonstrates that $E[L]$ is an increasing function of λ and decreasing function of μ. The explanation of this finding is very obvious and can be given as follows. If λ increases in the system, there will be more arrival of packets, and these packets which are not able to join the queue will enter to the orbit to retry later on. Therefore, it will increase the number of packets in the orbit. If the service rate is increased, the probability of these packets getting the service also increases, hence, $E[L]$ decreases in the system.

In the Fig. 3(b), the overall rate of retrial θ^* appears to be an increasing function of λ and increasing function of retrial rate θ. If the arrival of packets is increased in the system, more packets will be in the orbit and therefore, the overall retrial rate also increases. Similarly, if the retrial rate of these packets increases, the overall retrial rate of packets will also increase in the system.

Example 3: In this example, the probability of the server being in three different modes have been analyzed with respect to the discharging rate D_b and recharging rate R_b of the batteries.

Figure 4(a) shows the behaviour of the system being in normal mode P_{normal} with respect to discharging rate D_b and recharging rate R_b of batteries. It can be observed that P_{normal} exhibits a decreasing behaviour with respect to D_b and increasing behaviour with respect to R_b. This finding can be illustrated as follows. If D_b is increased in the system, the batteries will be discharged at a rapid speed and soon the charging level of the batteries will be below the threshold level K_1 (pre-defied). Hence, P_{normal} decreases in the system. Whereas, if the batteries are recharged with fast rate, P_{normal} increases as the system will be able to remain above the threshold level for some more time.

Similar behaviour can be observed from the Fig. 4(b) for the system being in power saving mode P_{power} with respect to D_b and R_b. Through the same explanation, this result can be explained. If the charging level of the batteries falls below the threshold value K_1 but remains above K_2, the system will remain in the power saving mode. If D_b increases, the batteries will be rapidly discharged, and the charging level of the batteries will soon fall below the K_1. Hence, P_{power} decreases in the system. If the

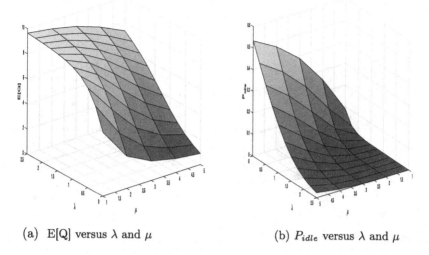

(a) E[Q] versus λ and μ (b) P_{idle} versus λ and μ

Fig. 2. Behaviour of expected number of packets in the system E[Q] and P_{idle} with respect to the arrival rate λ and service rate μ.

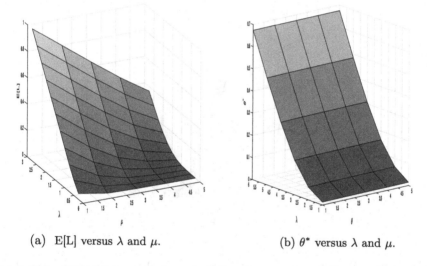

(a) E[L] versus λ and μ.   (b) θ^* versus λ and μ.

Fig. 3. Behaviour of the expected number of packets in orbit E[L] with respect to the arrival rate λ and service rate μ and overall rate of retrial θ^* with respect to the arrival rate λ and retrial rate θ.

batteries are recharged quickly, P_{power} increases since the system will be able to stay over the threshold level for a longer period of time.

On the similar track, Fig. 5 demonstrates the behaviour for the system being in ultra power saving mode P_{ultra} with respect to D_b and R_b. Here, it can be observed that P_{ultra} appears to be decreasing function of D_b and increasing function of R_b. The explanation for this finding remains the same as mentioned above for the P_{normal} and

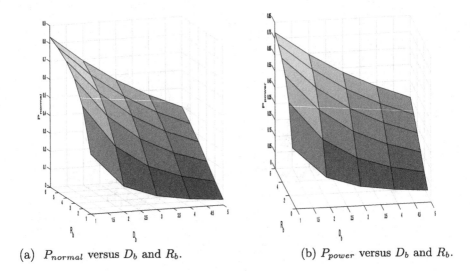

(a) P_{normal} versus D_b and R_b.   (b) P_{power} versus D_b and R_b.

Fig. 4. Behaviour of the probability of server being in normal mode P_{normal} and power saving model P_{power} with respect to battery discharging rate D_b and battery recharging rate R_b.

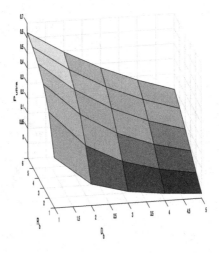

Fig. 5. Behaviour of the probability of server being in ultra power saving mode P_{ultra} with respect to battery discharging rate D_b and battery recharging rate R_b.

P_{power}. The value of P_{ultra} is less for the same value of D_b and R_b in comparison to P_{normal} and P_{power} which is obvious as when the system is in critically low charging level of batteries the probability value of being in the state also becomes less.

6   Conclusion and Future Directions

In the presented study, power management for the tethered HAP system is modeled by using a retrial queueing system. The power management in the tethered HAP system is explored through three different power saving modes, named as, normal mode, power saving mode and ultra power saving mode. The key concept behind introducing different modes is to keep the system active for a longer duration by saving its power while the service is degraded. It is assumed that the service rate is degraded from μ_1 to μ_2 when the system moves from normal mode to power saving mode, and the service rate is degraded from μ_2 to μ_3 when the system moves from power saving mode to ultra power saving mode, i.e., $\mu_1 > \mu_2 > \mu_3$. To handle the burstiness and correlation properties of the incoming traffic, this study considers more generalized *MAP* and *PH* distributions for the arrival and service processes of the system, respectively. The results of computing the stationary state probabilities and stability of the considered system are obtained using the matrix-geometric method. Through the numerical illustration, it has been shown that the probability of the system being in normal, power saving and ultra power saving mode are majorly affected by the discharging and recharging rate of the batteries. Also, the impact of arrival, service and retrial rates has been shown over a few selected performance measures of the system.

In the future, the proposed system can be thoroughly examined by taking into account multi server scenarios and different distributions, e.g., batch Markovian arrival process, general distribution, etc. The presented model can also be explored through various power saving optimization strategies in tethered HAP systems. Future research will focus heavily on a number of topics, particularly pertaining to the tethered HAP system's performance and availability.

References

1. Alam, M.S., Kurt, G.K., Yanikomeroglu, H., Zhu, P., Đào, N.D.: High altitude platform station based super macro base station constellations. IEEE Commun. Mag. **59**(1), 103–109 (2021)
2. Arum, S.C., Grace, D., Mitchell, P.D., Zakaria, M.D., Morozs, N.: Energy management of solar-powered aircraft-based high altitude platform for wireless communications. Electronics **9**(1), 179 (2020)
3. Dayar, T.: Analyzing Markov Chains Using Kronecker Products: Theory and Applications. Springer, New York (2012). https://doi.org/10.1007/978-1-4614-4190-8
4. Dudin, A., Kim, C., Dudin, S., Dudina, O.: Analysis and optimization of guard channel policy with buffering in cellular mobile networks. Comput. Netw. **107**, 258–269 (2016)
5. Dudin, A.N., Klimenok, V.I., Vishnevsky, V.M.: The Theory of Queuing Systems with Correlated Flows. Springer, Cham (2020). https://doi.org/10.1007/978-3-030-32072-0
6. Joo, C., Choi, J.: Low-delay broadband satellite communications with high-altitude unmanned aerial vehicles. J. Commun. Netw. **20**(1), 102–108 (2018)
7. Kozyrev, D.V., Phuong, N.D., Houankpo, H.G.K., Sokolov, A.: Reliability evaluation of a hexacopter-based flight module of a tethered unmanned high-altitude platform. In: Vishnevskiy, V.M., Samouylov, K.E., Kozyrev, D.V. (eds.) DCCN 2019. CCIS, vol. 1141, pp. 646–656. Springer, Cham (2019). https://doi.org/10.1007/978-3-030-36625-4_52

8. Latouche, G., Ramaswami, V.: Introduction to Matrix Analytic Methods in Stochastic Modeling. SIAM, Philadelphia (1999)
9. Li, S., Wang, L., David, G., Ma, D.: Cooperative directional inter-cell handover scheme in high altitude platform communications systems. J. Electron. **28**(2), 249–257 (2011)
10. Neuts, M.F.: Matrix-Geometric Solutions in Stochastic Models: An Algorithmic Approach. Courier Corporation (1994)
11. Qiu, J., Grace, D., Ding, G., Zakaria, M.D., Wu, Q.: Air-ground heterogeneous networks for 5G and beyond via integrating high and low altitude platforms. IEEE Wirel. Commun. **26**(6), 140–148 (2019)
12. Vishnevsky, V.M., Tereschenko, B.N., Tumchenok, D.A., Shirvanyan, A.M., Sokolov, A.: Principles of building a power transmission system for tethered unmanned telecommunication platforms. In: Vishnevskiy, V.M., Samouylov, K.E., Kozyrev, D.V. (eds.) DCCN 2019. LNCS, vol. 11965, pp. 94–110. Springer, Cham (2019). https://doi.org/10.1007/978-3-030-36614-8_8
13. Vishnevsky, V.M., Kozyrev, D.V., Rykov, V.V., Nguyen, Z.F.: Reliability modeling of an unmanned high-altitude module of a tethered telecommunication platform. Inform. Tekhnol. Vychslitel'nye Sist. (4), 26–38 (2020)
14. Widiawan, A.K., Tafazolli, R.: High altitude platform station (HAPS): a review of new infrastructure development for future wireless communications. Wirel. Pers. Commun. **42**(3), 387–404 (2007)

Investigation of a Finite-Source Retrial Queueing System with Two-Way Communication, Catastrophic Breakdown and Impatient Customers Using Simulation

János Sztrik⬡, Ádám Tóth$^{(\boxtimes)}$⬡, Ákos Pintér⬡, and Zoltán Bács⬡

University of Debrecen, University Square 1, Debrecen 4032, Hungary
{sztrik.janos,toth.adam}@inf.unideb.hu, apinter@science.unideb.hu,
bacs.zoltan@econ.unideb.hu

Abstract. An M/M/1//N retrial queueing system with two-way communication to the infinite source and impatient customers in the orbit is considered in the paper. There is a finite source in which the primary or regular customers are coming, while requests from the infinite source are the secondary customers. Because of not having waiting queues, the service of an arriving, primary customer begins immediately. Otherwise, in the case of a busy server, the primary customers are forwarded to the orbit waiting an exponentially distributed random time to try to reach the service unit. When the service unit is in an idle state, it may call a customer from the infinite source for service. All requests possess an impatience property resulting in an earlier departure from the system through the orbit if they wait too much for being served. Besides, the service unit is supposed to break down according to several distributions which have a specialty in removing all the customers located in the system. In the case of a faulty state, blocking is applied not allowing the customers into the system until the service unit fully recovers. This work concentrates on examining the effect of those distributions on several performance measures like the distribution of the number of customers in the system, and the probability of a primary customer departing because of catastrophe. The obtained results are graphically realized to show the differences and curiosities among the used parameter settings of the various distributions.

Keywords: Simulation · Catastrophic breakdown · Retrial queuing system · Collision · Impatience · Sensitivity analysis

1 Introduction

In many fields of our life, the phenomenon of waiting appears making inevitably creating queueing systems handling for example increasing network traffic in many info-communication systems. Throughout the years researchers have designed numerous tools and mechanisms which are suitable for modeling various organizations and one prime instance is retrial queueing systems that are capable of depicting arising real-life problems in telecommunication schemes like telephone switching systems, call centers, computer networks, and computer systems. Several publications exist where researchers exploit the advantages of retrial-queuing systems with repeated calls using for their models like in [2,7,8,12].

In the case of a retrial queueing system, a virtual waiting room is taken into consideration which is called the orbit meaning that whenever a service of a job can not start because of failure or occupation of the server it remains in the system. In the orbit, these customers have the opportunity to be at the service facility after a random time. The population of the customers is finite as the probability that a server calls a customer from the orbit is not very small and under such circumstances, it is more suitable to examine models with retrial queues. Exploring the available literature many papers have applied infinite and finite source queueing systems for example in [1,15,17,19].

It is also interesting to observe how the feature of two-way communication is used in papers in many fields of life. Its popularity originates from its usefulness to model applicable systems and creating real-life applications. One prime example can be mentioned in the topic of telecommunication, especially in call centres where agents may be occupied with other particular labor during an inactivity period like selling, advertising, and promoting products besides handling the calls of the customers. Optimizing the utilization of the service units or agents is always pivotal to increasing the efficiency of such systems. To mention some works about applying two-way communication schemes here are some instances [5,18].

Waiting is a natural occurrence in many aspects of our life and people experience the annoyance of having to wait in queues which is not a satisfying act. This may result in earlier departures of the requests leaving the system without being served which is called impatience. This behaviour is experienced in many territories like in healthcare applications, call centers, or telecommunication networks. This phenomenon is extensively studied in numerous articles and different types of impatience are distinguished: balking customers choose to avoid entering the system if the size of the queue is large, and jockeying customers can change their positions among the queues to get the service earlier or reneging customers decide to leave the queues if they have waited a specific long time there. About the investigation of this behaviour here are some examples: [10,14,16].

Random breakdowns and malfunctions occur in real-life scenarios caused by a power outage, human negligence, or other catastrophic activity. This greatly changes the operation of the system and the performance measures thus its investigation is necessary. In many cases, the service units are presumed to be accessible all the time which is not realistic. Systems with random failures have been investigated by many researchers for example in [6,13,20]. However, there are certain situations where the effect of breakdown is different types of failures can be investigated. For instance, power outages or mechanical failures may cause catastrophic events in which all the customers in the system are removed. This is known as a negative customer and it takes out every other request from service upon its arrival. This eventuates a disaster event because it also breaks down the service unit and in this case, every customer is forwarded back to the source. Papers in connection with negative customers can be found in [3,11].

The aim of this work is to realize a sensitivity analysis using various distributions of failure time on several performance measures while the departure of customers may happen. The results are obtained by our stochastic simulation program using the basics of SimPack [9] which contains the basic building blocks of a simulation model. This gives us the opportunity to model any type of queueing system to create any type of simulation model and we can calculate any performance measure using arbitrary random number generators for the desired random variable. The presented curves highlight both the effect of disaster events and the impatience of the customers applying various parameter settings and these figures concentrate on the interesting phenomena of these systems. The table of input parameters and graphical illustrations of the results are included demonstrating the influence of the used distributions on the main performance metrics.

2   System Model

A finite-source retrial queueing system of type $M/M/1//N$ is regarded as an unreliable service unit and impatient customers (see Fig. 1). This model has a service unit and exactly N individual resides in the finite-source in which request generation (primary request) is proceeded towards the system according to exponential law with parameter λ. This means that the inter-arrival times are exponentially distributed with mean λ. If an arriving job finds the server in an idle state then its service starts immediately which is an exponentially distributed variable with μ_1. Otherwise, in vain of a queue, jobs are not lost but remain in the system being forwarded to the orbit which is a virtual waiting space. From there these requests after an exponentially random time with parameter ν retry to reach the service unit. After spending futilely an exponentially distributed time with rate τ in the orbit a customer may choose to leave the system without being served so in other words, every request has an impatience characteristic. It is assumed random breakdowns take place according to various distributions like gamma, hyper-exponential, Pareto, hypo-exponential, and lognormal. The parameters are chosen in a way that a real sensitivity analysis would be accomplished. In these occurrences, disaster events develop resulting in interruption of the service of a job and with the customers, in the orbit, they all depart from the system. Blocking is applied during faulty periods so no customers are allowed by the system until the server fully recovers.

The repair process begins to be executed after the failure of the service unit which happens according to an exponential distribution with parameter γ_2. Two-way communication was also introduced in our model, when the server becomes free it may call a request from an infinite source (secondary customer) after an exponentially random time with parameter λ_2. That type of customer occupies instantly the service unit if it is not busy upon its arrival, otherwise, it is forwarded to a special buffer where it waits there until the server turns idle. At that moment a secondary customer automatically enters the service facility. In the case of a catastrophic event, every primary job returns to the finite-source, and every secondary customer exits from the system including the one who is under service. The service time of the secondary customer also follows an exponential distribution with the rate of μ_2. Every appeared arbitrary variable in the model construction is supposed to be independent of each other.

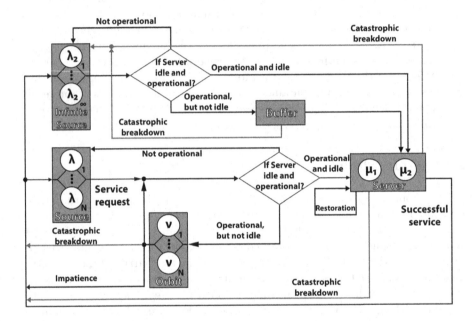

Fig. 1. System model.

3   Simulation Results

3.1   First Scenario

As mentioned earlier SimPack was used as a base of our program and we used a statistical package that was responsible for obtaining the desired performance measures. This utilizes the batch means method where the running period is split into batches (altogether S) and in every batch, $s = R - M/S$ are executed, in which M denotes the warm-up period observations at the beginning of the simulation which are rejected and R the length of the simulation. After the initial period, the sample average of the whole run is calculated, and what is really important here is to have long enough batches and approximately independent sample averages of the batches. Here is an article containing more information about the used process [4].

The confidence level of 99.9% is employed throughout the simulations and 0.00001 is the amount of the relative half-width of the confidence interval to pause the actual simulation sequence. The size of the batch in the initial transient period can not be too small therefore its value is set to 1000.

In Table 1 the used values of input parameters are presented.

Table 1. Numerical values of model parameters

N	ν	μ_1	μ_2	τ	γ_2	λ_2
100	0.01	1	2.5	0.05	1	0.5

The next table (Table 2) consists of the parameters of failure time, every chosen parameter is according to have the same mean and variance value in that way a valid comparison is achieved. The simulation program was tested by many parameter values and the reason for selecting these values to focus on interesting situations besides that it is worth mentioning that almost the same phenomenon appeared in this particular setting. The squared coefficient of variation is more than one in this scenario which is totally intentional to check the influence of peculiar random variables.

Table 2. Parameters of failure time

Distribution	Gamma	Hyper-exponential	Pareto	Lognormal
Parameters	$\alpha = 0.31225$	$p = 0.36197$	$\alpha = 2.1455$	$m = 1.00278$
	$\beta = 0.05588$	$\lambda_1 = 0.12955$	$k = 2.9835$	$\sigma = 1.19819$
		$\lambda_2 = 0.22835$		
Mean	5.558			
Variance	100			
Squared coefficient of variation	3.2024857438			

In Fig. 2 i represents the number of primary customers in the system on the X-axes, and $P(i)$ denotes the probability that exactly i primary customers are situated at the server and in the orbit altogether on the Y-axes. The distribution of the number of primary customers in the system is displayed when λ is 0.11 using various distributions of failure time. The mean number of primary customers in the system differs from each other greatly. In the case of the gamma distribution, customers tend to spend more time in the system compared to Pareto distribution. It is also noticeable that the highest probability is 0 and this can be explained by the fact that during faulty periods customers are not allowed to enter and for every catastrophic breakdown the system is emptied.

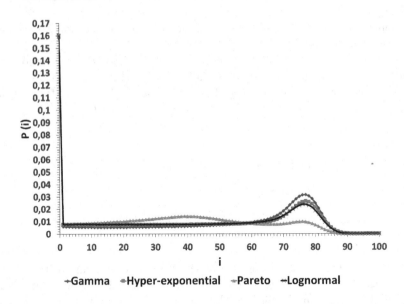

Fig. 2. Distribution of the number of primary customers in the system.

The expected response time of an arbitrary primary customer is presented in the function of the arrival intensity of incoming primary request in Figs. 3 and 4. Even though the mean and the variance value are equal to each other, huge gaps develop among the applied distributions, at gamma the highest average response times can be observed compared to the others (Fig. 3). Also with the increment of the arrival intensity, the expected response time of an arbitrary primary customer starts to increase then after a certain intensity arrival ($\lambda = 0.07$) it decreases except in the case of Pareto. The effect of impatience of the primary customer on the average response time is visible in Fig. 4 using the gamma distribution. Naturally, with the increment of rate τ the mean sojourn time decreases and this tendency is exactly the same for the other distributions. Although the mean operation time is 5.558 the values of the expected response times are higher than that which is quite interesting. Our intuition is that this can be explained by that the variance is quite high resulting in many small operation times and in most of them no customer can enter because they are so small. But there are several high operation periods in which it is very probable that many jobs enter and spend relatively a high amount of time.

Figure 5 demonstrates the development of the probability that a primary customer departs because of impatience besides increasing arrival intensity. At $\tau = 0.0001$ the probability of departure caused by impatience is basically 0 and as the rate of impatience increases this measure increases as well. At $\tau = 0.1$ almost half of the incoming primary customers leave the system earlier willingly.

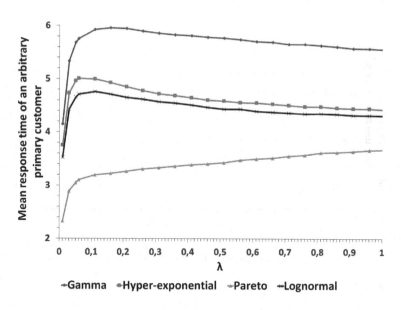

Fig. 3. Average response time of an arbitrary primary customer.

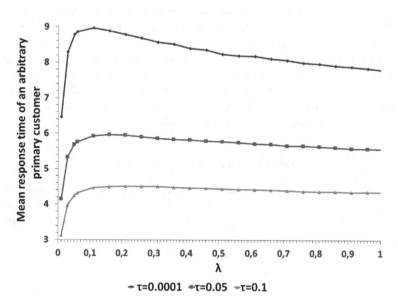

Fig. 4. Average response time of an arbitrary primary customer using different values of impatience.

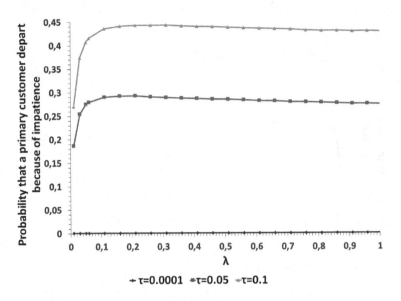

Fig. 5. Probability of departure of a primary customer due to impatience.

3.2 Second Scenario

Examining the results and phenomena of the first section we were excited to see how another parameter setting modifies the behaviour of the performance measures. In this second scenario, the parameters were chosen to have the squared coefficient of variation being less than one. Hypo-exponential replaces the hyperexponential distribution in which the squared coefficient of variation is always more than one. We will analyze the same measures as in the previous section but the new parameters of failure time which are presented in Table 3. The mean is the same but the variance is different in this case less than compared in the last section. All the remaining parameters are unchanged see Table 1.

Table 3. Parameters of failure time

Distribution	Gamma	Hypo-exponential	Pareto	Lognormal
Parameters	$\alpha = 1.2321$	$\mu_1 = 0.2$	$\alpha = 2.494$	$m = 1.4235$
	$\beta = 0.2205$	$\mu_2 = 1.7$	$k = 3.3478$	$\sigma = 0.7709$
Mean	5.588			
Variance	25.3460207612			
Squared coefficient of variation	0.811634349			

Figure 6 is in connection to the steady-state distribution of primary customers in the system. Analyzing the curves the obtained values are very near to each other, and the mean numbers are much closer than in the first section. However, there is a similarity as well, the most probable state is 0 and it is around 0.16 too. Likewise, these results were obtained besides the arrival intensity of 0.11.

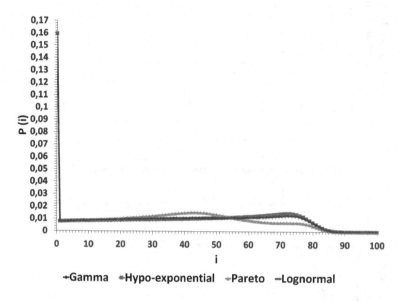

Fig. 6. Distribution of the number of customers in the system.

The next two figures (Fig. 7 and 8) present the average response time of an arbitrary primary customer versus the arrival intensity. On the first one (Fig. 7) the comparison of the used distribution can be seen where the difference is quite small among the values but the tendency remains the same, the highest amount is observed at gamma and the lowest at Pareto distribution. But none of them reach the mean value of operation time and based on our assumption this is because the variance is lower in this parameter setting. Figure 8 demonstrates the influence of impatience when the failure time follows hypo-exponential distribution however the same phenomenon is true for the others. Commonly, when rate τ begins to rise it makes the average response time lessens.

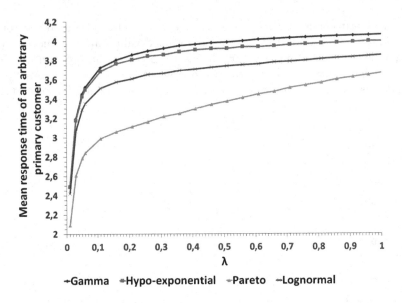

Fig. 7. Average response time of an arbitrary primary customer.

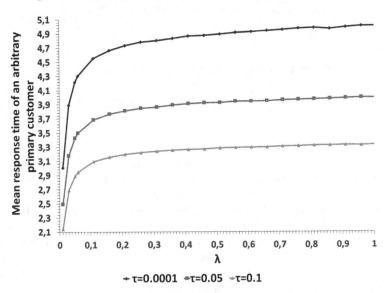

Fig. 8. Average response time of an arbitrary primary customer using different values of impatience.

4   Conclusion

A finite-source retrial queueing system is introduced with a non-reliable server, a two-way communication scheme, and impatient customers. We investigated several scenarios where different parameters are used to carry out a sensitivity analysis to figure out how the different performance measures develop. The results are obtained by our simulation program and several graphical figures depict the effect of using various distributions of failure time on the expected response time of primary and primary departed customers, on the probability of departure because of impatience, or on the distribution of customers in the system. In our figures, the differences were clearly seen among the values of several performance measures when the squared coefficient of variation is greater than one showing how pivotal applying a distribution can be and substantially slightly when it is less than one. The curves also reveal the impact of impatience on reducing the average response time of a primary customer. In the future we plan to continue our research work, examining other types of finite-source retrial queuing systems with two-way communication or adding another service unit for backup purposes.

References

1. Ali, A.A., Wei, S.: Modeling of coupled collision and congestion in finite source wireless access systems. In: 2015 IEEE Wireless Communications and Networking Conference (WCNC), pp. 1113–1118. IEEE (2015)
2. Artalejo, J., Corral, A.G.: Retrial Queueing Systems: A Computational Approach. Springer, Heidelberg (2008). https://doi.org/10.1007/978-3-540-78725-9
3. Chakravarthy, S.R., Subramanian, S.: A stochastic model for automated teller machines subject to catastrophic failures and repairs. SIMULATION **1**(1), 75–94 (2018)
4. Chen, E.J., Kelton, W.D.: A procedure for generating batch-means confidence intervals for simulation: checking independence and normality. SIMULATION **83**(10), 683–694 (2007)
5. Dragieva, V., Phung-Duc, T.: Two-way communication $M/M/1//N$ retrial queue. In: Thomas, N., Forshaw, M. (eds.) ASMTA 2017. LNCS, vol. 10378, pp. 81–94. Springer, Cham (2017). https://doi.org/10.1007/978-3-319-61428-1_6
6. Dragieva, V.I.: Number of retrials in a finite source retrial queue with unreliable server. Asia Pac. J. Oper. Res. **31**(2), 1–23 (2014). https://doi.org/10.1142/S0217595914400053
7. Falin, G., Artalejo, J.: A finite source retrial queue. Eur. J. Oper. Res. **108**, 409–424 (1998)
8. Fiems, D., Phung-Duc, T.: Light-traffic analysis of random access systems without collisions. Ann. Oper. Res. **277**(2), 311–327 (2019)
9. Fishwick, P.A.: SimPack: getting started with simulation programming in C and C++. In: 1992 Winter Simulation Conference, pp. 154–162 (1992)
10. Gupta, N.: Article: A view of queue analysis with customer behaviour and priorities. In: IJCA Proceedings on National Workshop-Cum-Conference on Recent Trends in Mathematics and Computing 2011, RTMC(4), May 2012
11. Gupta, U.C., Kumar, N., Barbhuiya, F.P.: A queueing system with batch renewal input and negative arrivals. In: Joshua, V.C., Varadhan, S.R.S., Vishnevsky, V.M. (eds.) Applied Probability and Stochastic Processes. ISFS, pp. 143–157. Springer, Singapore (2020). https://doi.org/10.1007/978-981-15-5951-8_10

12. Kim, J., Kim, B.: A survey of retrial queueing systems. Ann. Oper. Res. **247**(1), 3–36 (2016)
13. Krishnamoorthy, A., Pramod, P.K., Chakravarthy, S.R.: Queues with interruptions: a survey. TOP **22**(1), 290–320 (2014)
14. Kumar, R., Jain, N., Som, B.: Optimization of an $M/M/1/N$ feedback queue with retention of reneged customers. Oper. Res. Decis. **24**, 45–58 (2014). https://doi.org/10.5277/ord140303
15. Lakaour, L., Aïssani, D., Adel-Aissanou, K., Barkaoui, K.: $M/M/1$ retrial queue with collisions and transmission errors. Methodol. Computi. Appl. Probab. **21**(4), 1395–1406 (2018)
16. Panda, G., Goswami, V., Datta Banik, A., Guha, D.: Equilibrium balking strategies in renewal input queue with Bernoulli-schedule controlled vacation and vacation interruption. J. Ind. Manag. Optim. **12**, 851–878 (2015). https://doi.org/10.3934/jimo.2016.12.851
17. Peng, Y., Liu, Z., Wu, J.: An M/G/1 retrial G-queue with preemptive resume priority and collisions subject to the server breakdowns and delayed repairs. J. Appl. Math. Comput. **44**(1–2), 187–213 (2014). https://doi.org/10.1007/s12190-013-0688-7
18. Sztrik, J., Tóth, Á., Pintér, Á., Bács, Z.: The simulation of finite-source retrial queueing systems with two-way communication and impatient customers. In: Vishnevskiy, V.M., Samouylov, K.E., Kozyrev, D.V. (eds.) DCCN 2021. LNCS, vol. 13144, pp. 117–127. Springer, Cham (2021). https://doi.org/10.1007/978-3-030-92507-9_11
19. Takeda, T., Yoshihiro, T.: A distributed scheduling through queue-length exchange in CSMA-based wireless mesh networks. J. Inf. Process. **25**, 174–181 (2017)
20. Tóth, A., Sztrik, J., Pintér, A., Bács, Z.: Reliability analysis of finite-source retrial queuing system with collisions and impatient customers in the orbit using simulation. In: 2021 International Conference on Information and Digital Technologies (IDT), pp. 230–234 (2021). https://doi.org/10.1109/IDT52577.2021.9497567

Analysis of Retrial Queuing System with Limited Processor Sharing Discipline and Changing Effective Bandwidth

Alexander Dudin[1,2], Sergey Dudin[1], Olga Dudina[1], and Chesoong Kim[3(✉)]

[1] Department of Applied Mathematics and Computer Science,
Belarusian State University, 220030 Minsk, Belarus
dudina@bsu.by
[2] Peoples' Friendship University of Russia (RUDN University),
6 Miklukho-Maklaya Street, Moscow 117198, Russia
[3] Sangji University, Wonju, Kangwon 26339, Republic of Korea
dowoo@sangji.ac.kr

Abstract. In this paper, a queueing system with a flexible limited processor sharing discipline according to which a limited number of requests can receive service simultaneously is considered. Each request has the required service rate and when the number of requests in service is such that the bandwidth of the server is sufficient for providing this rate for all requests, the requests receive this service rate. We suggest that the requests on service can disturb (interfere) each other and the effective bandwidth of the server decreases with increase of the number of servicing requests. In the situation when the effective bandwidth is not enough, all requests receive service at a proportionally reduced rate. The arrival flow is defined by a Markovian arrival process. A request, which cannot enter the service immediately upon arrival due to a limitation on the number of requests that can receive service simultaneously, will make retrials according to the classical retrial strategy. The process of the system states is defined as the level-dependent multidimensional Markov process. The infinitesimal generator of this process is derived. The main performance measures of the system are obtained. Numerical illustrations are presented.

Keywords: Markovian arrival flow · Limiting processor sharing · Changing effective bandwidth · Retrials

1 Introduction

Queuing systems are effectively used to model and optimize various industrial, logistics and telecommunications systems and networks. In some of these systems, requests are serviced one at a time in the order specified by the service discipline. However, sometimes many requests can be serviced in the system at the same time. In this case, multi-server systems are often considered. That is,

V. M. Vishnevskiy et al. (Eds.): DCCN 2022, LNCS 13766, pp. 243–256, 2022.
https://doi.org/10.1007/978-3-031-23207-7_19

the system throughput (bandwidth) is divided into several parts, called servers, and each server can serve one request independently of other servers. Multi-server queuing systems are a popular subject for research, see, e.g., in [1]. It should be noted that multi-server systems have a drawback in terms of optimal use of the system resources. For example, in a situation when there is one request for service, and there are many available servers, then the bulk of the bandwidth is not used. As an alternative to multi-server systems, queuing systems with the discipline of processor sharing are considered in the literature. For a review of work on processor sharing systems, see, e.g., [2–5]. This discipline assumes that the entire resources of the system are aimed at simultaneously servicing all available requests. That is, even when one request is being serviced, the system resource is fully used. Because, sometimes classical processor sharing is not reasonable due to the necessity to guarantee users a certain minimum service rate and, therefore, to restrict the number of requests, which can receive service simultaneously, the discipline of limited processor sharing is often used, see, e.g., [6–9].

In [10], more references to the relevant research are presented and the opportunity of a combination of providing service via the fixed servers and via the limited processor sharing discipline is discussed.

This work is devoted to the analysis of a queuing model with the discipline of processor sharing. Compared with classical systems, this model has the following features that increase its adequacy to modern systems. First, we assume that the request has a required service rate that cannot be exceeded. In fact, if a user of a wireless communication network requires a certain amount of system bandwidth to operate, then it is not necessary to allocate all the system bandwidth to the user. He/she simply won't be able to use it and won't be served faster. However, if there are a lot of requests presenting in the system, and there is a lack of bandwidth to provide service to all requests at the required rate, then a decrease in the average service rate is allowed. Second, we assume that the number of requests that can receive a service simultaneously is limited by a given control parameter. The fact is that if we do not restrict access to the system, then a situation may arise in which the number of serviced requests will be so large that requests will be serviced at an unacceptably low service rate. Third, we assume that requests in service may interfere with each other. In fact, in wireless communication networks, in order to avoid interference and organize multiple access, part of the system's bandwidth can be spent on delimiting requests. In addition, in this paper, we assume that the arrival flow of requests is specified by a MAP (Markovian arrival process), see, for example, [11] and [12], which allows taking into account significant traffic fluctuations inherent in modern telecommunication networks. Also, for greater model adequacy, we assume that requests that are not allowed to enter the service upon arrival in the system can make repeated attempts to get serviced. In other words, we consider a system with repeated calls (retrials). The current state of the problem of studying the retrial systems, which are the adequate models of many telecommunication systems (including mobile cellular networks and local area networks), call centers, etc., is described, e.g., in [13]. As earlier surveys on this topic, we can mention [14,15].

The manuscript is organized as follows. The second section deals with the mathematical model of the system under study. The process describing the changes in the states of the system is given in the third section. In the same section, there is an infinitesimal generator of this process, and the issue of finding a stationary distribution of system states is solved. The fourth section presents the main performance characteristics of the system. In Sect. 5, the results of a numerical example are presented. Section 6 concludes the paper.

2 Model Description

We consider a retrial queuing system with a limited processor sharing service discipline.

The structure of the system is presented in Fig. 1.

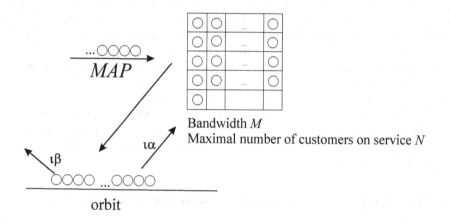

Fig. 1. Structure of the system

A single server can serve up to N requests at the same time. The parameter N is sometimes called multiprogramming level, see [16]. The buffer is missing. We suppose that the total bandwidth of the server is equal to M megabits per second. One request requires an average speed of service equal to X megabits per second. The average volume of one request is S megabits. Thus, if a request is serviced with the required bandwidth, then its average service time is defined as $b_1 = S/X$. We assume that the service time of one request has an exponential distribution. If the required bandwidth is available, the parameter of the exponential service time distribution is given as $\mu = \frac{1}{b_1}$. We suppose that during service requests can interfere with each other, that is, if there are n, $n = \overline{2, N}$, requests in service, then the effective throughput of the server is M_n, and $M \geq M_n \geq M_{n+1} > 0$ for any admissible n. If the number n of requests in service is such that $nX \leq M_n$, then all requests receive the required service rate and are served with intensity μ. Otherwise, each request is allocated bandwidth $X_n = \frac{M_n}{n}$ megabits and its service rate is $\mu_n = \frac{X_n}{S}$.

The MAP arrival process of requests enters the system. This flow is specified by the underlying process ν_t, $t \geq 0$. This process is an irreducible Markov chain with continuous-time and finite state space $\{1, 2, \ldots, W\}$, and matrices D_0 and D_1. Let us denote the mean intensity of incoming requests as λ.

A detailed description of the MAP process and its characteristics, as well as formulas for computation of the characteristics, can be found in [1,11], and [12].

If, during the arrival epoch of a request, the number of requests for service is less than the parameter N, then the request is accepted for service. Otherwise, the request goes to an orbit of unlimited capacity, from where it makes repeated attempts to get serviced at exponentially distributed intervals with the parameter α. An attempt is considered successful if, at the time of its execution, the number of requests in service is less than N. In this case, the request starts service, and the number of requests in the orbit reduces by one. If the attempt was unsuccessful, that is, at the time of its completion, the number of requests for service was equal to N, then the request returns to the orbit. Customers staying in orbit can be impatient. This means that each request can leave the orbit and permanently depart from the system after an exponentially distributed time with the parameter β, $\beta > 0$. Effect of impatience of requests staying in the orbit of a multi-server queue with MAP was earlier studied in [17]. Impatience of requests in multi-server queues with MAP arrivals and buffers was taken into account, e.g., in [18,19].

We suppose that N and M are the control parameters and the final purpose of analytical modeling is to provide a possibility to determine the optimal values of the parameters N and M, at which some fixed in advance cost criterion takes the optimal values.

3   The Process of System States and Its Analysis

Let us introduce the regular irreducible continuous-time Markov chain

$$\xi_t = \{i_t, n_t, \nu_t\}, \ t \geq 0,$$

where, at time t, $t \geq 0$,

- i_t is the number of requests in the orbit, $i_t \geq 0$;
- n_t is the number of requests on service, $n_t = \overline{0, N}$;
- ν_t is the state of the underlying process of the MAP, $\nu_t = \overline{1, W}$.

One can see that this process completely describes the behavior of the system under consideration.

Let us enumerate the states of the Markov chain ξ_t in the lexicographic order and call the set of $(N+1)W$ states having the same value i of the first component as the level i. Let $Q_{i,j}$, $|i - j| \leq 1$, be the matrices containing the intensities of transition of the Markov chain ξ_t from the states that belong to the level i to the states that belong to the level j. The diagonal entries of the matrix $Q_{i,i}$ are negative and define, up to the sign, the rate of the exit of the Markov chain ξ_t from the correspondent state.

Theorem 1. *The infinitesimal generator Q of Markov chain ξ_t, $t \geq 0$, has a block-tridiagonal structure*

$$
Q = \begin{pmatrix}
Q_{0,0} & Q_{0,1} & O & O & O & \ldots \\
Q_{1,0} & Q_{1,1} & Q_{1,2} & O & O & \ldots \\
O & Q_{2,1} & Q_{2,2} & Q_{2,3} & O & \ldots \\
\vdots & \vdots & \vdots & \vdots & \vdots & \ddots
\end{pmatrix}
$$

where the non-zero blocks $Q_{i,j}$, $|i - j| \leq 1$, of size $(N + 1)W$ are defined as follows:

$$
Q_{i,i} = \begin{pmatrix}
Q_{i,i}^{(0,0)} & Q_{i,i}^{(0,1)} & O & \ldots & O & O \\
Q_{i,i}^{(1,0)} & Q_{i,i}^{(1,1)} & Q_{i,i}^{(1,2)} & \ldots & O & O \\
\vdots & \vdots & \vdots & \ddots & \vdots & \vdots \\
O & O & O & \ldots & Q_{i,i}^{(N,N-1)} & Q_{i,i}^{(N,N)}
\end{pmatrix},
$$

$$
Q_{i,i}^{(0,0)} = D_0 - i(\alpha + \beta)I_W,
$$

$$
Q_{i,i}^{(n,n)} = D_0 - i(\alpha + \beta)I_W - n\mu I_W, \ n \leq \frac{M_n}{X},
$$

$$
Q_{i,i}^{(n,n)} = D_0 - i(\alpha + \beta)I_W - n\mu_n I_W, \ \frac{M_n}{X} < n < N,
$$

$$
Q_{i,i}^{(N,N)} = D_0 - i\beta I_W - N\mu_N I_W,
$$

$$
Q_{i,i}^{(n,n+1)} = D_1, \ 0 \leq n < N,
$$

$$
Q_{i,i}^{(n,n-1)} = n\mu I_W, \ n \leq \frac{M_n}{X},
$$

$$
Q_{i,i}^{(n,n-1)} = n\mu_n I_W, \ \frac{M_n}{X} < n \leq N,
$$

$$
Q_{i,i-1} = \begin{pmatrix}
Q_{i,i-1}^{(0,0)} & Q_{i,i-1}^{(0,1)} & O & \ldots & O & O \\
O & Q_{i,i-1}^{(1,1)} & Q_{i,i-1}^{(1,2)} & \ldots & O & O \\
\vdots & \vdots & \vdots & \ddots & \vdots & \vdots \\
O & O & O & \ldots & Q_{i,i-1}^{(N-1,N-1)} & Q_{i,i-1}^{(N-1,N)} \\
O & O & O & \ldots & O & Q_{i,i-1}^{(N,N)}
\end{pmatrix},
$$

$$
Q_{i,i-1}^{(n,n)} = i\beta I_W, \ 0 \leq n \leq N,
$$

$$
Q_{i,i-1}^{(n,n+1)} = i\alpha I_W, \ 0 \leq n \leq N - 1,
$$

$$
Q_{i,i+1} = \begin{pmatrix}
O & O & \ldots & O & O \\
\vdots & \vdots & \ddots & \vdots & \vdots \\
O & O & \ldots & O & O \\
O & O & \ldots & O & D_1
\end{pmatrix}.
$$

The proof of the theorem is carried out by means of careful analysis of all possible transitions of the Markov chain ξ_t and further grouping the intensities of these transitions into the blocks of the generator.

Note that since the requests in orbit are assumed to be impatient, it is easy to verify that the Markov chain ξ_t is ergodic for any values of the system parameters.

Let us denote by $\pi(i, n, \nu)$ the stationary probabilities of the states of the chain ξ_t:

$$\pi(i, n, \nu) = P\{i_t = i, n_t = n, \nu_t = \nu\}, \ i \geq 0, \ n = \overline{0, N}, \ \nu_t = \overline{1, W}.$$

Then, we form from these probabilities the row vectors

$$\boldsymbol{\pi}(i, n) = (\pi(i, n, 1), \dots, \pi(i, n, W)), \ i \geq 0, \ n = \overline{0, N},$$
$$\boldsymbol{\pi}_i = (\boldsymbol{\pi}(i, 0), \dots, \boldsymbol{\pi}(i, N)), \ i \geq 0,$$
$$\boldsymbol{\pi} = (\boldsymbol{\pi}_0, \boldsymbol{\pi}_1, \boldsymbol{\pi}_2, \dots)$$

It is well-known that the vector $\boldsymbol{\pi}$ satisfies the system of equations

$$\boldsymbol{\pi} Q = \mathbf{0}, \ \boldsymbol{\pi} \mathbf{e} = 1$$

where $\mathbf{0}$ is a row vector on 0's, $\mathbf{e}$ is column vector on 1's.

The problem of solving this infinite system of equations, to compute the vectors of stationary probabilities $\boldsymbol{\pi}_i$, $i \geq 0$, is not easy and often similar systems are solved in the literature via rough truncation of the system. Instead of this inaccurate way, the efficient and numerically stable algorithms developed in [20] and [21] are recommended.

4   Performance Measures

The average number of requests for service processing in the system at an arbitrary moment is computed as

$$N_{serv} = \sum_{i=0}^{\infty} \sum_{n=1}^{N} n \boldsymbol{\pi}(i, n) \mathbf{e}.$$

The average number of requests staying in the orbit at an arbitrary moment is computed as

$$N_{orbit} = \sum_{i=1}^{\infty} i \boldsymbol{\pi}_i \mathbf{e}.$$

The average number of requests in the system is computed as

$$L = \sum_{i=0}^{\infty} \sum_{n=0}^{N} (i + n) \boldsymbol{\pi}(i, n) \mathbf{e} = N_{serv} + N_{orbit}.$$

The probability that the system is empty at an arbitrary moment is calculated as

$$P_{empty} = \boldsymbol{\pi}(0,0)\mathbf{e}.$$

The probability that the server is idle at an arbitrary moment is computed as

$$P_{idle} = \sum_{i=0}^{\infty} \boldsymbol{\pi}(i,0)\mathbf{e}.$$

The probability that the orbit is empty at an arbitrary moment is calculated as

$$P_{idle} = \sum_{n=0}^{N} \boldsymbol{\pi}(0,n)\mathbf{e} = \boldsymbol{\pi}_0 \mathbf{e}.$$

The probability that an arbitrary request will enter the service immediately upon arrival is calculated as

$$P_{imm} = \sum_{i=0}^{\infty} \sum_{n=0}^{N-1} \boldsymbol{\pi}(i,n)\frac{D_1}{\lambda}\mathbf{e}.$$

The intensity of the flow of requests, which received service, is computed as

$$\lambda_{out} = \sum_{i=0}^{\infty} \sum_{n=1}^{N} \left(\delta_{n \le \frac{M_n}{\chi}} n\mu \boldsymbol{\pi}(i,n)\mathbf{e} + \delta_{\frac{M_n}{\chi} < n \le N} n\mu_n \boldsymbol{\pi}(i,n)\mathbf{e} \right),$$

where

$$\delta_a = \begin{cases} 1, & \text{if } a \text{ is true,} \\ 0, & \text{otherwise.} \end{cases}$$

The intensity of the flow of the lost requests is computed as

$$\lambda_{loss} = \lambda - \lambda_{out}.$$

The probability that an arbitrary request will be lost is calculated as

$$P_{loss} = \frac{1}{\lambda} \sum_{i=1}^{\infty} i\beta\pi_i = \frac{\lambda_{loss}}{\lambda} = 1 - \frac{\lambda_{out}}{\lambda}.$$

The probability that, at an arbitrary epoch, requests receive a reduced service rate is computed as

$$P_{shar} = \sum_{i=0}^{\infty} \sum_{n=1}^{N} \delta_{\frac{M_n}{\chi} < n \le N} \boldsymbol{\pi}(i,n)\mathbf{e}.$$

5   Numerical Example

In this numerical example, we assume that requests arrive at the system in the MAP that is defined by the matrices

$$D_0 = \begin{pmatrix} -22.3258798 & 0.1488581 \\ 0.3647022 & -5.5293771 \end{pmatrix}, \quad D_1 = \begin{pmatrix} 21.3230975 & 0.8539242 \\ 0.0334886 & 5.1311862 \end{pmatrix}.$$

The mean arrival intensity is $\lambda = 0.9$. The coefficient of variation of inter-arrival times is 1.894, and the coefficient of correlation of successive inter-arrival times is 0.209.

Let us assume that a request requires an average service speed of $X = 15$ megabits per second and the average volume of one request is $S = 30$ megabits. Therefore, the required service rate is equal to 0.5. Also, we assume that servicing requests can interfere with each other, and if there are n, $n = \overline{2, N}$, requests in service, then the effective throughput of the server is $M_n = M - cn$, where $c = 2$ megabits per second. The intensity of requests retrial is set $\alpha = 0.5$, and the intensity of requests' impatience from orbit is $\beta = 0.1$.

We assume that the quality of the system operation is defined by the following cost criterion:

$$E(M, N) = a\lambda_{out} - b\lambda_{loss} - dM$$

where λ_{out} and λ_{loss} are defined above as the rates of the flows of the served and lost requests, a is the profit gained by the system for service of one request, b is the charge paid by the system for one request loss, and d is the charge for using one megabit bandwidth per second.

Let us assume that the cost coefficients are defined as follows: $a = 1$, $b = 5$, $d = 0.015$.

The aim of the experiment is to define the values of the parameters M and N for which the value of the cost criterion is maximal.

To this end, let us vary the bandwidth of the server M in the interval [100,1000] with step 100 megabits per second, and the maximal admissible number N of requests in service over the interval $[1, \min\{100, \lceil M/a \rceil - 1\}]$ with the step 1, and investigate the dependence of the main performance measures of the system on the parameters M and N. Note that the limitation $N < \lceil M/a \rceil$ is required to guarantee that the effective throughput of the server is greater than zero.

The computations were implemented on the notebook Lenovo Gaming 3 with Intel(r) Core(TM) i7-10750H CPU 2.60 GHz, 16 GB RAM, Wolfram Mathematica 12.2. The computation time is a bit less than 80 min for all 948 different pairs of parameters (M, N) or, on average, 12 pairs per minute.

Figure 2 illustrates the dependence of the average number N_{orbit} of requests in the orbit on the parameters M and N.

It is evidently seen that the number N_{orbit} quickly decreases with growth of N and bandwidth M.

Figure 3 illustrates the dependence of the average number N_{serv} of requests in the service on the parameters M and N.

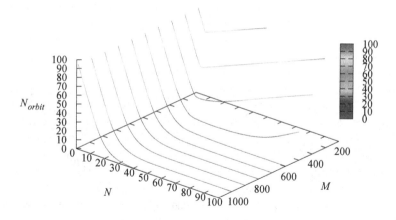

Fig. 2. Dependence of the average number N_{orbit} of requests in the orbit on the parameters M and N

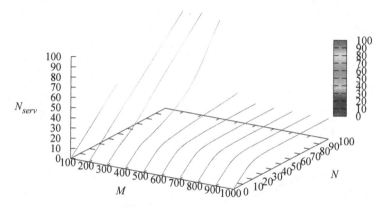

Fig. 3. Dependence of the average number N_{serv} of requests in the service on the parameters M and N

For small values of bandwidth M, the increase of the threshold N leads to a quick increase in the number N_{serv} of requests in the service up to the value N. For large values of bandwidth M, the increase of the threshold N initially leads to the increase of N_{serv}. But for enough large N the increase of N_{serv} becomes very small because a large value of M allows requests quickly receive system and depart.

Figures 4 and 5 illustrate the dependence of the average intensities of the flow of served requests λ_{out} and the average intensities of the flow of lost requests λ_{loss} on the parameters M and N.

For a large values of bandwidth M, the increase of the threshold N leads to the increase of λ_{out}. For small values of bandwidth M, the increase of the threshold N initially leads to the increase of λ_{out}. But then the increase of N leads to higher interference of requests and a decrease in effective bandwidth.

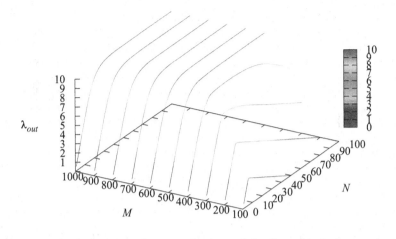

Fig. 4. Dependence of the average intensity λ_{out} of the flow of served requests on the parameters M and N

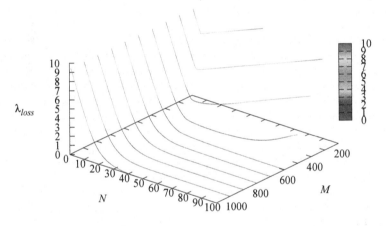

Fig. 5. Dependence of the average intensity λ_{loss} of the flow of lost requests on the parameters M and N

In turn, this implies a decrease of λ_{out}. Because the sum of λ_{out} and λ_{loss} is the constant equal to λ, behavior of λ_{out} illustrated in Fig. 4 easy explains the curves for λ_{loss} presented on Fig. 5.

Figure 6 illustrates the dependence of the probability P_{shar} that, at an arbitrary moment, requests receive a reduced service rate on the parameters M and N.

For large values of bandwidth M, server quickly provides service and requests succeed to receive service without exploiting the reduced service rate. For smaller values of bandwidth M, value P_{shar} quickly increases with the increase of N which implies higher competition for requests and the use uced service rates.

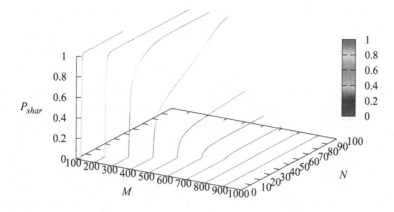

Fig. 6. Dependence of the probability P_{shar} on the parameters M and N

Figure 7 illustrates the dependence of the loss probability P_{loss} on the parameters M and N.

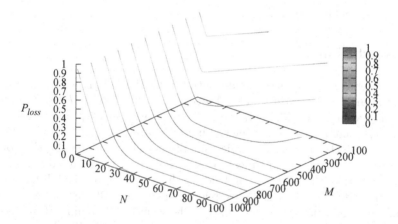

Fig. 7. Dependence of the loss probability P_{loss} on the parameters M and N

This figure matches Fig. 5 for λ_{loss} because the probability P_{loss} is equal to λ_{loss} divided by the constant λ.

The dependence of the values of the cost criterion E on the parameters M and N is illustrated in Fig. 8.

It is clear from this figure that for the majority of values of N and M the values of criterion $E(M, N)$ defining profit of the system are negative. However, it appears that it is possible to find values of N and M such that the profit is positive.

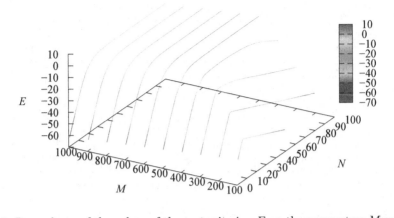

Fig. 8. Dependence of the values of the cost criterion E on the parameters M and N

The optimal value of the cost criterion is reached for $M = 500$ megabits per second and value $N = 82$ of the number of requests that can be allowed to receive service simultaneously (multiprogramming level). The optimal value of the criterion is equal to 2.22191.

6   Conclusion

In this paper, a retrial queuing model with flexible limited processor sharing discipline is considered. Customer arrivals occur according to the MAP what allows taking into account possible burstiness of arrival processes typical for many real modern telecommunication systems. Limited sharing means that the number of requests which can receive service simultaneously is limited by the control parameter N. Flexible sharing means that if the sum of the required bandwidths becomes less than the effective bandwidth, which is less than the full bandwidth of the server due to the interference phenomenon, then all requests receive service at the proportionally reduced rate. The requests that did not succeed in entering the service immediately upon arrival due to limitation defined by the parameter N retry for entering service after exponentially distributed intervals of time while staying on so-called orbit. Customers staying in orbit are impatient and can depart from the system without service. The effects of requests retrials and impatience imply state inhomogeneous behavior of the process defining the dynamics of the system. This essentially complicated its analysis. We derived the generator of this process, which is the multidimensional Markov chain. This allows to compute its stationary distribution and obtain dependence of performance measures of the system on the value of bandwidth of the server and the threshold N. Numerical illustrations are presented. The obtained results can be expanded to the case when requests are impatient also during service and service at a reduced rate in particular.

Acknowledgments. This work was supported by the Basic Science Research Program through the National Research Foundation of Korea (NRF) funded by the Ministry of Education (NRF-2022R1I1A3067543) and by the RUDN University Strategic Academic Leadership Program.

References

1. Dudin, A.N., Klimenok, V.I., Vishnevsky, V.M.: The Theory of Queuing Systems with Correlated Flows. Springer, Cham (2020). https://doi.org/10.1007/978-3-030-32072-0. 410 p
2. Yashkov, S.F.: Processor-sharing queues: some progress in analysis. Queueing Syst. **2**(1), 1–17 (1987)
3. Yashkov, S.F., Yashkova, A.S.: Processor sharing: A survey of the mathematical theory. Autom. Remote Control **68**(9), 1662–1731 (2007)
4. Altman, E., Avrachenkov, K., Ayesta, U.: A survey on discriminatory processor sharing. Queueing Syst. **53**, 53–63 (2006)
5. Kim, C., Dudin, S.A., Dudina, O.S., Dudin, A.N.: Mathematical models for the operation of a cell with bandwidth sharing and moving users. IEEE Trans. Wirel. Commun. **19**(2), 744–755 (2019)
6. Samouylov, K.E., Sopin, E.S., Gudkova, I.A.: Sojourn time analysis for processor sharing loss queuing system with service interruptions and MAP arrivals. Commun. Comput. Inf. Sci. **678**, 406–417 (2016)
7. Dudin, S., Dudin, A., Dudina, O., Samouylov, K.: Analysis of a retrial queue with limited processor sharing operating in the random environment. In: Koucheryavy, Y., Mamatas, L., Matta, I., Ometov, A., Papadimitriou, P. (eds.) WWIC 2017. LNCS, vol. 10372, pp. 38–49. Springer, Cham (2017). https://doi.org/10.1007/978-3-319-61382-6_4
8. Telek, M., van Houdt, B.: Response time distribution of a class of limited processor sharing queues. In: Proceedings of IFIP WG. 7.3 Performance conference, New York City (2017)
9. Dudin, A.N., Dudin, S.A., Dudina, O.S., Samouylov, K.E.: Analysis of queueing model with processor sharing discipline and customers impatience. Oper. Res. Perspect. **5**, 245–255 (2018)
10. D'Apice, C., Dudin, A., Dudin, S., Manzo, R.: Priority queueing system with many types of requests and restricted processor sharing. J. Amb. Intell. Humaniz. Comput., 1–12 (2022). https://doi.org/10.1007/s12652-022-04233-w
11. Chakravarthy, S.R.: The batch Markovian arrival process: a review and future work. In: Advances in Probability Theory and Stochastic Processes, pp. 21–29. Notable Publications Inc., New Jersey (2001)
12. Lucantoni, D.: New results on the single server queue with a batch Markovian arrival process. Commun. Stat. Stoch. Model. **7**, 1–46 (1991)
13. Kim, J., Kim, B.: A survey of retrial queueing systems. Ann. Oper. Res. **247**, 3–36 (2016)
14. Falin, G.: A survey of retrial queues. Queueing Syst. **7**(2), 127–167 (1990)
15. Falin, G., Templeton, J.G.: Retrial Queues, vol. 75. CRC Press, Boca Raton (1997)
16. Nair, J., Wierman, A., Zwart, B.: Tail-robust scheduling via limited processor sharing. Perform. Eval. **67**(11), 978–995 (2010)
17. Klimenok, V.I., Orlovsky, D.S., Dudin, A.N.: A $BMAP/PH/N$ system with impatient repeated calls. Asia Pac. J. Oper. Res. **24**, 293–312 (2007)

18. Dudin, S.A., Dudina, O.S.: Call center operation model as a $MAP/PH/N/R-N$ system with impatient customers. Probl. Inf. Transm. **47**(4), 364–377 (2011)
19. Dudin, S., Kim, C., Dudina, O.: $MMAP/M/N$ queueing system with impatient heterogeneous customers as a model of a contact center. Comput. Oper. Res. **40**(7), 1790–1803 (2013)
20. Dudin, S., Dudina, O.: Retrial multi-server queuing system with PHF service time distribution as a model of a channel with unreliable transmission of information. Appl. Math. Model. **65**, 676–695 (2019)
21. Dudin, S., Dudin, A., Kostyukova, O., Dudina, O.: Effective algorithm for computation of the stationary distribution of multi-dimensional level-dependent Markov chains with upper block-Hessenberg structure of the generator. J. Comput. Appl. Math. **366**, 112425 (2020)

On the Distribution of the Number of Consecutively Lost Customers in the $BMAP/PH/1/N$ System

Valentina Klimenok$^{(\boxtimes)}$ and Alexander Dudin

Department of Applied Mathematics and Computer Science,
Belarusian State University, 220030 Minsk, Belarus
{klimenok,dudin}@bsu.by

Abstract. In this paper, we propose method for calculating the distribution of the number of consecutively lost customers in the single-server queueing system with a finite buffer, batch Markovian arrival process and phase type distribution of service time. The most well-known and important performance measure of finite capacity systems is the probability of losing an arbitrary customer. Loss probability is the subject of research in the literature under various assumptions about the nature of the input flow and the distribution of service time. At the same time, this characteristic may be not always a good estimate of the quality of service in queuing systems that arise in the mathematical modeling of telecommunication networks. More indicative in this case is the probability of losing several customers in a row caused by an overflowing buffer. We propose explicit formulas that characterize the distribution and mathematical expectation of the number of consecutively lost customers in the system under consideration.

Keywords: Single-server queue · Finite buffer · Batch Markovian arrival process · Phase type distribution of service time · Consecutively lost customers

1 Introduction

Queueing systems with losses form a practically important and mathematically interesting class of systems in queueing theory. An overview of early work on such systems can be found in [1]. A significant number of results in this area were obtained at the Peoples' Friendship University of Russia, see, for example, the book by P.P. Bocharov et al. [2] and numerous papers. Most of early works on queueing systems with losses was devoted to systems with stationary Poisson and renewal flows. However, it is already well known that the flows in the modern telecommunication networks are bursty and correlated. Therefore, another model of arrival process should be considered to get an adequate mathematical model of real life processes. As a result of research efforts in searching for convenient descriptors for real life flows, a mathematical model of $BMAP$ (Batch

© The Author(s), under exclusive license to Springer Nature Switzerland AG 2022
V. M. Vishnevskiy et al. (Eds.): DCCN 2022, LNCS 13766, pp. 257–269, 2022.
https://doi.org/10.1007/978-3-031-23207-7_20

Markovian Arrival Process) has been developed and widely used. The $BMAP$ well describes the correlated burst traffic and, at the same time, allows a transparent mathematical interpretation. This arrival process is the generalization of versatile point process introduced by Marcel Neuts in [3]. Approximately at the same time the similar flows called as Markov Chain arrival flows were introduced in the scientific school by G. P. Basharin in Peoples' Friendship University of Russia. Later the name of versatile flow was changed to N-flow (Neuts flow). N-flows were defined by a large number of parameters. At the same time, they did not have a convenient mathematical description, which significantly hampered their perception and work with them. In 1991, the article by D. Lucantoni [4] appeared, in which a convenient form of parameter storage was proposed, which makes it possible to mathematically transparently treat the batch analogue of the N-flow. This analog was called as a $BMAP$.

Queueing systems with the $BMAP$ and a finite buffer have been considered in a number of papers under different assumptions about the service discipline and distribution of service times, see, for example, the papers [5–22] and references therein.

The most well-known and important performance indicator of systems with a finite buffer is the probability of a loss of an arbitrary customer that enters the system and meets the buffer fully occupied. Most researchers, when estimating losses in such systems, confine themselves to calculating just this probability. However, as it is emphasized in [23], the loss probability alone cannot fully characterize the quality of service in telecommunication networks. A more efficient loss estimate is the probability of losing several customers in a row caused by an overflowed buffer. But, as it is shown in [24], the probability of losing several customers in a row cannot be inferred from the loss probability. Therefore, the topic of calculating this probability is actual in the literature. A number of works can be indicated as relevant ones, see, for example, [24–27] where the problem of finding the burst ratio caused by an overflowed buffer is under consideration. This problem has been studied in [24–26] for the stationary Poisson arrival process and general distribution of service time. In [27] the results was extended to the case of $MMPP$ (Markov modulated Poisson process). In just cited works, the authors, among other things, find the conditional probability of losing several customers in a row using the transition probabilities of the embedded Markov chain and some recursive procedures.

A number of papers are also known, where the problems of calculating the probability of losing more than k customers in a row during the busy period of the system are considered. Such problems for the systems $M/G/1/N$, $GI/M/1/N$ and their group analogs were considered in the works [28–31] and were studied by compiling and solving a system of linear algebraic equations for the probabilities that during the busy period generated by i customers in the system, there will be no loss of more than k customers in a row. Using the idea developed in the cited papers, the authors of [22] construct a similar system of linear algebraic equations for the $BMAP/G(a;b)/1/N$ and $BMAP/MSP(a;b)/1/N$ queueing systems with group service. In this case, the system of linear algebraic equations has a complex form, since the coefficients are represented by the matrices the calculation of which is an independent task.

The main contribution of our work is that we derive explicit formulas for the stationary probabilities and the average number of consecutive lost customers in the $BMAP/PH/1/N$ system. Note that the mentioned average number, known in the telecommunications literature as the average packet gap is widely used in the study of telecommunications networks, see, for example, [32,33].

2   The System Description

Customers arrive to the system in a $BMAP$ under the control of an irreducible Markov chain with continuous time $\nu_t, t \geq 0$, which admits values in a finite state space $\{0, 1, \dots, W\}$ and is called as underlying process of the $BMAP$. Time spent by the process ν_t in the state ν has an exponential distribution with the parameter $\lambda_\nu, \nu = \overline{0, W}$. After the sojourn time of the process ν_t in this state expired, with probability $p_k(\nu, \nu')$ the process goes into the state ν' and a batch consisting of $k, k \geq 0$, customers is generated. It is assumed that transition from the ν state to the same state without generating customers impossible, i.e. $p_0(\nu, \nu) = 0$, and the probabilities $p_k(\nu, \nu')$ satisfy the normalization condition $\sum_{k=0}^{\infty} \sum_{\nu'=0}^{W} p_k(\nu, \nu') = 1$, $\nu = \overline{0, W}$. Thus, a $BMAP$ is completely characterized by the dimension $W + 1$ of the state space of the underlying process ν_t, the intensities $\lambda_\nu, \nu = \overline{0, W}$, and a set of probabilities $p_k(\nu, \nu')$, $\nu, \nu' = \overline{0, W}, k \geq 0$.

D. Lucantoni in [4] proposed to store this information in a form of a set of square matrices $D_k, k \geq 0$, of size $(W + 1)$ or their matrix generating function

$$D(z) = \sum_{k=0}^{\infty} D_k z^k, |z| \leq 1.$$ The (ν, ν')th entry of the matrix D_k has the form:

$$(D_k)_{\nu,\nu'} = \lambda_\nu p_k(\nu, \nu'), \nu, \nu' = \overline{0, W}, k \geq 1, (D_0)_{\nu,\nu'} = \begin{cases} \lambda_\nu p_0(\nu, \nu'), \nu \neq \nu', \nu, \nu' = \overline{0, W}, \\ -\lambda_\nu, \nu = \nu', \nu = \overline{0, W}. \end{cases}$$

The entries of the matrices $D_k, k \geq 1$, are the transition rates of the process $\nu_t, t \geq 0$, accompanied by the generation of a batch consisting of k customers. The off-diagonal entries of the matrix D_0 have a similar meaning and the diagonal entries of this matrix are the transition rates of the process ν_t from its states taken with the opposite sign.

The matrix $D(1)$ is an irreducible infinitesimal generator of the Markov chain $\nu_t, t \geq 0$. The stationary distribution of this chain is defined by the row vector $\boldsymbol{\theta}$ which is the unique solution of the system of equations $\boldsymbol{\theta} D(1) = \mathbf{0}, \boldsymbol{\theta} \mathbf{e} = 1$. Here, $\mathbf{e}$ is a column vector consisting of 1's and $\mathbf{0}$ is a row vector consisting of 0's.

The fundamental rate λ of arrivals in the $BMAP$ is defined as $\lambda = \boldsymbol{\theta} D'(z)|_{z=1} \mathbf{e}$. The rate $\lambda^{(b)}$ of groups arrival is defined as $\lambda^{(b)} = \boldsymbol{\theta}(-D_0)\mathbf{e}$. Coefficient of variation, c_{var}, of the inter-arrival interval is determined by the formula $c_{var}^2 = 2\lambda_b \boldsymbol{\theta}(-D_0)^{-1}\mathbf{e} - 1$. Coefficient of correlation, c_{cor}, of the lengths of two adjacent intervals is calculated as follows: $c_{cor} = (\lambda_b \boldsymbol{\theta}(-D_0)^{-1}(D(1) - D_0)(-D_0)^{-1}\mathbf{e} - 1)/c_{var}^2$. The $BMAP$ includes the following special cases: stationary Poisson flow ($W = 0, D_0 = -\lambda, D_1 = \lambda, D_k = 0, k > 1$), flows with the PH distribution of inter-arrival intervals (for example, Erlang distribution, hyperexponential distribution, etc.), an ordinary Markov flow (Markovian Arrival Process – MAP), $MMPP$ (Markov modulated Poisson process) as well as their superpositions.

The most important characteristic of a $BMAP$, which we will use in what follows, is the counting function, that is characterized by a square matrices $P(k,t)$, $k \geq 0$, of order $W + 1$. The (ν, ν')th entries of this matrix is the probability that k customers arrive during the time interval of the length t and the underlying process transits to the state ν' provided that at the beginning of this interval it has been staying in the state ν. The generating function of these matrices has the form $P(z) = \sum_{n=0}^{\infty} P(n,t)z^n = e^{D(z)t}$. A more detailed description of the $BMAP$ can be found in [4,34,36].

The service time of a customer has a PH distribution given by an irreducible representation (β, S) and the underlying process (Markov chain) $m_t, t \geq 0$, with phase space $\{1, \ldots, M, M + 1\}$, where $\{1, \ldots, M\}$ are the transient phases and $M + 1$ is an absorbing phase. At the service completion epoch, the phase for the next service is selected according to the stochastic row vector β. After start of the service, the random walk begins in the space of transient phases in accordance with the $M \times M$ transition rates matrix S. The delay in the phase m has an exponential distribution with the parameter $(-S)_{m,m}$. The probability of the underlying process transition from the phase m to the phase m' in time t is defined as $(e^{St})_{m,m'}$. After entering the absorbing phase, the service of the customer ends. The rates of transition to the absorbing phase are determined by the entries of the column vector $S_0 = -Se$. A more detailed description of the PH distribution can be found in the monographs [35] and [36].

Let at time t, $t \geq 0$,

- n_t be the number of customers in the system;
- ν_t be the state of underlying process of the $BMAP$;
- m_t be the phase of underlying process of service time.

Then the operation of the system is described by the Markov chain $\xi_t = \{n_t, \nu_t, m_t\}, t \geq 0$, whose state space depends on the discipline of customers acceptance to the system. This discipline determines the rule by which customers are admitted to the system if they arrive in a batch whose size exceeds the number of free at the arrival moment places in the system.

In such situations, three disciplines of admission are usually considered:

- partial admission (only a part of the group corresponding to a number of free places in the system is admitted while the rest of the group is lost),
- complete rejection (a whole group leaves the system without service if the number of free places is less than the number of customers in the group),
- complete admission (all group is accepted. A part of the group corresponding to a number of free places joins the system while the rest of the group waits in some extended memory for entering the system until the corresponding number of earlier accepted customers depart after receiving service in the system).

The discipline, which is most popular in literature (due to relative easiness of its analysis), is partial admission one. The next section is devoted to finding the desired characteristics under this admission discipline.

3 Stationary Distribution and the Average Number of Consecutive Lost Customers in the System $BMAP/PH/1/N$. The Case of Partial Admission Discipline

In the case of partial admission discipline, the Markov chain $\xi_t, t \geq 0$, has the state space $\{(n, \nu, m), n = \overline{0, N}, \nu = \overline{0, W}, m = \overline{1, M}\}$. Denote by $\boldsymbol{\pi}_n$ the vector of stationary probabilities of the states of this chain corresponding to the presence of n customers in the system, $n = \overline{0, N}$. We assume that the components of this vector are ordered in lexicographic order.

The algorithm for calculating the vectors $\boldsymbol{\pi}_n, n = \overline{0, N}$, is known in the literature, see, for example, [2,36] and references therein. For the convenience of the reader, we present this algorithm below. Denote by Q the infinitesimal generator of the chain under consideration. This generator has the block form $Q = (Q_{n,n'})_{n,n'=\overline{0,N}}$ where blocks $Q_{n,n'}$ of order $(W+1)M$ are composed of the transition rates from the states corresponding to the state n of the first component to the state n' of this component.

Algorithm 1. Stationary probability vectors $\boldsymbol{\pi}_n$, $n = \overline{0, N}$ are calculated as follows:

$$\boldsymbol{\pi}_n = \boldsymbol{\pi}_0 F_n,\ n = \overline{1, N},$$

where the matrices F_n are calculated recursively:

$$F_n = (\bar{Q}_{0,n} + \sum_{i=1}^{n-1} F_i \bar{Q}_{i,n})(-\bar{Q}_{n,n})^{-1},\ n = \overline{1, N-1},$$

$$F_N = (\bar{Q}_{0,N} + \sum_{i=1}^{N-1} F_i \bar{Q}_{i,N})(-\bar{Q}_{N,N})^{-1},$$

the matrices $\bar{Q}_{i,N}$ are computed using the reverse recursion

$$\bar{Q}_{i,N} = Q_{i,N},\ i = \overline{0, N},$$

$$\bar{Q}_{i,n} = Q_{i,n} + \bar{Q}_{i,n+1} G_n,\ i = \overline{0, n}, n = N-1, N-2, \ldots, 0,$$

the matrices $G_n, n = \overline{0, N-1}$, are also calculated using the reverse recursion

$$G_n = (-Q_{n+1,n+1} - \sum_{l=1}^{N-n-1} Q_{n+1,n+1+l} G_{n+l} G_{n+l-1} \ldots G_{n+1})^{-1} Q_{n+1,n},$$

$$n = N-1, N-2, \ldots, 0,$$

the vector π_0 is calculated as the unique solution of the system

$$\pi_0 \bar{Q}_{0,0} = \mathbf{0}, \quad \pi_0(\mathbf{e} + \sum_{n=1}^{N} F_n \mathbf{e}) = 1.$$

Let us denote by q_k the joint probability of occurrence of two events: at an arbitrary time of the system operation, an overflow period begins (i.e., the buffer becomes full) and $k, k \geq 0$, customers in a row will be lost during this period. The following assertion is true.

Theorem 1. The probabilities $q_k, k \geq 0$, for the $BMAP/PH/1$ system with partial admission discipline are calculated by the following formula:

$$q_k = \lambda^{-1} \sum_{i=0}^{N-1} \pi_i \sum_{l=N-i}^{N-i+k} (D_l \otimes I_M) \int_0^\infty P(N-i+k-l,t) \otimes e^{St} dt (\mathbf{e}_{W+1} \otimes \boldsymbol{S}_0)$$

$$(1)$$

where $\otimes$ is the symbol of Kronecker product of matrices, see, e.g. [37].

Proof. Analyzing the situation associated with the occurrence of the above two events, we can see that such a situation is possible if and only if a) there are i customers in the system, $i = \overline{0, N-1}$; b) a batch of customers of size l arrives, $l = \overline{N-i, N-i+k}$, $N-i$ customers of the batch occupy free places in the system, the rest are lost; c) during the residual service time (during the service time, if $N=1$) $N-i+k-l$ customers arrive in the $BMAP$, which will also be lost, since all places in the system are already occupied. Formula (1) follows from what has been said.

Note that the entry of the vector $\lambda^{-1} \sum_{i=0}^{N-1} \pi_i \sum_{l=N-i}^{N-i+k} (D_l \otimes I_M)$ corresponding to the state ν of the underlying process of the $BMAP$ and the phase m of the PH service process defines the joint probability that at an arbitrary time an overflow period begins and at the beginning of this period the underlying process of the $BMAP$ is in the state ν and the service process PH is in the phase m.

Corollary 1. The generating function of probabilities $q_k, k \geq 0$, for the $BMAP/PH/1$ system with partial admission discipline has the following form:

$$Q(z) = \lambda^{-1} \sum_{i=0}^{N-1} \pi_i z^{i-N} \sum_{l=N-i}^{\infty} (D_l \otimes I_M)[-(D(z) \oplus S)]^{-1}(\mathbf{e}_{W+1} \otimes \boldsymbol{S}_0), \quad (2)$$

where $\oplus$ is the symbol of Kronecker sum of matrices, see, e.g. [37].

The main problem in calculating the probabilities q_k is the calculation of the integrals $Y_n = \int_0^\infty P(n,t) \otimes e^{St} dt, n \geq 0$. We can suggest the following methods for calculating these integrals.

1) Method of uniformization of the matrix exponent, see [36], pp. 66–67. Using this method, we obtain the following expansion for the matrix $P(n,t)$:

$$P(n,t) = \sum_{j=0}^{\infty} e^{-\gamma t} \frac{(\gamma t)^j}{j!} K_n^{(j)}, \quad n \geq 1, \tag{3}$$

where $\gamma = \max_{i=\overline{0,W}} (-D_0)_{ii}$ and the matrices $K_n^{(j)}$ satisfy the following system of recurrence relations:

$$K_0^{(0)} = I, \quad K_n^{(0)} = O, \quad n \geq 1, \quad K_0^{(j+1)} = K_0^{(j)}(I + \gamma^{-1}D_0),$$

$$K_n^{(j+1)} = \gamma^{-1} \sum_{i=0}^{n-1} K_i^{(j)} D_{n-i} + K_n^{(j)}(I + \gamma^{-1}D_0), \quad n \geq 1, \; j \geq 0.$$

Using expansion (3), the integrals Y_n can be calculated explicitly.

2) In this method the recursive calculation of Y_n similar to that proposed in [36], pp. 80–81 is used. The recursive procedure is as follows:

$$Y_n = Z_n(I_{\bar{W}} \otimes S_0), \quad n \geq 0,$$

where

$$Z_0 = -(I_{\bar{W}} \otimes \beta)(D_0 \oplus S)^{-1},$$

$$Z_n = -\sum_{i=0}^{n-1} Z_i(D_{n-i} \otimes I_M)(D_0 \oplus S)^{-1}, \quad n \geq 1.$$

3) Theoretically, to calculate the integrals Y_n, one can use the expansion of the matrix generating function

$$Y(z) = \sum_{n=0}^{\infty} Y_n z^n = \int_0^{\infty} e^{D(z)t} \otimes e^{St} dt = [-(D(z) \oplus S)]^{-1}.$$

Having calculated the probabilities q_k, $k \geq 0$, we can calculate the average number of consecutively lost customers in the overflow period according to the formula $\sum_{k=1}^{\infty} kq_k$. As noted above, this average (average packet gap) is widely used in the study of telecommunication networks. Therefore, it seems very important to obtain an explicit formula for this characteristic. Such a formula is presented in the following statement.

Corollary 2. The average number, $\bar{q}$, of lost customers in a row in the $BMAP/PH/1/N$ system with partial admission discipline is calculated by the following formula:

$$\bar{q} = \lambda^{-1} \left\{ \sum_{i=0}^{N-1} \pi_i \sum_{l=N-i}^{\infty} (l - N + i)(D_l \otimes I_M)\mathbf{e} + \sum_{i=0}^{N-1} \pi_i \sum_{l=N-i}^{\infty} (D_l \otimes I_M)\mathbf{g} \right\} \tag{4}$$

where the column vector $\mathbf{g}$ is computed as

$$\mathbf{g} = -\Big\{ I_{W+1} \otimes S^{-1} + [(\mathbf{e}\boldsymbol{\theta} - D(1))^{-1} \otimes I_M] \times$$

$$[(D(1) \oplus S)^{-1}(I_{\bar{W}} \otimes S) - D(1) \otimes S^{-1} - I]\Big\}(D'(1) \otimes I_M)\mathbf{e}. \tag{5}$$

Proof. The desired average $\bar{q}$ will be calculated as the derivative of the generating function (2) with respect to z at the point $z = 1$. The main difficulty in differentiation is taking the derivative of the term $[-(D(z) \oplus S)]^{-1}(\mathbf{e}_{W+1} \otimes \boldsymbol{S}_0)$ on the right hand side of (2). We will present the corresponding calculations in the form of a series of transformations, where we will indicate the sources along the way, from which the validity of nontrivial transitions follows.

$$\mathbf{g} = \{[-(D(z) \oplus S)]^{-1}(\mathbf{e}_{W+1} \otimes \boldsymbol{S}_0)\}'|_{z=1} = \int_0^\infty (e^{D(z)t})'|_{z=1} \otimes e^{St} dt(\mathbf{e}_{W+1} \otimes \boldsymbol{S}_0) =$$

At this stage of the transformations, we will use the formula for derivative $(e^{D(z)t})'|_{z=1}$, given in [36], p. 68.

$$= \int_0^\infty [tI_{W+1} - (\mathbf{e}\boldsymbol{\theta} - D(1))^{-1}(e^{D(1)t} - tD(1) - I)]D'(1) \otimes e^{St} dt(\mathbf{e}_{W+1} \otimes \boldsymbol{S}_0)$$

$$= \Big\{ I_{W+1} \otimes S^{-2} - [(\mathbf{e}\boldsymbol{\theta} - D(1))^{-1} \otimes I_M] \int_0^\infty (e^{D(1)t} - tD(1) - I) \otimes e^{St} dt] \Big\}$$

$$\times (D'(1) \otimes (-S))\mathbf{e} = -\Big\{ I_{W+1} \otimes S^{-1} + [(\mathbf{e}\boldsymbol{\theta} - D(1))^{-1} \otimes I_M][(D(1) \oplus S)^{-1}(I_{\bar{W}} \otimes S)$$

$$- D(1) \otimes S^{-1} - I]\Big\}(D_1 \otimes I_M)\mathbf{e}.$$

4   Stationary Distribution and the Average Number of Consecutive Lost Customers in the System $BMAP/PH/1/N$. The Case of Complete Rejection and Complete Admission Disciplines

In the case of complete rejection discipline, the state space of the Markov chain ξ_t, describing the operation of the system is the same as in the case of partial admission discipline. Therefore, the stationary distribution of the system states is still given by a finite number of vectors $\boldsymbol{\pi}_n, n = \overline{0, N}$.

   In the case of a complete admission discipline, the state space of the system consists of a countable number of states and has the form $\{(n, \nu, m), n \geq 0, \nu = \overline{0, W}, m = \overline{1, M}\}$. Accordingly, the stationary distribution is given by countably many vectors $\boldsymbol{\pi}_n, n \geq 0$.

In both cases, the algorithms for calculating the vectors π_n involve operation with blocks $Q_{n,n'}$ of the generator $Q = (Q_{n,n'})_{n,n'}$ of the chain under consideration. In the case of complete rejection discipline, this algorithm is the same as Algorithm 1 presented above for the case of partial admission discipline. In the case of a complete admission discipline, a description of the algorithm for calculating the $\pi_n, n \geq 0$, vectors can be found in [38]. Considering some differences compared to the [38], we present our algorithm below.

Algorithm 2. Stationary probability vectors $\pi_i, i \geq 0$ are calculated in the following way:

$$\pi_i = \pi_0 F_i, i \geq 1,$$

where the matrices F_i are calculated recurrently

$$F_i = (\bar{Q}_{0,i} + \sum_{l=1}^{i-1} F_l \bar{Q}_{l,i})(-\bar{Q}_{i,i})^{-1}, i \geq 1,$$

the matrices $\bar{Q}_{l,i}$ are calculated as:

$$\bar{Q}_{l,i} = Q_{l,i} + \sum_{k=1}^{\infty} Q_{l,i+k} G^{max\{0,i+k-N\}} \cdot G_{min\{N,i+k\}-1} \cdot G_{min\{N,i+k\}-2} \cdots G_i,$$

$$l = 0, \ldots, i, i \geq 1,$$

the matrix G has a form

$$G = (-D(1) \oplus S)^{-1}(I_{\bar{W}} \otimes S_0 \beta),$$

the matrices $G_l, l = \overline{0, N-1}$ are calculated from the backward recursion;

$$G_l = -(Q_{l+1,l+1} + \sum_{i=l+2}^{\infty} Q_{l+1,i} G^{max\{0,i-N\}} \cdot G_{min\{N,l\}-1} \times$$

$$G_{min\{N,i\}-2} \cdots G_{l+1})^{-1} Q_{l+1,l}, l = N-1, N-2, .., 0,$$

the vector π_0 is the unique solution of the system:

$$\pi_0 \bar{Q}_{0,0} = 0, \quad \pi_0(e + \sum_{i=1}^{\infty} F_i e) = 1.$$

Having calculated the vectors π_n and using the same reasoning as above for the partial admission discipline, we can find the stationary distribution and the average number of lost customers in a row for the case of complete rejection and complete admission disciplines. The corresponding results are presented in the following assertions.

Theorem 2. The probabilities $q_k, k \geq 0$, for the $BMAP/PH/1$ system with complete rejection discipline and complete admission discipline are calculated by the following formulas:

(i) in the case of complete rejection discipline

$$q_k = \lambda^{-1} \sum_{i=0}^{N-1} \pi_i (D_{N-i} \otimes I_M) \int_0^\infty P(k,t) \otimes e^{St} dt (\mathbf{e}_{W+1} \otimes \mathbf{S}_0), k \geq 0.$$

(ii) in the case of complete admission discipline

$$q_k = \lambda^{-1} \sum_{i=0}^{N-1} \pi_i \sum_{l=N-i}^\infty (D_l \otimes I_M) \int_0^\infty P(k,t) \otimes e^{St} dt (\mathbf{e}_{W+1} \otimes \mathbf{S}_0), k \geq 0.$$

Corollary 3. The generating function of probabilities $q_k, k \geq 0$, for the system $BMAP/PH/1$ with complete rejection discipline and complete admission discipline has the following form:

(i) in the case of complete rejection discipline

$$Q(z) = \lambda^{-1} \sum_{i=0}^{N-1} \pi_i (D_{N-i} \otimes I_M)[-(D(z) \oplus S)]^{-1} (\mathbf{e}_{W+1} \otimes \mathbf{S}_0),$$

(ii) in the case of complete admission discipline

$$Q(z) = \lambda^{-1} \sum_{i=0}^{N-1} \pi_i \sum_{l=N-i}^\infty (D_l \otimes I_M)[-(D(z) \oplus S)]^{-1} (\mathbf{e}_{W+1} \otimes \mathbf{S}_0).$$

Corollary 4. The average value $\bar{q}$ of the number of consecutive lost customers in the system $BMAP/PH/1/N$ with complete rejection discipline and complete admission discipline is calculated by the following formulas:

(i) in the case of complete rejection discipline:

$$\bar{q} = \lambda^{-1} \sum_{i=0}^{N-1} \pi_i (D_{N-i} \otimes I_M) \mathbf{g}.$$

(ii) in the case of complete admission discipline:

$$\bar{q} = \lambda^{-1} \sum_{i=0}^{N-1} \pi_i \sum_{l=N-i}^\infty (D_l \otimes I_M) \mathbf{g},$$

where the vector $\mathbf{g}$ is calculated by formula (5).

5 Conclusion

In this paper, explicit formulas are obtained for the joint probabilities q_k, $k \geq 0$, that at an arbitrary moment of time an overflow period in the $BMAP/PH/1/N$ system begins and k customers will be lost in a row during this period. Three well-known disciplines for accepting customers arriving in a batch flow to a system with a finite number of waiting places are considered: partial admission, complete rejection and complete admission.

For each discipline, formulas are obtained for the generating function of the probabilities q_k, $k \geq 0$, and the average number of customers lost in a row in the overflow period. These formulas can serve for adequate evaluation of the quality of service in telecommunication networks for various purposes. In continuation of the work, it is planned to extend the results to the case of a batch marked Markovian arrival process ($BMMAP$, see [39]), which well describes the batch correlated flow of heterogeneous customers. In this case, explicit formulas for the average number of customers of each type that are lost in a row in the overflow period are of practical interest.

References

1. Chaudhry, M.L., Templeton, J.G.C.: First Course in Bulk Queues. Wiley, New York (1983)
2. Bocharov, P.P., D'Apice, C., Pechinkin, A.V.: Queueing Theory. Walter de Gruyter, Berlin (2011)
3. Neuts, M.F.: A versatile Markovian point process. J. Appl. Probab. **14**, 764–779 (1979)
4. Lucantoni, D.M.: New results on the single server queue with a batch Markovian arrival process. Commun. Stat. Stoch. Model. **7**, 1–46 (1991)
5. Bocharov, P.P., D'Apice, C., Phong, N.H.: On a retrial single-server queueing system with finite buffer and Poisson flow. Probl. Inf. Transm. **37**, 248–262 (2001)
6. Bocharov, P.P.: A $MAP/G/1/r$ system with a large service time variation coefficient. Autom. Remote Control **66**, 1782–1790 (2005)
7. D'Apiche, C., Manzo, R.: A finite capacity $BMAP_K/G_k/1$ queue with the generalized foreground-background processor-sharing discipline. Autom. Remote Control **67**, 428–434 (2006)
8. Pechinkin, A.V.: Markov queueing system with finite buffer and negative customers affecting the queue end. Autom. Remote Control **68**, 1104–1117 (2007)
9. Baiocchi, A.: Analysis of the loss probability of the $MAP/G/1/K$ queue. Part I: asymptotic theory. Commun. Stat. Stoch. Model. **10**, 867–893 (1994)
10. Baiocchi, A., Blefari-Melezzi, N.: Analysis of the loss probability of the $MAP/G/1/K$ queue. Part 2: approximations and numerical results. Commun. Stat. Stoch. Model. **10**, 895–925 (1994)
11. Blondia, C.: The $N/G/1$ finite capacity queue. Part 2: approximations and numerical results. Commun. Stat. Stoch. Model. **5**, 273–294 (1989)
12. Dudin, A.N., Klimenok, V.I., Tsarenkov, G.V.: A single-server queueing system with batch Markov arrivals, semi-Markov service, and finite buffer: its characteristics. Autom. Remote Control **63**, 1285–1297 (2002)

13. Dudin, A.N., Nishimura, S.: Optimal hysteretic control for a $BMAP/SM/1/N$ queue with two operation modes. Math. Probl. Eng. **5**, 397–419 (2000)
14. Dudin, A.N., Shaban, A.A., Klimenok, V.I.: Analysis of a $BMAP|G|1|N$ queue. Int. J. Simul. Syst. Sci. Technol. **6**, 13–23 (2005)
15. Dudin, A., Shaban, A.: Analysis of the $BMAP/SM/1/N$ type system with randomized choice of customers admission discipline. In: Dudin, A., Gortsev, A., Nazarov, A., Yakupov, R. (eds.) ITMM 2016. CCIS, vol. 638, pp. 44–56. Springer, Cham (2016). https://doi.org/10.1007/978-3-319-44615-8_4
16. Dudin, A.N., Kazimirsky, A.V., Klimenok, V.I., Breuer, L., Krieger, U.: The queuing model MAP/PH/1/N with feedback operating in a Markovian random environment. Austrian J. Stat. **34**, 101–110 (2005)
17. Yoshigoe, K., Cuyt, A.: Computing packet loss probabilities of $D - BMAP/PH/1/N$ queues with group services. Perform. Eval. **67**, 160–173 (2010)
18. Banerjee, A., Gupta, U.C., Chakravarthy, S.R.: Analysis of a finite-buffer bulk-service queue under Markovian arrival process with batch-size-dependent service. Comput. Oper. Res. **60**, 138–149 (2015)
19. Banik, A.D.: Queueing analysis and optimal control of $BMAP/G(a;b)/1/N$ and $BMAP/MSP(a;b)/1/N$ systems. Comput. Ind. Eng. **57**, 748–761 (2009)
20. Banik, A.D., Chaudhry, M.L.: Efficient computational analysis of stationary probabilities for the queueing system $BMAP/G/1/N$ with or without vacation(s). INFORMS J. Comput. **29**, 140–151 (2017)
21. Banik, A.D., Ghosh, S.: Efficient computational analysis of non-exhaustive service vacation queues: $BMAP/R/1/N(1)$ under gated-limited discipline. Appl. Math. Model. **68**, 540–562 (2019)
22. Ghosh, S., Banik, A.D., Walraevens, J., Bruneel, H.: A detailed note on the finite-buffer capacity queueing system with correlated batch arrivals and batch-size-phase-dependent bulk-service. 4OR **20**, 241–272 (2022)
23. Kant, L., Sanders, W.H.: Analysis of the distribution of consecutive cell losses in an ATM switch using stochastic activity networks. Comput. Syst. Sci. Eng. **12**, 117–129 (1997)
24. Chydzinski, A.: On the distribution of consecutive losses in a finite capacity queue. WSEAS Trans. Circuits Syst **4**, 117–124 (2005)
25. Chydzinski, A., Samociuk, D., Adamczyk, B.: Burst ratio in the finite-buffer queue with batch Poisson arrivals. Appl. Math. Comput. **330**, 225–238 (2018)
26. Chydzinski, A., Samociuk, D.: Burst ratio in a single-server queue. Telecommun. Syst. **70**, 263–276 (2019)
27. Chydzinski, A.: On the structure of data losses induced by an overflowed buffer. Appl. Math. Comput. **415**, 1–12 (2022)
28. De Boer, P.T.: Analysis and efficient simulation of queueing models of telecommunication systems. Ph.D. thesis, Centre for Telematics and Information Technology University of Twente (2000)
29. Pacheco, A., Ribeiro, H.: Consecutive customer loss probabilities in $M/G/1/n$ and $GI/M(m)/1/n$. In: Proceedings of Workshop on Tools for Solving Structured Markov Chains (SMCtools 2006), Pisa, Italy (2006)
30. Pacheco, A., Ribeiro, H.: Consecutive customer losses in oscillating $GI^X/M/1/n$ systems with state dependent services rates. Ann. Oper. Res. **162**, 143–158 (2008)
31. Pacheco, A., Ribeiro, H.: Consecutive customer losses in regular and oscillating $M^X/G/1/n$ systems. Queueing Systems **58**, 121–136 (2008)
32. Ferrandiz, J.M., Lazar, A.A.: Monitoring the packet gap of real-time packet traffic. Queueing Syst. **12**, 231–242 (1992)

33. Lee, C.W., Andersland, M.S.: Consecutive cell loss controls for leaky-bucket admission systems. In: Proceedings of Global Telecommunications Conference (GLOBE-COM 1996), vol. 3, pp. 1732–1738 (1996)
34. Chakravarthy, S.R.: The batch Markovian arrival process: a review and future work. In: Advances in Probability Theory and Stochastic Process, pp. 21–49. Notable Publications (2001)
35. Neuts, M.F.: Matrix-Geometric Solutions in Stochastic Models. The Johns Hopkins University Press, Baltimore (1981)
36. Dudin, A.N., Klimenok, V.I., Vishnevsky, V.M.: The Theory of Queuing Systems with Correlated Flows. Springer, Cham (2020). https://doi.org/10.1007/978-3-030-32072-0
37. Graham, A.: Kronecker Products and Matrix Calculus with Applications. Ellis Horwood, Cichester (1981)
38. Klimenok, V.I., Kim, C.S., Orlovsky, D.S., Dudin, A.N.: Lack of invariant property of Erlang $BMAP/PH/N/0$ model. Queueing Syst. **49**, 187–213 (2005)
39. He, Q.M.: Queues with marked customers. Adv. Appl. Probab. **28**, 567–587 (1996)

Analysis of Tandem Retrial Queue with Common Orbit and MMPP Incoming Flow

Anatoly Nazarov[1] ⓘ, Svetlana Paul[1], Tuan Phung-Duc[2] ⓘ,
and Mariya Morozova[1](✉) ⓘ

[1] National Research Tomsk State University, 36 Lenina Avenue,
634050 Tomsk, Russia
morozova_mariya_a@mail.ru
[2] University of Tsukuba, 1-1-1 Tennodai, Tsukuba, Ibaraki 305-8573, Japan
tuan@sk.tsukuba.ac.jp
https://en.tsu.ru, https://www.tsukuba.ac.jp/en

Abstract. In this paper, we consider a tandem queueing system with one orbit, MMPP incoming flow and two sequentially connected servers by the method of asymptotic diffusion analysis. We obtain the necessary condition for the existence of the steady-state regime and the limiting probability distribution of the number of calls there under the scaling regime when the retrial rate is extremely small. Then we evaluate the applicability of the asymptotic results by simulation.

Keywords: Tandem system · Retrial queue system · MMPP incoming flow · Asymptotic diffusion analysis

1 Introduction

Retrial phenomenon naturally arises in various systems such as communication systems and service systems. For example, in call centers, if an operator is not available upon the arrival of a customer, the customer may hear some message such that "the system is currently congested, please wait or make a phone call later". In computer systems, if a request is not processed in a period of time, some automatic program tries to repeat the request in some fixed intervals. From a queueing model point of view, in these situations customers who cannot receive service immediately upon arrival due to the unavailability of the servers join the orbit and retry to enter the server after some random time. During the retrial interval, customers stay in a virtual waiting room called the orbit.

Retrial queues characterized by the phenomenon above have been extensively studied in the literature [1]. The analysis of these models is more difficult than that of their counterpart models with a buffer in front of the server(s) because the underlying Markovian chain of retrial queues is spatially inhomogeneous due to the total retrial rate of customers that depends on the number of customers in the orbit. As a reason, analytical results for retrial queues are available in only a few special cases with one or two servers [5].

ⓒ The Author(s), under exclusive license to Springer Nature Switzerland AG 2022
V. M. Vishnevskiy et al. (Eds.): DCCN 2022, LNCS 13766, pp. 270–283, 2022.
https://doi.org/10.1007/978-3-031-23207-7_21

The analysis of queueing network with retrial is even more challenging because these models do not possess a product form solution. There are several related works on tandem queue with two connected servers in which only blocked customers at the first server join orbit while those who find the second server busy upon service completion at the first server are lost. For such a model, in the cases with exponential service time distributions in both servers, explicit solution is derived [4]. For matrix-analytic solutions, some more general solutions are available [2].

However, for the models with retrials from any server, analytical solutions have not been obtained yet for even the simplest model with Poisson input. In our recent study [3], we have obtained a scaling limit for the exponential distribution setting at both servers. In this paper, we extend our previous work to present the solution for such a model with MMPP input. It should be noted that explicit solution for the joint stationary distribution of the number of customers in orbit and the state of the servers is challenging if not impossible.

Here we focus on a special regime when the retrial rate is extremely small. In this regime, the number of customers in orbit explodes. However, after a proper scaling, we prove that two scaled versions of the number of customers in the orbit converge to a deterministic process and a diffusion process respectively. The later result is then used to build an approximation to the distribution of the number of customers in orbit.

The rest of the paper is organized as follows. Section 2 presents the model in detail. Section 3 shows the set of Kolmogorov equations for the underlying model while. Section 4 is devoted to the first order asymptotic analysis of the distribution of the number of customers in the orbit. Section 5 is for the necessary stability condition. Section 6 is devoted to the second order asymptotic. Section 7 presents the asymptotic diffusion approach where we obtain an approximation for the distribution of the number of customers in the orbit in the stationary regime. Section 8 demonstrates an approximation for the model and conclusion is presented in Sect. 9.

2   Description of the Mathematical Model and Problem Statement

We consider a retrial queueing system with MMPP arrivals and two sequentially connected servers (Fig. 1). Upon the arrival of a call, if the first server is free, the call occupies it. The call is served for a random time exponentially distributed with parameter μ_1 and then tries to go to the second server. If the second server is free, the call moves to it for a random time exponentially distributed with parameter μ_2 and after that departs from the system. When a call arrives, if the first server is busy, the call instantly goes to the orbit. Upon service completion at the first server if the second server is busy the call also goes to the orbit. Calls stay in the orbit for a random time exponentially distributed with parameter σ and then retries to occupy the first server again. The behavior of a call from the orbit is the same as that of a new one.

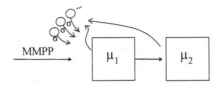

Fig. 1. Tandem Retrial Queue with common orbit and incoming MMPP-flow.

Let us denote: process $k(t)$ – Markovian chain which regulates the MMPP-flow, defined by infinitesimal generator – matrix $\mathbf{Q}$ with elements q_{vk}, v, $k = \overline{1, K}$; the arrival rate is given by $\lambda_k \geq 0$; process $n_1(t)$ – the state of the first server at time t: 0, if the server is free; 1, if the server is busy; process $n_2(t)$ – the state of the second server at time t: 0, if the server is free; 1, if the server is busy; process $i(t)$ – number of calls in the orbit at the time t.

The goal of the study is to obtain a probability distribution number of calls in the orbit $i(t)$ in the considered system.

3   The System of Differential Kolmogorov Equations

We define probabilities

$$P_{n_1 n_2}(k, i, t) = P\{n_1(t) = n_1, n_2(t) = n_2, k(t) = k, i(t) = i\};$$
$$n_1 = 0, 1; \ n_2 = 0, 1. \tag{1}$$

The four-dimensional process $\{n_1(t), n_2(t), k(t), i(t)\}$ is a Markovian chain. For probability distribution (1), we can write the system of differential Kolmogorov equations.

$$\frac{\partial P_{00}(k, i, t)}{\partial t} = -(\lambda_k + i\sigma)P_{00}(k, i, t) + \mu_2 P_{01}(k, i, t) + \sum_v P_{00}(v, i, t)q_{vk},$$

$$\frac{\partial P_{10}(k, i, t)}{\partial t} = \lambda_k P_{00}(k, i, t) + (i+1)\sigma P_{00}(k, i+1, t) - (\lambda_k + \mu_1)P_{10}(k, i, t)$$
$$+ \lambda_k P_{10}(k, i-1, t) + \mu_2 P_{11}(k, i, t) + \sum_v P_{10}(v, i, t)q_{vk},$$

$$\frac{\partial P_{01}(k, i, t)}{\partial t} = \mu_1 P_{10}(k, i, t) - (\lambda_k + \mu_2 + i\sigma)P_{01}(k, i, t)$$
$$+ \mu_1 P_{11}(k, i-1, t) + \sum_v P_{01}(v, i, t)q_{vk},$$

$$\frac{\partial P_{11}(k, i, t)}{\partial t} = -(\lambda_k + \mu_1 + \mu_2 + i\sigma)P_{11}(k, i, t)$$
$$+ \lambda_k P_{11}(k, i, t) + \sum_v P_{11}(v, i, t)q_{vk}, \tag{2}$$

We introduce partial characteristic functions, denoting $j = \sqrt{-1}$

$$H_{n_1 n_2}(k, u, t) = \sum_{i=0}^{\infty} e^{jui} P_{n_1 n_2}(k, i, t). \tag{3}$$

So, we have

$$\frac{\partial H_{00}(k, u, t)}{\partial t} = -\lambda_k H_{00}(k, u, t) + \mu_2 H_{01}(k, u, t)$$
$$+ \sum_v H_{00}(v, u, t) q_{vk} + j\sigma \frac{\partial H_{00}(k, u, t)}{\partial u},$$

$$\frac{\partial H_{10}(k, u, t)}{\partial t} = \left(\lambda_k \left(e^{ju} - 1\right) - \mu_1\right) H_{10}(k, u, t) + \lambda_k H_{00}(k, u, t)$$
$$+ \mu_2 H_{11}(k, u, t) + \sum_v H_{01}(v, u, t) q_{vk} - j\sigma e^{-ju} \frac{\partial H_{00}(u, t)}{\partial u},$$

$$\frac{\partial H_{01}(k, u, t)}{\partial t} = \mu_1 H_{10}(k, u, t) - (\lambda_k + \mu_2) H_{01}(k, u, t) + \mu_1 e^{ju} H_{11}(k, u, t)$$
$$+ \sum_v H_{10}(v, u, t) q_{vk} + j\sigma \frac{\partial H_{01}(k, u, t)}{\partial u},$$

$$\frac{\partial H_{11}(k, u, t)}{\partial t} = \left(\lambda_k \left(e^{ju} - 1\right) - \mu_1 - \mu_2\right) H_{11}(k, u, t)$$
$$+ \sum_v H_{11}(v, u, t) q_{vk} + j\sigma e^{-ju} \frac{\partial H_{01}(k, u, t)}{\partial u}. \tag{4}$$

Let us denote row vectors to remain the compactness of further computation

$$\mathbf{H}_{n_1 n_2}(u, t) = \{H_{n_1 n_2}(1, u, t), H_{n_1 n_2}(2, u, t), \dots, H_{n_1 n_2}(K, u, t),\}$$
$$\mathbf{H}(u, t) = \{\mathbf{H}_{00}(u, t), \mathbf{H}_{10}(u, t), \mathbf{H}_{01}(u, t), \mathbf{H}_{11}(u, t)\} \tag{5}$$

Furthermore, we define the following block structured matrices

$$\mathbf{A} = \begin{bmatrix} \mathbf{Q} - \mathbf{\Lambda} & \mathbf{\Lambda} & \mathbf{O} & \mathbf{O} \\ \mathbf{O} & \mathbf{Q} - (\mathbf{\Lambda} + \mu_1 \mathbf{I}) & \mu_1 \mathbf{I} & \mathbf{O} \\ \mu_2 \mathbf{I} & \mathbf{O} & \mathbf{Q} - (\mathbf{\Lambda} + \mu_2 \mathbf{I}) & \mathbf{\Lambda} \\ \mathbf{O} & \mu_2 \mathbf{I} & \mathbf{O} & \mathbf{Q} - (\mathbf{\Lambda} + \mu_1 \mathbf{I} + \mu_2 \mathbf{I}) \end{bmatrix},$$

$$\mathbf{B} = \begin{bmatrix} \mathbf{O} & \mathbf{O} & \mathbf{O} & \mathbf{O} \\ \mathbf{O} & \mathbf{\Lambda} & \mathbf{O} & \mathbf{O} \\ \mathbf{O} & \mathbf{O} & \mathbf{O} & \mathbf{O} \\ \mathbf{O} & \mathbf{O} & \mu_2 \mathbf{I} & \mathbf{\Lambda} \end{bmatrix}, \mathbf{I}_0 = \begin{bmatrix} \mathbf{I} & \mathbf{O} & \mathbf{O} & \mathbf{O} \\ \mathbf{O} & \mathbf{O} & \mathbf{O} & \mathbf{O} \\ \mathbf{O} & \mathbf{O} & \mathbf{I} & \mathbf{O} \\ \mathbf{O} & \mathbf{O} & \mathbf{O} & \mathbf{O} \end{bmatrix}, \mathbf{I}_1 = \begin{bmatrix} \mathbf{O} & \mathbf{I} & \mathbf{O} & \mathbf{O} \\ \mathbf{O} & \mathbf{O} & \mathbf{O} & \mathbf{O} \\ \mathbf{O} & \mathbf{O} & \mathbf{O} & \mathbf{I} \\ \mathbf{O} & \mathbf{O} & \mathbf{O} & \mathbf{O} \end{bmatrix}. \tag{6}$$

There are all blocks with the dimension $K \times K$, $\mathbf{O}$ is a zero block with the dimension $K \times K$, $\mathbf{I}$ is identity matrix. Using these matrices and multiplying all equations of system of differential Kolmogorov equations by identity column vector $\mathbf{e}$ with dimension 4K, combining the matrix equation and the scalar equation we have the system

$$\frac{\partial \mathbf{H}(u,t)}{\partial t} = \mathbf{H}(u,t)\{\mathbf{A} + e^{ju}\mathbf{B}\} + j\sigma \frac{\partial \mathbf{H}(u,t)}{\partial u}\{\mathbf{I}_0 - e^{-ju}\mathbf{I}_1\},$$

$$\frac{\partial \mathbf{H}(u,t)}{\partial t}\mathbf{e} = (e^{ju} - 1)\left\{\mathbf{H}(u,t)\mathbf{B} + j\sigma e^{-ju}\frac{\partial \mathbf{H}(u,t)}{\partial u}\mathbf{I}_1\right\}\mathbf{e}. \qquad (7)$$

This system of equations is the basis in further research. We solved it by a method of asymptotic diffusion analysis under the asymptotic condition $\sigma \to 0$ for similar tandem queueing system with Poisson arrival process in [3].

4   The First Order Asymptotic

By denoting $\sigma = \varepsilon$ and performing substitution in the system (7)

$$\tau = t\varepsilon, \ u = w\varepsilon, \ \mathbf{H}(u,t) = \mathbf{F}(w,\tau,\varepsilon) \qquad (8)$$

we rewrite the system (7) as

$$\varepsilon\frac{\partial \mathbf{F}(w,\tau,\varepsilon)}{\partial t} = \mathbf{F}(w,\tau,\varepsilon)(\mathbf{A} + e^{j\varepsilon w}\mathbf{B}) + j\frac{\partial \mathbf{F}(w,\tau,\varepsilon)}{\partial w}(\mathbf{I}_0 - e^{-j\varepsilon w}\mathbf{I}_1),$$

$$\varepsilon\frac{\partial \mathbf{F}(w,\tau,\varepsilon)}{\partial t}\mathbf{e} = (e^{j\varepsilon w} - 1)\left\{\mathbf{F}(w,\tau,\varepsilon)\mathbf{B} + je^{-j\varepsilon w}\frac{\partial \mathbf{F}(w,\tau,\varepsilon)}{\partial w}\mathbf{I}_0\right\}\mathbf{e}. \qquad (9)$$

Let us denote

$$\lim_{\varepsilon \to 0} \mathbf{F}(w,\tau,\varepsilon) = \mathbf{F}(w,\tau), \ \mathbf{F}(0,\tau) = \mathbf{r}. \qquad (10)$$

The row vector $\mathbf{r}$ defines two-dimensional probability distribution of the states of servers (n_1, n_2). It will be seen below that the row vector $\mathbf{r}$, that satisfies the normalization condition $\mathbf{re} = 1$, is a solution of the matrix equation

$$\mathbf{r}\{\mathbf{A} + \mathbf{B} - x(\mathbf{I}_0 - \mathbf{I}_1)\} = 0. \qquad (11)$$

Coefficients of this equation depend on variable x, so, solution $\mathbf{r}$ depends on value of x, therefore denote $\mathbf{r} = \mathbf{r}(x)$. To solve the system (9) under asymptotic condition $\varepsilon \to 0$ $(\sigma \to 0)$ we will use the following statement.

Theorem 1. *Under the limit condition $\sigma \to 0$, the following equality is true*

$$\lim_{\sigma \to 0} \mathbb{E}e^{jw\sigma i(\frac{\tau}{\sigma})} = e^{jwx(\tau)}. \qquad (12)$$

Here the scalar function $x = x(\tau)$ is a solution of differential equation

$$x'(\tau) = \mathbf{r}(x)(\mathbf{B} - x\mathbf{I}_0)\mathbf{e}, \qquad (13)$$

where the vector $\mathbf{r}(x)$ *satisfies the normalization condition*

$$\mathbf{r}(x)\mathbf{e} = 1 \tag{14}$$

and is a solution of the matrix equation

$$\mathbf{r}(x)\{\mathbf{A} + \mathbf{B} - x(\mathbf{I}_0 - \mathbf{I}_1)\} = 0. \tag{15}$$

Let's substitute the solution $\mathbf{r}(x)$ of the system of Eqs. (15) in the scalar equation (13) and we will get

$$a(x) = \mathbf{r}(x)(\mathbf{B} - x\mathbf{I}_0)\mathbf{e}, \tag{16}$$

Function $a(x)$ is very important for study retrial queuing systems by the method of asymptotic diffusion analysis. Firstly, as we have shown in Theorem 1, $a(x) = x'(\tau)$, therefore, function $a(x)$ characterizes dynamic of the process $x(\tau)$, the limit under $\sigma \to 0$ for normalized number of calls in the orbit $\sigma i(\tau/\sigma)$. Then we will show that function $a(x)$ is a drift coefficient for diffusion process which determines the asymptotic number of customers in the orbit $i(t)$. Using $a(x)$ we will get necessary condition for the existence of steady-state regime in the retrial queuing system under consideration.

5 Existence of Steady-State Regime

The inequality $\lim_{x \to \infty} a(x) < 0$ is the necessary equation for the existence of the steady-state regime. Let us prove the following statement. We will use the following statement.

Theorem 2. *A necessary condition for the existence of steady-state regime in RQ-system under consideration is inequality*

$$\mathbf{r}_1 \Lambda \mathbf{e}_1 < \frac{\mu_1 \mu_2}{\mu_1 + \mu_2}. \tag{17}$$

Vector $\mathbf{r}_1$ *is the vector of steady-state distribution of the control process* $k(t)$ *for which* $\mathbf{r}_1 \mathbf{Q} = 0$, $\mathbf{r}_1 \mathbf{e}_1 = 1$, Λ *is the diagonal matrix with elements* λ_k, $k = \overline{1, K}$ *and vector* $\mathbf{e}_1$ *is vector of units with dimension* K.

Proof. In order to find the condition for the existence of the steady-state, we find the limit of function $a(x)$ when $x \to \infty$. Using (16), let's rewrite this equality for components $\mathbf{r}_{n_1 n_2}(x)$

$$a(x) = -(\mathbf{r}_{00}(x) + \mathbf{r}_{01}(x))x\mathbf{e}_1 + (\mathbf{r}_{10}(x) + \mathbf{r}_{11}(x))\Lambda\mathbf{e}_1 + \mathbf{r}_{11}(x)\mu_1\mathbf{e}_1.$$

Let's define vector's components $\mathbf{r}_{n_1 n_2}(x)$ of vector $\mathbf{r}(x) = \mathbf{r}_{00}(x), \mathbf{r}_{10}(x), \mathbf{r}_{01}(x),$ $\mathbf{r}_{11}(x)$ defined by equalities (14) and (15) to find $\lim\limits_{x \to \infty} a(x)$.

$$\mathbf{r}_{00}(x)(\mathbf{Q} - \Lambda - x\mathbf{I}) + \mu_2 \mathbf{r}_{01}(x)\mathbf{I} = 0,$$
$$\mathbf{r}_{00}(x)(\Lambda + x\mathbf{I}) + \mathbf{r}_{10}(x)(\mathbf{Q} - \mu_1 \mathbf{I}) + \mu_2 \mathbf{r}_{11}(x)\mathbf{I} = 0,$$
$$\mu_1 \mathbf{r}_{10}(x) + \mathbf{r}_{01}(x)(\mathbf{Q} - (\Lambda + \mu_2 \mathbf{I} + x\mathbf{I})) + \mu_1 \mathbf{r}_{11}(x) = 0,$$
$$\mathbf{r}_{01}(\Lambda + x\mathbf{I}) + \mathbf{r}_{11}(x)(\mathbf{Q} - \mu_1 \mathbf{I} - \mu_2 \mathbf{I}) = 0,$$
$$\mathbf{r}_{00}(x) + \mathbf{r}_{10}(x) + \mathbf{r}_{01}(x) + \mathbf{r}_{11}(x) = \mathbf{r}_1$$

Here vector $\mathbf{r}_1$ is the vector of steady-state probabilities of the control process $k(t)$, $k = \overline{1, K}$. The solution of the system is as follows:

$$\mathbf{r}_{01}(x) = \frac{1}{\mu_2} \mathbf{r}_{00}(x)(\Lambda + x\mathbf{I} - \mathbf{Q}) + \frac{x}{\mu_2} \mathbf{r}_{00}(x) \left(\mathbf{I} + O\left(\frac{1}{x} \right) \right),$$

$$\mathbf{r}_{10}(x) = \frac{x^2}{\mu_2} \mathbf{r}_{00}(x) \left(\frac{1}{\mu_1} \mathbf{I} + (\mu_1 \mathbf{I} + \mu_2 \mathbf{I} - \mathbf{Q})^{-1} \right) \left(\mathbf{I} + O\left(\frac{1}{x} \right) \right),$$

$$\mathbf{r}_{11}(x) = \frac{x^2}{\mu_2} \mathbf{r}_{00}(x)(\mu_1 \mathbf{I} + \mu_2 \mathbf{I} - \mathbf{Q})^{-1} \left(\mathbf{I} + O\left(\frac{1}{x} \right) \right),$$

$$\mathbf{r}_{00}(x) + \mathbf{r}_{10}(x) + \mathbf{r}_{01}(x) + \mathbf{r}_{11}(x) = \mathbf{r}_1 = \frac{x^2}{\mu_1 \mu_2} \mathbf{r}_{00}(x) \left(\mathbf{I} + O\left(\frac{1}{x} \right) \right).$$

From the last equality we obtain

$$x^2 \mathbf{r}_{00}(x) = \mu_1 \mu_2 \mathbf{r}_1 \left(\mathbf{I} + O\left(\frac{1}{x} \right) \right). \tag{18}$$

Let us come back to the function $a(x)$ and let us to find the limit of the function $a(x)$ from (16) under the condition $x \to \infty$

$$a(x) = \{(\mathbf{r}_{10}(x) + \mathbf{r}_{11}(x))\Lambda + \mathbf{r}_{11}(x)\mu_1 - (\mathbf{r}_{00}(x) + \mathbf{r}_{01}(x))x\}\mathbf{e}_1$$
$$= \frac{x^2}{\mu_2} \mathbf{r}_{00}(x) \left\{ \frac{1}{\mu_1} \Lambda + \mu_1(\mu_1 \mathbf{I} + \mu_2 \mathbf{I} - \mathbf{Q})^{-1} - \mathbf{I} \right\} \mathbf{e}_1 \left(1 + O\left(\frac{1}{x} \right) \right),$$

Let's consider the expression individually

$$\mu_1(\mu_1 \mathbf{I} + \mu_2 \mathbf{I} - \mathbf{Q})^{-1} \mathbf{e}_1 = \frac{\mu_1}{\mu_1 + \mu_2} \left(\mathbf{I} - \frac{1}{\mu_1 + \mu_2} \mathbf{Q} \right)^{-1} \mathbf{e}_1.$$

We denote $\left(\mathbf{I} - \frac{1}{\mu_1 + \mu_2} \mathbf{Q} \right)^{-1} \mathbf{e}_1 = \mathbf{v}$, multiply the left and right parts of the last equation by the matrix $\left(\mathbf{I} - \frac{1}{\mu_1 + \mu_2} \mathbf{Q} \right)$ and get

$$\mathbf{e}_1 = \left(\mathbf{I} - \frac{1}{\mu_1 + \mu_2} \mathbf{Q} \right) \mathbf{v}.$$

It is obvious that $\mathbf{v} = \mathbf{e}_1$. So, we can rewrite the expression for $a(x)$ as

$$a(x) = \frac{x^2}{\mu_2} \mathbf{r}_{00}(x) \left\{ \frac{1}{\mu_1} \boldsymbol{\Lambda} - \frac{\mu_2}{\mu_1 + \mu_2} \mathbf{I} \right\} \left(1 + O\left(\frac{1}{x} \right) \right).$$

Let's substitute to it the expression (18)

$$a(x) = \left\{ \mathbf{r}_1 \boldsymbol{\Lambda} \mathbf{e}_1 - \frac{\mu_1 \mu_2}{\mu_1 + \mu_2} \right\} \left(1 + O\left(\frac{1}{x} \right) \right).$$

The inequality $\lim_{x \to \infty} a(x) < 0$ that defines the necessary condition for the existence of the steady-state regime in the RQ-system under consideration is giving by

$$\mathbf{r}_1 \boldsymbol{\Lambda} \mathbf{e}_1 < \frac{\mu_1 \mu_2}{\mu_1 + \mu_2}.$$

So, the Theorem 2 is proved.

6   The Second Order Asymptotic

Substituting the following in the system (7)

$$\mathbf{H}(u, t) = e^{j\frac{u}{\sigma} x(\sigma t)} \mathbf{H}^{(1)}(u, t) \tag{19}$$

and taking into account the Eq. (16), we obtain

$$\frac{\partial \mathbf{H}^{(1)}(u, t)}{\partial t} + jua(x)\mathbf{H}^{(1)}(u, t) = \mathbf{H}^{(1)}(u, t) \left(\mathbf{A} + e^{ju}\mathbf{B} - x \left(\mathbf{I}_0 - e^{-ju}\mathbf{I}_1 \right) \right)$$

$$+ j\sigma \frac{\partial \mathbf{H}^{(1)}(u, t)}{\partial u} (\mathbf{I}_0 - e^{-ju}\mathbf{I}_1),$$

$$\frac{\partial \mathbf{H}^{(1)}(u, t)}{\partial t} \mathbf{e} + jua(x)\mathbf{H}^{(1)}(u, t)\mathbf{e} = (e^{ju} - 1)$$

$$\times \left\{ \mathbf{H}^{(1)}(u, t)(\mathbf{B} - e^{-ju}x\mathbf{I}_0) + j\sigma e^{-ju} \frac{\partial \mathbf{H}^{(1)}(u, t)}{\partial u} \mathbf{I}_0 \right\} \mathbf{e}. \tag{20}$$

We make a substitute (19) with a view to asymptotic centring of random process $i(t)$ because $\mathbf{H}^{(1)}(u, t)$ is the vector characteristic function of the centring random process, where the function $x(\tau)$ was receiving in the first stage of asymptotic analysis.

By denoting $\sigma = \varepsilon^2$ in the system (20) and making substitutions

$$\tau = t\varepsilon^2, \quad u = w\varepsilon, \quad \mathbf{H}^{(1)}(u, t) = \mathbf{F}^{(1)}(w, \tau, \varepsilon), \tag{21}$$

we can rewrite the system in the following form

$$\varepsilon^2 \frac{\partial \mathbf{F}^{(1)}(w, \tau, \varepsilon)}{\partial t} + j\varepsilon w a(x) \mathbf{F}^{(1)}(w, \tau, \varepsilon)$$

$$= \mathbf{F}^{(1)}(w, \tau, \varepsilon) \left(\mathbf{A} + e^{j\varepsilon w} \mathbf{B} - x(\mathbf{I}_0 - e^{-j\varepsilon w} \mathbf{I}_1) \right)$$

$$+ j\varepsilon \frac{\partial \mathbf{F}^{(1)}(w, \tau, \varepsilon)}{\partial w} (\mathbf{I}_0 - e^{-j\varepsilon w} \mathbf{I}_1),$$

$$\varepsilon^2 \frac{\partial \mathbf{F}^{(1)}(w, \tau, \varepsilon)}{\partial \tau} \mathbf{e} + j\varepsilon w a(x) \mathbf{F}^{(1)}(w, \tau, \varepsilon) \mathbf{e} = (e^{j\varepsilon w} - 1)$$

$$\times \left(\mathbf{F}^{(1)}(w, \tau, \varepsilon) \left(\mathbf{B} - e^{-j\varepsilon w} x \mathbf{I}_0 \right) + e^{-j\varepsilon w} j\varepsilon \frac{\partial \mathbf{F}^{(1)}(w, \tau, \varepsilon)}{\partial w} \mathbf{I}_0 \right) \mathbf{e}. \quad (22)$$

Denote

$$\lim_{\varepsilon \to 0} \mathbf{F}^{(1)}(w, \tau, \varepsilon) = \mathbf{F}^{(1)}(w, \tau), \quad \lim_{\varepsilon \to 0} \frac{\partial \mathbf{F}^{(1)}(w, \tau, \varepsilon)}{\partial \tau} = \frac{\partial \mathbf{F}^{(1)}(w, \tau)}{\partial \tau},$$

$$\lim_{\varepsilon \to 0} \frac{\partial \mathbf{F}^{(1)}(w, \tau, \varepsilon)}{\partial \tau} = \frac{\partial \mathbf{F}^{(1)}(w, \tau)}{\partial \tau}. \quad (23)$$

Theorem 3. *Function* $\mathbf{F}^{(1)}(w, \tau)$ *has the following form*

$$\mathbf{F}^{(1)}(w, \tau) = \Phi(w, \tau) \mathbf{r}(x), \quad (24)$$

where the row vector $\mathbf{r}(x)$ *depends on variable* x. *The vector* $\mathbf{r}(x)$ *is determined by Theorem 1, in vector's components* $\mathbf{r}_{n_1 n_2}(x)$, *and the scalar function* $\Phi(w, \tau)$ *is the solution of the partial differential equation*

$$\frac{\partial \Phi(w, \tau)}{\partial \tau} = a'(x) w \frac{\partial \Phi(w, \tau)}{\partial w} + b(x) \frac{(jw)^2}{2} \Phi(w, \tau). \quad (25)$$

Here the function $a(x)$ *is determined by (16) and the scalar function* $b(x)$ *has the form*

$$b(x) = a(x) + 2\mathbf{g}(x)(\mathbf{B} - x\mathbf{I}_0)\mathbf{e} + 2x\mathbf{r}(x)\mathbf{e}, \quad (26)$$

where vector $\mathbf{g}(x)$ *is determined by the system of equations*

$$\mathbf{g}(x) \left(\mathbf{A} + \mathbf{B} + x(\mathbf{I}_1 - \mathbf{I}_0) \right) = a(x)\mathbf{r}(x) + \mathbf{r}(x)(x\mathbf{I}_0 - \mathbf{B}),$$

$$\mathbf{g}(x)\mathbf{e} = 0. \quad (27)$$

Later we will show that function $b(x)$ is the diffusion coefficient of a diffusion process which has the function $a(x)$ as the coefficient of drift given by (16). Thus, we have defined functions $a(x)$ by the Eq. (16) and $b(x)$ by the Eq. (26). Later we will show their application in the method of asymptotic diffusion analysis in the considered system.

7   Method of Asymptotic Diffusion Analysis

In this section of the paper, we will consider the implementation of the method of asymptotic diffusion analysis for finding the probability distribution number of calls in the orbit $i(t)$ under the asymptotic condition $\sigma \to 0$ in the considered RQ-system.

Lemma 1. *Asymptotic stochastic process under the condition $\sigma \to 0$*

$$y(\tau) = \lim_{\sigma \to 0} \sqrt{\sigma} \left\{ i(\tau) - \frac{1}{\sigma} x(\tau) \right\},$$

is a solution of the stochastic differential equation

$$dy(\tau) = a'(x)y d\tau + \sqrt{b(x)} dw(\tau), \tag{28}$$

that depends on continuous parameter x.

Let us consider the following stochastic process

$$z(\tau) = x(\tau) + \varepsilon y(\tau),$$

where $\varepsilon = \sqrt{\sigma}$ as before. This process is the sum of the normalized mean and the centered number of calls in the orbit.

Lemma 2. *Stochastic process $z(\tau)$ is a solution to stochastic differential equation*

$$dz(\tau) = a(z)d\tau + \sqrt{\sigma b(z)} dw(\tau) \tag{29}$$

with a precision up to an infinitesimal of order ε^2

Suppose that the system is in steady-state regime. We consider the stationary probability density function for the process $z(\tau)$

$$s(z, \tau) = s(z) = \frac{\partial P\{z(\tau) < z\}}{\partial z}. \tag{30}$$

Theorem 4. *Stationary probability density $s(z)$ of the stochastic process $z(\tau)$ has the form*

$$s(z) = \frac{C}{b(z)} \exp\left\{ \frac{2}{\sigma} \int_0^z \frac{a(x)}{b(x)} dx \right\}, \tag{31}$$

where C is a normalizing constant.

8   Approximations Accuracy

We have constructed an approximation for discrete probability distribution $P(i)$ in [3]. We have written a non-negative function $G(i)$ of the discrete argument i in the form

$$G(i) = \frac{C}{b(\sigma i)} \exp \left\{ \frac{2}{\sigma} \int_0^{\sigma i} \frac{a(x)}{b(x)} dx \right\}, \tag{32}$$

where function $a(x)$ as a coefficient of drift defined by the equation

$$a(x) = \mathbf{r}(x)(\mathbf{B} - x\mathbf{I}_0)\mathbf{e}, \tag{33}$$

and where $x = x(\tau)$ is a solution of differential equation

$$x'(\tau) = \mathbf{r}(x)(\mathbf{B} - x\mathbf{I}_0)\mathbf{e}. \tag{34}$$

The vector $\mathbf{r}(x)$ satisfies the normalization condition $\mathbf{r}(x)\mathbf{e} = 1$ and is a solution of the matrix equation

$$\mathbf{r}(x)(\mathbf{A} + \mathbf{B} - x(\mathbf{I}_0 - \mathbf{I}_1)) = 0, \tag{35}$$

and function $b(x)$ is the diffusion coefficient of a diffusion process defined by the equation

$$b(x) = a(x) + 2\mathbf{g}(x)(\mathbf{B} - x\mathbf{I}_0)\mathbf{e} + 2x\mathbf{r}(x)\mathbf{e}, \tag{36}$$

where vector $\mathbf{g}(x)$ is determined by system of equations

$$\mathbf{g}(x)\,(\mathbf{A} + \mathbf{B} + x(\mathbf{I}_1 - \mathbf{I}_0)) = a(x)\mathbf{r}(x) + \mathbf{r}(x)(x\mathbf{I}_0 - \mathbf{B}),$$
$$\mathbf{g}(x)\mathbf{e} = 0. \tag{37}$$

Using the normalization condition, we can write

$$Pdif_i = \frac{G(i)}{\sum\limits_{i=0}^{\infty} G(i)}. \tag{38}$$

This probability distribution $Pdif_i$ we will use as an approximation for the probability distribution $Pdif_i = P\{i(t) = i\}$ for the number of calls in the orbit.

Approximations accuracy will be defined and compare by using Kolmogorov range

$$\Delta = \max_{k \geq 0} \left| \sum_{i=0}^{k} (Pdif_i - P_i) \right|, \tag{39}$$

where P_i is an empirical probability distribution of the number i of calls in the orbit obtained by the simulation.

Let's denote matrices

$$\mathbf{Q} = \begin{bmatrix} -1 & 0.3 & 0.7 \\ 0.1 & -0.6 & 0.5 \\ 0.4 & 0.3 & -0.7 \end{bmatrix}, \mathbf{\Lambda}_1 = \begin{bmatrix} 1 & 0 & 0 \\ 0 & 2 & 0 \\ 0 & 0 & 3 \end{bmatrix}.$$

Load of the system by ρ $(0 < \rho < 1)$ and intensity values of the incoming flow - elements of matrix $\mathbf{\Lambda}$ (under the condition for the existence of steady-state regime (17)) set by the equation

$$\mathbf{\Lambda} = \rho \frac{\mathbf{\Lambda}_1}{\mathbf{r}_1 \mathbf{\Lambda}_1 \mathbf{e}_1} \frac{\mu_1 \mu_2}{\mu_1 + \mu_2}. \tag{40}$$

We consider $\mu_1 = 1$, $\mu_2 = 2$ and $\rho = 0.5$ for different parameters σ (Table 1).

Table 1. Kolmogorov range.

$\sigma = 2$	$\sigma = 1.5$	$\sigma = 1$	$\sigma = 0.5$	$\sigma = 0.1$	$\sigma = 0.05$	$\sigma = 0.01$
0.057	0.049	0.033	0.011	0.033	0.019	0.007

Density diagrams of probability distributions are shown in Figs. 2, 3 and 4. The dotted line represents the probability distribution of the number i of calls in the orbit obtained by the simulation, the solid - approximation obtained by method of asymptotic diffusion and analysis (Pdif).

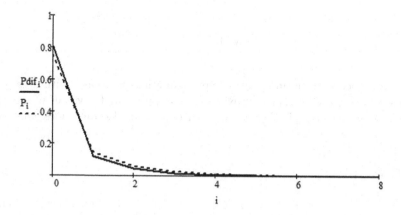

Fig. 2. $\sigma = 2$.

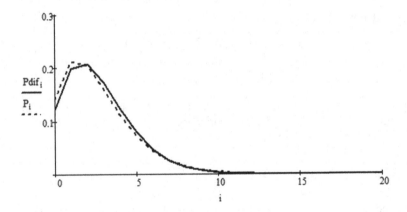

Fig. 3. $\sigma = 0.1$.

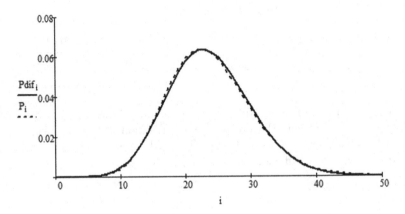

Fig. 4. $\sigma = 0.01$.

It can be seen, the accuracy of the approximations in-creases with decreasing parameter σ. The Gaussian approximation is applicable for values of $\sigma < 1.5$, where the relative error, in the form of the Kolmogorov distance, does not exceed 0.05.

9   Conclusion

In this paper, we consider the tandem retrial queueing system with incoming MMPP-flow. For this system, we define the condition for the existence of the steady-state regime. Using the method of asymptotic diffusion analysis under the asymptotic condition of the long delay in the orbit, we obtain parameters of the diffusion process. Probability density distribution of this process has enabled us to construct an approximation for probability distribution number of calls in the orbit in the considered retrial queueing system. Comparing with the results of simulation, it is shown that the accuracy of the approximations increases with the decrease in the retrial rate σ.

References

1. Falin, G.I., Templeton, J.: Retrial Queues, 1st edn. Chapman & Hall, London (1997)
2. Kim, C.S., Klimenok, V., Taramin, O.: A tandem retrial queueing system with two Markovian flows and reservation of channels. Comput. Oper. Res. **37**(7), 1238–1246 (2010)
3. Nazarov, A., Paul, S., Phung-Duc, T., Morozova, M.: Analysis of tandem retrial queue with common orbit and Poisson arrival process. In: Ballarini, P., Castel, H., Dimitriou, I., Iacono, M., Phung-Duc, T., Walraevens, J. (eds.) EPEW/ASMTA -2021. LNCS, vol. 13104, pp. 441–456. Springer, Cham (2021). https://doi.org/10. 1007/978-3-030-91825-5_27
4. Phung-Duc, T.: An explicit solution for a tandem queue with retrials and losses. Oper. Res. Int. J. **12**(2), 189–207 (2012). https://doi.org/10.1007/s12351-011-0113-7
5. Phung-Duc, T., Masuyama, H., Kasahara, S., Takahashi, Y.: State-dependent $M/M/c/c + r$ retrial queues with Bernoulli abandonment. J. Ind. Manag. Optim. **6**(3), 517–540 (2010)

Average Cost Minimization in a Multi-server Retrial Queueing System with a Controllable Reserve Group of Servers

Dmitry Efrosinin[1,2]([⊠]) [iD] and Natalia Stepanova[3] [iD]

[1] Johannes Kepler University Linz, Altenbergerstrasse 69, 4040 Linz, Austria
dmitry.efrosinin@jku.at
[2] Peoples' Friendship University of Russia (RUDN University),
Miklukho-Maklaya St. 6, 117198 Moscow, Russia
[3] AO NPF INSET, Zvezdniy b-r 19-1, 129085 Moscow, Russia
natalia0410@rambler.ru
http://www.jku.at, http://www.eng.rudn.ru, http://www.incet.ru

Abstract. The paper deals with a dynamic optimal activation and deactivation of a reserve group of servers in a multi-server retrial queueing system. The repeated attempts to occupy the server from the orbit occur due to a classical retrial discipline, and the activation of reserves requires a random set-up time. The optimal policy minimizes the average cost per unit of time which consists of the holding and usage cost components. The optimal control policy belongs to a class of hysteretic control policies defined by two threshold levels for the orbit size and prescribes the necessity to activate and deactivate a standby reserve group of servers. A dynamic-programming approach for the Markov decision process is used to show that the optimal thresholds depend on the number of busy servers as well. In order to provide a less computationally expensive method for calculating optimal thresholds, explicit heuristic expressions are proposed to guarantee at least a quasi-optimal solution to the optimization problem. Numerical examples confirm the efficiency of the proposed approach.

Keywords: Retrial queueing system · Markov decision process · Hysteretic policy · Heuristic solution

1 Introduction

Modern telecommunications and computer systems are now inconceivable without the parallel use of a large number of service devices or servers. The number

This paper has been supported by the RUDN University Strategic Academic Leadership Program (recipient D. Efrosinin). The reported study was funded by the Russian Science Foundation within scientific project No. 22-49-02023 "Development and study of methods for reliability enhancement of tethered high-altitude unmanned telecommunication platforms".

of servers must be sufficient to provide uninterrupted service for customers, but increasing the capacity of the service area is often limited by increasing operating costs associated with service, maintenance, power consumption, etc. The solution to this problem of keeping some balance between the number and capacity of servers and the economic costs of operation is the use of a standby service devices that can be activated and deactivated as required. As it is well known, if the activation or deactivation of a server or a group of servers is subject to a delay or some switching cost must be taken into account, and the frequency of such switchings is relatively high, then it is reasonable to apply hysteretic control policy, see e.g. [10], which has two different thresholds depending on current service level as well as on the number of waiting customers respectively for the activation and deactivation. Although the optimality of this class of policy is still unproven, the numerical results show that this policy reduces the activation/deactivation frequency and hence it avoids a ping-pong effect, which in turn reduces the long-run average cost. There are large number of papers on the problem of optimal activation and deactivation of servers using a hysteretic control policy. We will only mention a few of them here. In [14] and [15] the authors studied a model with set-up costs using a hysteretic control rule, thereby stressing the algorithmic aspects of the optimal control structure. The same system has been discussed in [13], where it was proposed a direct method that provided a closed-form expression for the stationary occupancy distribution. The optimal hysteretic policy for the ordinary queueing system with a unreliable reserve server pool was studied in [5], where the authors for these proposes have implemented a genetic optimization algorithm. The problem of choosing the optimal hysteresis strategy to control the number of active servers was studied in [11], where the interested reader can consult the paper for further references to results on hysteretic control for the systems with different attributes.

In retrial queueing system the customer who finds in the service area all servers busy becomes a blocked one and goes to an orbit, where it repeats the attempt to occupy the server in a random time. Most retrial queues assume to involve a level-dependent retrial intensity or so-called classical retrial discipline, i.e. each customer in orbit seeks service independently. A comprehensive overview of the literature on such retrial queues can be found in the classified bibliographies [1,2,8], in surveys [4,7,18], and in book [3]. Obtaining analytical and numerical results in the area of multi-server retrial queueing systems as is known is a non-trivial task. Therefore, in many cases one tries to find an approximation for the original model or some heuristic solution in the case of controllable systems. The paper [12], where results were obtained for a heuristic estimation of optimal thresholds for hysteresis policy which minimizes the average cost in case of an ordinary queueing system, motivated us to try to obtain similar results but for a multi-server retrial system. Note that a derivation of analytical expressions for the average cost as a function of unknown thresholds is a complex task. Therefore, we formulate the problem as a Markov decision process (MDP) and use algorithmic calculation methods such as dynamic programming to calculate the optimal control policy. Then we confirm the fact that this policy

286 D. Efrosinin and N. Stepanova

has also a hysteresis nature but the thresholds depend on the number of busy servers as well. Moreover, we need these results to confirm the effectiveness of the heuristic thresholds we derived in this paper.

The rest of the paper is organized as follows. In Sect. 2, we discuss the MDP formalism and structural properties of the optimal control policy. In Sect. 3, we describe a heuristic approach to calculate quasi-optimal thresholds of the hysteretic policy. In Sect. 4, we verify numerically the heuristic results based on the real values obtained by means of a policy-iteration algorithm.

For use in sequel, let $\mathbf{e}_j$ denote the (column) of dimension 3 with 1 in the jth position beginning from 0th and 0 elsewhere. The notation $\text{sgn}(x)$ means a modified sign function defined for an non-negative integer number x and is equal to 0 for $x = 0$ and 1 for $x > 0$. At last, the notation $1_{\{A\}}$ specifies the indicator function, which takes the value 1 if the event A holds, and 0 otherwise.

2 The MDP Formalism and Structural Properties

We consider a multi-server retrial queueing system illustrated in Fig. 1, where primary customers form a Poisson input flow with a rate λ. The service area consists of $N = n_1 + n_2$ identical servers with service rate μ, of which the main group of n_1 servers is always available, while another group of n_2 servers is a standby reserve with controllable activation and deactivation. If a primary customer finds an idle server, it occupies immediately the server and starts receive a service. When upon an arrival a customer finds all servers busy, it is blocked. In this case the blocked customer is transmitted to an orbit and becomes a retrial one. In the orbit a retrial customer repeats the attempt to get a service according to a classical retrial discipline, i.e. when the customer makes retrials independently of other orbiting customers in exponentially distributed time with a rate γ.

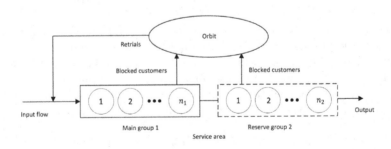

Fig. 1. Structure of a retrial queue with a reserve group of servers

The reserve group of servers has in a service area a controllable availability. The decision maker can activate the reserves at an arrival epoch of a primary customer. After the activation the exponential distributed set-up time with a rate τ to make the reserves on is required. The reserve group can be deactivated at service completion times. After the deactivation the reserve servers becomes immediately off. The customers served at this point of time by the reserve group of servers leave the service area and join the set of retrial customers. We assume the mutual independence of the inter-arrival, service, retrial and set-up times.

Denote by $Q(t)$ the number of customers in the orbit, $D(t)$ – the number of busy servers in the system and by $S(t)$ – the state of a reserve group of servers at time t, where

$$S(t) = \begin{cases} 0 & \text{the reserve group is deactivated,} \\ 1 & \text{the reserve group is activated and is not operational,} \\ 2 & \text{the reserve group is activated and is operational.} \end{cases}$$

The Markov-decision process (MDP) associated with a continuous-time Markov chain

$$\{X(t)\}_{t \geq 0} = \{Q(t), D(t), S(t)\}_{t \geq 0} \tag{1}$$

is described by a five-tuple

$$\{E, A_0, A_1, \lambda_{xy}(a), c(x)\}$$

with the following components.

1. E is a state space,

$$E = \bigcup_{s=0}^{2} E_s, \tag{2}$$

where

$$E_s = \{(q, d, s) : q \in \mathbb{N}_0, d \in [0, n_1] \cap \mathbb{N}_0\}, \ s \in \{0, 1\},$$
$$E_2 = \{(q, d, 2) : q \in \mathbb{N}_0, d \in [0, N] \cap \mathbb{N}_0\}.$$

Further in the paper the notations $q(x)$, $d(x)$ and $s(x)$ will be used to specify the certain elements of the vector state $x \in E$.

2. $A_0 = \{0, 1\}$ is a set of control actions at arrival epochs defined for states $x \in E_0$. The control action $a = 0$ and $a = 1$ with $a \in A_0$ means respectively to keep the reserve group in a deactivated mode and to activate it.

3. $A_1 = \{0, 1\}$ is a set of control actions at service completion epochs defined for states $x \in E_2$. The control actions $a = 0$ and $a = 1$ with $a \in A_1$ means respectively to keep the reserve group in an activated mode and to switch it off.

4. $\lambda_{xy}(a)$ is a transition rate to go from state x to state y given the control action is a which is given for $y \neq x$ by (3), where $\mu_1(x) = \mu \min\{d(x), n_1\}$ and $\mu_2(x) = \mu \min\{d(x), N\}$. The first row with λ represents the transition due to arrival of a primary customer to a state $x \in E_0$, where the decision maker chooses an action $a \in A_0$. The next two rows with λ describe the transitions due to arrivals in states $x \in E_1$ and $x \in E_2$, respectively. The rows with $\mu_1(x)$ and $\mu_2(x)$ stand for service completion epochs with possible selection of an action in state $x \in E_2$. The rows with γ are associated with the retrial arrivals and the second last row with τ describes the transition in an activation set-up time.

$$
\lambda_{xy}(a) = \begin{cases}
\lambda & y = x + \mathbf{e}_{\mathrm{sgn}(n_1 - d(x))} + a\mathbf{e}_2, \ x \in E_0, \ a \in A_0, \\
\lambda & y = x + \mathbf{e}_{\mathrm{sgn}(n_1 - d(x))}, \ x \in E_1, \\
\lambda & y = x + \mathbf{e}_{\mathrm{sgn}(N - d(x))}, \ x \in E_2, \\
\mu_1(x) & y = x - \mathbf{e}_1, \ d(x) > 0, \ x \in E_0 \cup E_1, \\
\mu_2(x) & y = x - \mathbf{e}_1 - 2a\mathbf{e}_2, \ d(x) > 0, \ x \in E_2, \ a \in A_1, \\
\gamma q(x) & y = x - \mathbf{e}_0 + \mathbf{e}_{\mathrm{sgn}(n_1 - d(x))}, \ q(x) > 0, \ x \in E_0 \cup E_1, \\
\gamma q(x) & y = x - \mathbf{e}_0 + \mathbf{e}_{\mathrm{sgn}(N - d(x))}, \ q(x) > 0, \ x \in E_2, \\
\tau & y = x + \mathbf{e}_2, \ x \in E_1, \\
0 & \text{otherwise,}
\end{cases}
\tag{3}
$$

5. $c(x)$ is an immediate cost in state $x \in E$,

$$
c(x) = c_0(d(x) + q(x)) + c_1(n_1 + \mathrm{sgn}(s(x))n_2),
\tag{4}
$$

where c_1 is the holding cost per unit of time for each customer in the system and c_2 the usage cost per unit of time per activated server.

A stationary policy $f = (f_0, f_1)$ is a mapping from states to actions, where the component $f_0 : E_0 \to A_0$ and $f_1 : E_2 \to A_1$ specifies a control action that must be chosen at an arrival epoch of a primary customer and at a service completion epoch. The system performance is characterized by the average cost g per unit of time which depends on the policy f and for the introduced cost structure can be expressed as

$$
g = c_0 L + c_1 C,
\tag{5}
$$

where L is the mean number of customers in the system and C is the mean number of activated servers.

It is difficult to calculate the average cost g in explicit form using steady-state probabilities, since the optimal control policy f is unknown and enumerating all possible policies is computationally a very complex task. Hence we use the dynamic-programming approach here, see e.g. [9, 16, 17]. The optimality equation of the MDP can be expressed in form

$$v(x) = \frac{1}{\lambda_x} \Big[c(x) - g + \lambda \min_{a \in A_0} v(x + \mathbf{e}_{\mathrm{sgn}(n_1 - d(x))} + a\mathbf{e}_2) 1_{\{x \in E_0\}} \tag{6}$$

$$+ \lambda [v(x + \mathbf{e}_{\mathrm{sgn}(n_1 - d(x))}) 1_{\{x \in E_1\}} + v(x + \mathbf{e}_{\mathrm{sgn}(N - d(x))}) 1_{\{x \in E_2\}}]$$

$$+ \mu_1(x) v(x - \mathbf{e}_1) 1_{\{d(x) > 0, x \in E_0 \cup E_1\}} + \mu_2(x) \min_{a \in A_1} v(x - \mathbf{e}_1 - 2a\mathbf{e}_2) 1_{\{d(x) > 0, x \in E_2\}}$$

$$+ \gamma q(x) [v(x - \mathbf{e}_0 + \mathbf{e}_{\mathrm{sgn}(n_1 - d(x))}) 1_{\{x \in E_0 \cup E_1\}} + v(x - \mathbf{e}_0 + \mathbf{e}_{\mathrm{sgn}(N - d(x))}) 1_{\{x \in E_2\}}]$$

$$\times 1_{\{q(x) > 0\}} + \tau v(x + \mathbf{e}_2) 1_{\{x \in E_1\}} \Big], \quad x \in E,$$

where

$$\lambda_x = \lambda + \mu_1(x) 1_{\{x \in E_0 \cup E_1\}} + \mu_2(x) 1_{\{x \in E_2\}} + \gamma q(x) 1_{\{q(x) > 0\}} + \tau 1_{\{x \in E_1\}}$$

is a total transition rate from the state $x \in E$, $v : E \to \mathbb{R}$ is the optimal value function and g is the optimal long-run average cost per unit of time defined in (5). The meanings of terms in the right-hand side of the Eq. (6) can easily be read from the description of the transition intensities (3) of the corresponding Markov chain. The optimal control actions for the activation and deactivation can be defined then respectively as

$$f_0(x) = \underset{a \in A_0}{\mathrm{argmin}}\, v(x + \mathbf{e}_{\mathrm{sgn}(n_1 - d(x))} + a\mathbf{e}_2), \quad x \in E_0,$$

$$f_1(x) = \underset{a \in A_1}{\mathrm{argmin}}\, v(x - \mathbf{e}_1 - 2a\mathbf{e}_2), \quad x \in E_2.$$

The policy-iteration algorithm [9] is used to calculate the optimal control policy f. According to this algorithm we start by choosing an arbitrary policy and we iteratively evaluate and improve the policy until convergence.

Example 1. Consider the queueing system with $N = 20$ servers consisting of $n_1 = 10$ main and $n_2 = 10$ reserve servers. The system parameters take the following values:

$$(\lambda, \mu, \gamma, \tau, c_0, c_1) = (8, 1, 2, 0.1, 1, 2). \tag{7}$$

The orbit size is chosen to be substantially large in order to minimize the probability of losing a customer and to approximate the calculation results to a system with an infinite orbit. The optimal control actions $f_0(x), x \in E_0$, and $f_1(x), x \in E_2$, are summarized in Tables 1 and 2. Note that in states x with $13 \le d(x) \le N$ and $s(x) = 2$ the optimal action $f_1(x) = 0$.

The optimal value of the average cost is $g = 31.0693$. We can observe in presented tables as well as in numerous experiments for other values of system parameters the hysteretic nature of the optimal control policy, i.e. for two types of states $x = (q, d, 0)$ and $x = (q, d, 2)$ there are thresholds $q_A(\cdot)$ and $q_D(\cdot)$ of the orbit sizes $q(x)$ required respectively to activate and deactivate the reserve group of servers. Moreover, these thresholds depend on the number of busy servers, $q_A(d), d \in [0, n_1] \cap \mathbb{N}_0$, and $q_D(d), d \in [1, N] \cap \mathbb{N}_0$, e.g. $q_A(0) = 54, q_A(10) = 45$ and $q_D(1) = 15, q_D(12) = 1$.

Table 1. Optimal policy f_0

System state x	Orbit $q(x)$																
(d,s)	...	42	43	44	45	46	47	48	49	50	51	52	53	54	55	56	...
(0, 0)	0	0	0	0	0	0	0	0	0	0	0	0	0	1	1	1	1
(1,0)	0	0	0	0	0	0	0	0	0	0	0	0	1	1	1	1	1
(2, 0)	0	0	0	0	0	0	0	0	0	0	0	1	1	1	1	1	1
(3, 0)	0	0	0	0	0	0	0	0	0	0	0	1	1	1	1	1	1
(4, 0)	0	0	0	0	0	0	0	0	0	0	1	1	1	1	1	1	1
(5, 0)	0	0	0	0	0	0	0	0	0	1	1	1	1	1	1	1	1
(6, 0)	0	0	0	0	0	0	0	0	1	1	1	1	1	1	1	1	1
(7, 0)	0	0	0	0	0	0	0	1	1	1	1	1	1	1	1	1	1
(8, 0)	0	0	0	0	0	0	1	1	1	1	1	1	1	1	1	1	1
(9, 0)	0	0	0	0	0	1	1	1	1	1	1	1	1	1	1	1	1
(10, 0)	0	0	0	0	1	1	1	1	1	1	1	1	1	1	1	1	1

Table 2. Optimal policy f_1

System state x	Orbit $q(x)$																
(d,s)	0	1	2	3	4	5	6	7	8	9	10	11	12	13	14	15	...
(1, 2)	1	1	1	1	1	1	1	1	1	1	1	1	1	1	1	1	0
(2, 2)	1	1	1	1	1	1	1	1	1	1	1	1	1	1	1	0	0
(3, 2)	1	1	1	1	1	1	1	1	1	1	1	1	1	1	0	0	0
(4, 2)	1	1	1	1	1	1	1	1	1	1	1	1	1	0	0	0	0
(5, 2)	1	1	1	1	1	1	1	1	1	1	1	1	0	0	0	0	0
(6, 2)	1	1	1	1	1	1	1	1	1	1	1	0	0	0	0	0	0
(7, 2)	1	1	1	1	1	1	1	1	1	1	0	0	0	0	0	0	0
(8, 2)	1	1	1	1	1	1	1	1	1	0	0	0	0	0	0	0	0
(9, 2)	1	1	1	1	1	1	1	0	0	0	0	0	0	0	0	0	0
(10, 2)	1	1	1	1	1	1	0	0	0	0	0	0	0	0	0	0	0
(11, 2)	1	1	1	1	0	0	0	0	0	0	0	0	0	0	0	0	0
(12, 2)	1	1	0	0	0	0	0	0	0	0	0	0	0	0	0	0	0
(13, 2)	0	0	0	0	0	0	0	0	0	0	0	0	0	0	0	0	0

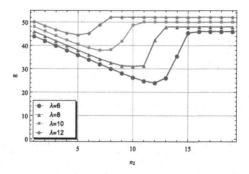

Fig. 2. Average cost g versus n_2 and λ

Example 2. The size n_2 of a reserve group of servers can also be optimized with a constraint $N = n_1 + n_2$. Figure 2 illustrates the average cost g as a function of a reserve group size n_2 for $N = 20$, system parameters (7) and varied arrival rate λ. The optimal pair (n_2^*, g^*) for $\lambda \in \{6, 8, 10, 12\}$ takes the following values:

$$(n_2^*, g^*) = (12, 24.0289), (10, 31.0692), (7, 37.9294), (5, 44.5595). \qquad (8)$$

As we can see from our experiments, as the server load $\rho = \frac{\lambda}{\mu}$ increases, the optimal size n_2^* of the reserve group decreases and the optimal number of servers $n_1^* = N - n_2^*$ in the main group increases accordingly.

The above example illustrates the fact that an optimal control policy although having structural properties is a dynamic one, i.e. it depends on specific states of the system, which makes the practical use of such a policy difficult. Thus, there is an obvious question of finding a suitable heuristic control policy given explicitly, but which proposes a quasi-optimal result for a given average cost function (5).

3 Heuristic Solution

According to proposed numerical results, the optimal thresholds depend on the number of busy servers, but we can observe that the values

$$q_A := q_A(d(x)) + d(x), \; x \in E_0, \text{ and } q_D := q_D(d(x)) + d(x), \; x \in E_2,$$

describing the number of customers in the system in state x, remain almost constant. This fact is used to derive a heuristic solution for the optimal threshold levels. We assume that the deactivation of the reserve group of servers must be performed when utilization of an additional server becomes more costly than servicing customers through the main group of servers. Therefore, the corresponding threshold q_D should coincide with the optimal size n_1^* of the main server group.

Proposition 1. *The optimal threshold q_D for deactivation of the reserve group of servers in states $x \in E_2$ can be set to*

$$q_D = n_1^* = \min\{N - 1, \max\{n : c_0(L(n - 1) - L(n)) - c_1 > 0\}\} \qquad (9)$$

where

$$L(n) = \qquad\qquad\qquad\qquad\qquad\qquad\qquad\qquad\qquad\qquad\qquad (10)$$
$$\frac{\rho\gamma + \lambda - (n(n - 1)\mu + \lambda(\rho - B_\infty(n) - 2(n - 1)))\log(1 - \frac{\rho}{n})}{\gamma(n - \rho)} B_\infty(n) + \rho,$$

and $B_\infty(n)$ is a mean busy period of the ordinary $M/M/n$ queueing system,

$$B_\infty(n) = \frac{\rho^n}{\rho^n + (n - 1)!(n - \rho)\sum_{i=0}^{n-1} \frac{\rho^i}{i!}}, \quad \rho = \frac{\lambda}{\mu}. \qquad (11)$$

Proof. Denote by $L(n)$ the mean number of customers in the retrial queueing system with n servers. The $n - 1$ servers have an advantage over the n servers with respect to the introduced cost structure if increase in the average holding cost $c_1(L(n-1) - L(n))$ will be higher than the usage cost c_2 of a reserve server. The mean value $L(n)$ can be represented as $L(n) = Q(n) + C$, i.e. as a sum of the mean number of customers in the orbit $Q(n)$ and the mean number of busy servers C which is equal to ρ and independent of the number of servers n. According to [3,6], the mean value $Q(n)$ satisfies the following asymptotic relation,

$$Q(n) = (1 + Z(n))Q_\infty(n), \qquad (12)$$

where $Z(n)$ is given by expression

$$Z(n) = \frac{\lambda - (n(n - 1)\mu + \lambda(\rho - B_\infty(n) - 2(n - 1)))\log(1 - \frac{\rho}{n})}{\rho\gamma},$$

$B_\infty(n)$ defined as (11) stands for the mean busy period of the ordinary $M/M/n$ queueing system when $\gamma \to \infty$ and

$$Q_\infty(n) = \frac{\rho}{n - \rho} B_\infty(n)$$

is the standard counterpart of the mean number of the queueing customers. After simple algebra we get the relation (10).

Corollary 1. *In case of an ordinary queueing system with $\gamma \to \infty$ the mean number of customers due to $\lim_{\gamma \to \infty} Z(n) = 0$ is of the form,*

$$L_\infty(n) = \frac{\rho}{n - \rho} B_\infty(n) + \rho.$$

The mean value $L_\infty(n)$ can be alternatively approximated by the mean number of customers in the $M/M/1$ queue with arrival rate λ and service rate $n\mu$,

$$L_\infty(n) = \frac{\rho}{n - \rho}. \tag{13}$$

In this case the Eq. (9) has an explicit solution,

$$q_D = \min\left\{N - 1, \left\lfloor \rho + \frac{1}{2}\left(1 + \sqrt{1 + 4\rho\frac{c_0}{c_1}}\right)\right\rfloor\right\}.$$

Proposition 2. *The optimal threshold q_A for activation of the reserve group of servers in states $x \in E_0$ can be set to*

$$q_A = \max\left\{n_1 - 1, \left\lfloor \frac{1}{2} + \frac{1}{\tau T(n_1)} + \frac{T(N)n_2 c_1}{(T(n_1) - T(N))c_0} + \right.\right. \tag{14}$$

$$\left.\left.\sqrt{\frac{2n_2 c_1}{\tau(T(n_1) - T(N))c_0} + \left(\frac{1}{2} - \frac{T(N)n_2 c_1}{(T(n_1) - T(N))c_0}\right)^2}\right\rfloor\right\},$$

where

$$T(n) = \frac{1 + Z(n)}{n\mu - \lambda}. \tag{15}$$

Proof. The proof is based on a comparison of the total average costs incurred by serving the q_A customers without and with the usage of the reserve group of servers. When n_1 servers are operational in a ordinary queueing system, then we could assume the effective rate for the queue decreasing at rate $\frac{1}{n_1\mu - \lambda}$. For the mean sojourn time of a customer in a retrial queueing system due to the Little's formula and relation (12) we have $T(n_1) = (1 + Z(n_1))T_\infty(n_1)$. The mean sojourn time $T_\infty(n_1)$ can be approximated using the relation (13), i.e.

$$T(n_1) = \frac{(1 + Z(n_1))L_\infty(n_1)}{\lambda} = \frac{1 + Z(n_1)}{n_1\mu - \lambda}. \tag{16}$$

Therefore, the overall average cost required to handle q_A customers using solely the main group of n_1 servers consists only of the holding cost part and hence it can be represented as

$$c_0 T(n_1) \sum_{i=0}^{q_A}(q_A - i) + c_1 n_1 q_A T(n_1) = c_0 T(n_1)\frac{q_A(q_A + 1)}{2}. \tag{17}$$

Denote by $l = \left\lfloor \frac{1}{\tau T(n_1)} \right\rfloor$ the mean number of customers served during a set-up time. When the number of customers is q_A and the decision maker chooses a control action to use a reserve group of servers then the average holding cost is of the form

$$c_0 T(n_1) \sum_{i=0}^{l}(q_A - i) + c_0 T(N) \sum_{i=l+1}^{q_A}(q_A - i) = \tag{18}$$

$$c_0 T(n_1)\frac{(l + 1)(2q_A - l)}{2} + c_0 T(N)\frac{(q_A - l)(q_A - l - 1)}{2},$$

where the first term specifies an average holding cost of l customers served during a set-up time when only n_1 servers are available and the second term is an average holding cost for the remaining $q_A - l - 1$ customers in a system with N available servers. The usage of fewer server system can be more profitable than the system with a reserve group of servers when the difference of (17) and (18) in the average holding cost must be higher than the average usage cost of the reserve n_2 servers. Thus q_A must be the smallest integer value that satisfies the inequality

$$c_0(T(n_1) - T(N))\frac{(q_A - l)(q_A - l - 1)}{2} > c_1 n_2\left(\frac{1}{\tau} + T(N)(q_A - l - 1)\right). \quad (19)$$

The term $T(N)(q_A - l - 1)$ specifies the average time required to serve $q_A - l - 1$ customers remaining in the system after the set-up time. Solving the last inequality and taking into account that q_A must be larger as $n_1 - 1$ we get (14).

Remark 1. Note that formula (14) can be easily transformed for the special cases of an ordinary queueing system without retrials when $\gamma \to \infty$, and/or there is no set-up time for the activation of the reserves when $\tau \to \infty$.

4   Numerical Analysis

In this section we verify explicitly derived heuristic results for the optimal control policy. We consider the system with the same parameter values as given in Example 1. Numerical solution of the inequality (9) is given by $n_1^* = 8, 10, 13, 15$ respectively for the arrival rates $\lambda = 6, 8, 10, 12$, which coincides with those results (8) obtained by a policy-iteration algorithm. Numerical results for thresholds of the optimal policy (OP) and heuristic policy (HS) are reported in Table 3 for the activation and in Table 4 for the deactivation of the reserves. As we can see, specific values of the optimal thresholds and the resulting heuristic thresholds are although different but nevertheless are relative close. But as further experiments show, this difference does not affect the value of the cost functional, so the heuristic policy can be assumed to be quasi-optimal.

Table 3. Threshold levels $q_A(d(x)), x \in E_0$

f_0	$q_A(0)$	$q_A(1)$	$q_A(2)$	$q_A(3)$	$q_A(4)$	$q_A(5)$	$q_A(6)$	$q_A(7)$	$q_A(8)$	$q_A(9)$	$q_A(10)$
OP	54	53	52	52	51	50	49	48	47	46	45
HP	57	56	55	54	53	52	51	50	49	48	47

Table 4. Threshold levels $q_D(d(x)), x \in E_2$

f_1	$q_D(1)$	$q_D(2)$	$q_D(3)$	$q_D(4)$	$q_D(5)$	$q_D(6)$	$q_D(7)$	$q_D(8)$	$q_D(9)$	$q_D(10)$	$q_D(11)$	$q_D(12)$	$q_D(13)$
OP	15	14	13	12	11	10	9	8	6	5	3	1	0
HP	11	10	9	8	7	6	5	4	3	2	1	0	0

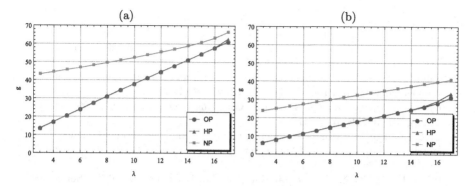

Fig. 3. Average cost g versus λ, policy f, $c_0 = 1, c_1 = 2$ (a) and $c_0 = 1, c_1 = 0.5$ (b)

We keep now the values $(\mu, \gamma, \tau) = (1, 2, 0.1)$ fixed as in (7). Now we compare the systems operating under the optimal policy (OP) evaluated by solving the optimality Eqs. (6), heuristic policy (HP) defined by thresholds (9), (14) and the system with a null policy (NP) $n_2 = 0$, i.e. the system has no reserves and the average cost is of the form $g = c_0 L(N) + c_1 N$, where $L(N)$ is the mean number of customers in the system with N servers (10).

The results of a comparison analysis of the functions g by varying λ for different control policies are summarized in Fig. 3(a, b), where (a) illustrates the curves for the costs $(c_0, c_1) = (1, 2)$ and (b) – for the costs $(c_0, c_1) = (1, 0.5)$. The graphs show that investigation of the model with controllable reserve group of servers for a given cost structure is justified, as we see that this model significantly outperforms its non-controllable counterpart on the average cost criterion. Further, we can note the fact that the heuristic relations we obtained for the optimal hysteretic policy in form of two thresholds can indeed be considered as a quasi-optimal one, since the corresponding average cost functions are indistinguishable in the graphs for both of cases (a) and (b). The small deviations of the g function values for large arrival rate are due to the impact of the final buffer required for the numerical calculations. Finally, while reducing the usage cost of a server, it reduces both the average cost and the difference between the controllable and non-controllable models.

References

1. Artalejo, J.R.: A classified bibliography of research on retrial queues: progress in 1990–1999. TOP **7**, 187–211 (1999). https://doi.org/10.1007/BF02564721
2. Artalejo, J.R.: Accessible bibliography on retrial queues. Math. Comput. Model. **30**, 1–6 (1999)
3. Artalejo, J.R., Gomez-Corral, A.: Retrial Queueing Systems: A Computational Approach. Springer, Berlin (2008). https://doi.org/10.1007/978-3-540-78725-9
4. Choi, B.D., Chang, Y.: Single server retrial queues with priority calls. Math. Comput. Model. **30**, 7–32 (1999)

5. Efrosinin, D., Gudkova, I., Stepanova, N.: Performance analysis and optimal control for queueing system with a reserve unreliable server pool. In: Dudin, A., Nazarov, A., Moiseev, A. (eds.) ITMM 2019. CCIS, vol. 1109, pp. 109–120. Springer, Cham (2019). https://doi.org/10.1007/978-3-030-33388-1_10
6. Falin, G.: Multichannel queueing systems with repeated calls under high intensity of repetition. J. Inf. Process. Cybern. **23**, 37–47 (1987)
7. Falin, G.: A survey of retrial queues. Queueing Syst. **7**, 127–167 (1990). https://doi.org/10.1007/BF01158472
8. Gomez-Corral, A.: A bibliographical guide to the analysis of retrial queues through matrix analytic techniques. Ann. Oper. Res. **141**, 163–191 (2006). https://doi.org/10.1007/s10479-006-5298-4
9. Howard, R.A.: Dynamic Programming and Markov Processes. Wiley Series (1960)
10. Hipp, S.K., Holzbaur, U.D.: Decision processes with monotone hysteretic policies. Oper. Res. **36**(4), 585–588 (1988)
11. Kim, C., Dudin, A., Dudin, S., Dudina, O.: Hysteresis control by the number of active servers in queueing system MMAP/PH/N with priority service. Perform. Eval. **101**, 20–33 (2016)
12. Mitrani, I.: Managing performance and power consumption in a server farm. Ann. Oper. Res. **202**, 121–134 (2013). https://doi.org/10.1007/s10479-011-0932-1
13. Le Ny, L.M., Tuffin, B.: A simple analysis of heterogeneous multi-server threshold queues with hysteresis. Institut National de Recherche en Informatique, France (2000)
14. Nobel, R.: Hysteretic and heuristic control of queueing systems. Ph.D. dissertation, Vrije University (1998)
15. Nobel, R., Tijms, H.C.: Optimal control of a queueing system with heterogeneous servers and set-up costs. IEEE Trans. Autom. Control **45**(4), 780–784 (2000)
16. Puterman, M.L.: Markov Decision Processes: Discrete Stochastic Dynamic Programming. Wiley, Hoboken (2005)
17. Sennott, L.: Stochastic Dynamic Programming and Control of Queueing Systems. Wiley, New York (1999)
18. Yang, T., Templeton, J.G.C.: A survey on retrial queues. Queueing Syst. **2**, 201–233 (1987). https://doi.org/10.1007/BF01158899

Verification of Stability Condition in Unreliable Two-Class Retrial System with Constant Retrial Rates

Ruslana Nekrasova[1,2]([✉])[ID], Evsey Morozov[1,2,3][ID], and Dmitry Efrosinin[4,5][ID]

[1] IAMR Karelian Research Centre RAS, Petrozavodsk, Russia
ruslana.nekrasova@mail.ru, emorozov@karelia.ru
[2] Petrozavodsk State University, Petrozavodsk, Russia
[3] Moscow Center for Fundamental and Applied Mathematics, Moscow State University, Moscow 119991, Russia
[4] Johannes Kepler University, Linz, Austria
[5] Peoples Friendship University of Russia, Moscow, Russia

Abstract. A two-class single-server retrial system with Poisson inputs is considered. In this system, unlike conventional retrial systems, each new ith class customer joins the 'end' of a virtual ith class orbit, and the 'oldest' customer from each orbit is only allowed to make an attempt to occupy server after a class-dependent exponential retrial time. Moreover, the server is assumed to be not reliable, and a customer whose service is interrupted joins the 'top' of class-i orbit queue. Thus FIFO discipline is applied in both orbits. Using regenerative methodology and Markov Chain approach we derive stability conditions of this system relying on analysis for less-complicated model with reliable server. Obtained conditions are verified by simulation. Additionally, we analyze a controllable variant of the main model operating under a $c\mu$-rule. For that case the system becomes less stable comparing to the non-controllable counterpart.

Keywords: Retrial system · Unreliable server · Regenerative stability analysis · Markov Chain approach · $c\mu$-rule

1 Introduction

It is well known that the models of the queueing theory are widely used to analyze the operational properties of telecommunication and computer service

The research of RN was prepared with the support of Russian Science Foundation according to the research project No. 21-71-10135 https://rscf.ru/en/project/21-71-10135/. The research of EM was published with the financial support of the Ministry of Education and Science of the Russian Federation as part of the program of the Moscow Center for Fundamental and Applied Mathematics under the agreement № 075-15-2022-284.

systems, and also help in solving various optimization problems. One of the most important element of such analysis is the study of the steady-state behaviour of the system, when the performance and reliability metrics as well as the optimal control policies in controllable models are independent of time. However in order to obtain results it is necessary to guarantee the existence of a stationary mode, i.e. to obtain some ergodicity or stability condition for the corresponding queueing systems. A large variety of queues with derived ergodic conditions can be found in numerous papers and monographs, e.g. in [1–4] and references cited therein.

Although many papers have been devoted to the stability of queues, there are still gaps for certain classes of systems which are characterized by different additional features like reliability attributes of a server, retrial phenomenon and different classes of customers. The server is unreliable when it is subject to breakdowns and repairs. The retrial effect occurs in systems where a primary customer is blocked when the server is busy. In this case the customer joins the queue of retrial customers where it repeats the attempt to occupy the server either independently of other orbiting customers due to the classical retrial discipline or only if it is located at the head of the retrial queue with respect to the constant retrial discipline. Different classes of customers arise in situations where a certain group of customers is willing to incur additional costs to reduce waiting times and improve service quality. The existence of two classes of customers is the motivation for considering the optimal scheduling problem, which cannot be solved without an accurate understanding of the stability conditions of a given system. For example, there are still relatively few works on ergodicity condition of single-server systems combined with double retrial queues for different classes of customers. The stability of the system consisting of ordinary and executive orbit with classical retrial discipline was investigated in [5]. In [6], the authors proposed ergodicity condition for the retrial queueing system with two orbits operating under a constant retrial discipline under specified symmetric constraints when arrival, retrial and service rates are equal for different classes of customers. The stability analysis under the same retrial discipline but with asymmetric assumptions on the rates was provided in [7]. Stability conditions for some other double orbit retrial queues with classic retrial discipline with different modifications were established in [8,9], where the service process was accompanied with interruptions.

The model discussed in this article differs from those previously studied. The queueing system combines not only the unreliable server, two classes of customers and the repeated customers with constant retrial rate, but also the condition that all customers without exception are served in order of arrival time, i.e. according to the FIFO rule. In other words, the primary arrival has not an access to the server independently of his state and becomes immediately blocked upon arrival. This customer must be sent to the end of the retrial queue where he waits until becomes in the head of the queue, and then he attempts to occupy the server. The results presented in this paper can be applied to both symmetric and asymmetric systems in which the arrival, service, retrial and failure rates are unequal for different classes of customers.

Moreover, in the paper we discuss the fact that an equivalent controllable system, in which the controller can prioritise customers of a certain class according to some control policy in order to minimise the mean cost function, has a smaller stability region than a non-controllable system for which a stability condition is obtained. This fact is illustrated by the example of a system with a control policy $c\mu$ which is independent of the arrival flow, where the queue with the highest product of the holding cost and the overall service rate is emptied in the priority order.

The remainder of the paper is organized as follows. Section 2 provides the formal queueing model description. Section 3 presents stability analysis of the system with a derivation of the stability condition. Section 4 presents simulation of the queueing system with some results on queue dynamics in stable and non-stable regimes. Section 5 examines a controllable variant of the main model operating under the $c\mu$-rule in terms of its stability. Finally, some concluding remarks are provided in Sect. 6.

2   Description of the Model

We consider a two-class single-server retrial system with constant retrial rate in which the arrivals of class i (class-i) follow a Poisson input with rate λ_i, $i = 1, 2$. We denote the system by Σ, $\lambda = \lambda_1 + \lambda_2$ denotes the total input rate. Thus class-i arrival joins the system with a probability $p_i = \lambda_i/\lambda$, $i = 1, 2$. The model Σ has the following important property. Each new class-i customer joins the 'end' of a virtual class-i orbit (even if the server is idle upon his arrival). The customers in orbit i form a FIFO-type virtual queue. In other words, only the top ('oldest') customer in orbit i makes attempts to occupy server after exponential retrial time ξ_i. We denote the rate of class-i orbit by $\theta_i = 1/\mathsf{E}\xi_i$. Thus the retrial rate of the orbital customers stays constant independently on orbit size (the number of customers in orbit), and the model belongs to a class of the retrial queuing system with a *constant retrial rate*. The orbital customer makes the attempts until he finds the server idle and ready for service. It is assumed that the server is unreliable. Namely, during service of a class-i customer a failure of the server happens after exponential time A_i with rate $\alpha_i = 1/\mathsf{E}A_i$. (If $\alpha_1 = \alpha_2 = 0$ then the server is reliable.) After a failure happens, the server is ready to work again after exponential repair time B with rate $\beta = 1/\mathsf{E}B$ which is assumed to be independent of the class number. If the service of a customer is interrupted by a failure then the customer returns to the 'top' of the corresponding virtual orbit-queue, and thus the orbits follow FIFO service discipline. The service times are assumed to be exponential and class-dependent with the service rate μ_i, $i = 1, 2$.

3   Stability Analysis

Our approach to stability analysis of this system is based on the technique developed in [10] for stability analysis of a two-dimensional Markov Chain (MC).

We outline the approach. Let $Y^{(i)}(t)$ be the number of the customers in orbit i at instant t^-. Consider the sequence $\{D_n, n \geq 1\}$ of instants, when the server become idle and ready for the service, either after the competition of previous service or because the repair is finished. Introduce the embedded (discrete-time) orbit size process,

$$Y_n^{(i)} := Y^{(i)}(D_n), \quad i = 1, 2, n \geq 1,$$

which describes the state of orbit i just after the instant D_n. Because the governing distributions are assumed to be exponential, then it is easy to see that the process

$$\mathbf{Y} = \{Y_n^{(1)}, Y_n^{(2)}\}, \quad n \geq 1,$$

defines a two-dimensional MC. The ergodicity of the MC $\mathbf{Y}$ means the stability of the system under consideration. In the book [10] the ergodicity criterion of a two-dimensional MC in an explicit form have been obtained. More precisely, this criterion is expressed by the following set of the *negative drift conditions* written via the mean increments of the marginals of the MC $\mathbf{Y}$:

$$\mathsf{M}_i^{11} = \mathsf{E}\big[Y_{n+1}^{(i)} - Y_n^{(i)}\big|Y_n^{(1)} > 0, \, Y_n^{(2)} > 0\big], \, i = 1, 2, \tag{1}$$

$$\mathsf{M}_i^{01} = \mathsf{E}\big[Y_{n+1}^{(i)} - Y_n^{(i)}\big|Y_n^{(1)} = 0, \, Y_n^{(2)} > 0\big], \, i = 1, 2, \tag{2}$$

$$\mathsf{M}_i^{10} = \mathsf{E}\big[Y_{n+1}^{(i)} - Y_n^{(i)}\big|Y_n^{(1)} > 0, \, Y_n^{(2)} = 0\big], \, i = 1, 2. \tag{3}$$

In the recent paper [7] this approach has applied to establish stability criterion of a conventional two-class retrial system with constant retrial rate and *reliable* server, where the incoming customers follow Poisson inputs and receive the service immediately if meet server idle. Based on [10], the authors of [7] show that the stability (ergodicity) of the MC describing the orbit-queue process is reduced to the following inequalities:

$$\mathsf{M}_1^{11}\mathsf{M}_2^{10} - \mathsf{M}_2^{11}\mathsf{M}_1^{10} < 0, \tag{4}$$

$$\mathsf{M}_2^{11}\mathsf{M}_1^{01} - \mathsf{M}_1^{11}\mathsf{M}_2^{01} < 0. \tag{5}$$

In this work we develop this approach for the (more complicated) system Σ with exponential governing distributions. In particular (after tedious and routine computation) the following explicit expressions for the mean increments in the model Σ are obtained:

$$\mathsf{M}_1^{11} = \frac{\lambda_1}{\theta_1 + \theta_2}\Big(1 + \frac{\theta_1}{\alpha_1 + \mu_1}\big(1 + \frac{\alpha_1}{\beta}\big) + \frac{\theta_2}{\alpha_2 + \mu_2}\big(1 + \frac{\alpha_2}{\beta}\big) - \frac{\theta_1\mu_1}{\lambda_1(\alpha_1 + \mu_1)}\Big),$$

$$\mathsf{M}_2^{11} = \frac{\lambda_2}{\theta_1 + \theta_2}\Big(1 + \frac{\theta_1}{\alpha_1 + \mu_1}\big(1 + \frac{\alpha_1}{\beta}\big) + \frac{\theta_2}{\alpha_2 + \mu_2}\big(1 + \frac{\alpha_2}{\beta}\big) - \frac{\theta_2\mu_2}{\lambda_2(\alpha_2 + \mu_2)}\Big),$$

$$\mathsf{M}_1^{01} = \frac{\lambda_1}{(\lambda_1 + \theta_2)(\theta_1 + \theta_2)}\Big(\frac{\lambda_1\theta_1}{\mu_1 + \alpha_1}\big(1 + \frac{\alpha_1}{\beta}\big) + \frac{(\lambda_1 + \theta_1 + \theta_2)\theta_2}{\mu_2 + \alpha_2}\big(1 + \frac{\alpha_2}{\beta}\big)$$

$$+ \frac{\theta_1\alpha_1}{\mu_1 + \alpha_1} + (\lambda_1 + \theta_2) + \frac{\theta_1\theta_2}{\lambda_1 + \theta_2}\Big),$$

$$M_2^{01} = \frac{\lambda_2}{(\lambda_1 + \theta_2)(\theta_1 + \theta_2)}\left(\frac{\lambda_1\theta_1}{\mu_1 + \alpha_1}\left(1 + \frac{\alpha_1}{\beta}\right) + \frac{(\lambda_1 + \theta_1 + \theta_2)\theta_2}{\mu_2 + \alpha_2}\left(1 + \frac{\alpha_2}{\beta}\right)\right.$$
$$\left. + (\lambda_1 + \theta_1 + \theta_2) - \frac{(\lambda_1 + \theta_1 + \theta_2)\theta_2\mu_2}{\lambda_2(\mu_2 + \alpha_2)}\right),$$

$$M_1^{10} = \frac{\lambda_1}{(\lambda_2 + \theta_1)(\theta_1 + \theta_2)}\left(\frac{(\lambda_1 + \theta_1 + \theta_2)\theta_1}{\mu_1 + \alpha_1}\left(1 + \frac{\alpha_1}{\beta}\right) + \frac{\lambda_2\theta_2}{\mu_2 + \alpha_2}\left(1 + \frac{\alpha_2}{\beta}\right)\right.$$
$$\left. + (\lambda_2 + \theta_1 + \theta_2) - \frac{(\lambda_2 + \theta_1 + \theta_2)\theta_1\mu_1}{\lambda_1(\mu_1 + \alpha_1)}\right),$$

$$M_2^{10} = \frac{\lambda_2}{(\lambda_2 + \theta_1)(\theta_1 + \theta_2)}\left(\frac{(\lambda_2 + \theta_1 + \theta_2)\theta_1}{\mu_1 + \alpha_1}\left(1 + \frac{\alpha_1}{\beta}\right) + \frac{\lambda_2\theta_2}{\mu_2 + \alpha_2}\left(1 + \frac{\alpha_2}{\beta}\right)\right.$$
$$\left. + \frac{\theta_2\alpha_2}{\mu_2 + \alpha_2} + (\lambda_2 + \theta_1) + \frac{\theta_1\theta_2}{\lambda_2 + \theta_1}\right). \tag{6}$$

Denote
$$\rho_1 = \lambda_1/\mu_1, \quad \rho_2 = \lambda_2/\mu_2,$$
and assume that the following condition holds:
$$\rho_1\left(1 + \alpha_1/\beta\right) + \rho_2\left(1 + \alpha_2/\beta\right) < 1. \tag{7}$$

In this case, the inequalities (4), (5), rewritten (after a simple but tedious and routine algebra) with the use of the explicit expressions (6) and *replaced by the equations*, have the following positive solution (θ_1, θ_2):

$$\theta_1^* = \frac{\rho_1(\alpha_1 + \mu_1)}{1 - \left(\rho_1\left(1 + \alpha_1/\beta\right) + \rho_2\left(1 + \alpha_2/\beta\right)\right)} > 0, \tag{8}$$

$$\theta_2^* = \frac{\beta\rho_2(\alpha_2 + \mu_2)}{1 - \left(\rho_1\left(1 + \alpha_1/\beta\right) + \rho_2\left(1 + \alpha_2/\beta\right)\right)} > 0. \tag{9}$$

Thus, provided assumption (7) holds, inequalities (4), (5) have a non-empty solution set of the parameters θ_1, θ_2, which represents the stability zone of the system.

Now we illustrate this theoretical finding by the numerical analysis of the stability zone of the model for a particular case of the symmetric classes with the following parameters

$$\lambda_1 = \lambda_2 = 1, \ \mu_1 = \mu_2 = 3, \ \alpha_1 = \alpha_2 = 2, \ \beta = 5, \tag{10}$$

in which case $\theta_1^* = \theta_2^* = 25 > 0$. Moreover in this case inequalities (4), (5) take the following form:

$$\theta_2(-3\theta_1^2 + 107\theta_1 + 75) + \theta_1(-3\theta_1^2 + 32\theta_1 + 220) < 0, \tag{11}$$
$$\theta_1(-3\theta_2^2 + 107\theta_2 + 75) + \theta_2(-3\theta_2^2 + 32\theta_2 + 220) < 0. \tag{12}$$

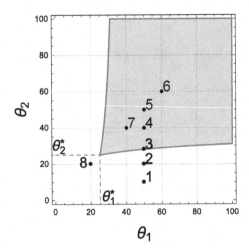

Fig. 1. Stability zone for retrial rates in symmetrical class model. (Color figure online)

The stability zone, that is the set of parameters θ_1, θ_2 satisfying (11), (12) is presented on Fig. 1 (colored by blue). Note that other parameters were chosen according to (7).

Next our goal in this research is to verify the obtained theoretical results by simulation. To do this, we analyze the mean dynamics of both orbits sizes for the different values of the retrial rates θ_1, θ_2.

4   Simulation

We present simulation results for the symmetric class model discussed in previous section. We varied the values of retrial rates and estimated the mean orbit dynamics

$$X_m^{(i)}(j) := \frac{1}{m} \sum_{k=1}^{m} \hat{Y}_k^{(i)}(j), \quad i = 1, 2, j = 1, \ldots, n, \tag{13}$$

for both classes with $m = 100$ independent replications (see Table 1). Here $\hat{Y}_k^{(i)}(j)$ is the number of customers in orbit i in the k-th replication just before the j-th arrival. Each trajectory includes $n = 10\,000$ arrivals. All experiments are implemented in *RStudio development environment*. Particular values of the retrial rates according to Table 1 are illustrated on Fig. 1.

Table 1. Retrial rates for symmetric class model.

Case	1	2	3	4	5	6	7	8
θ_1	50.0	50.0	50.0	50.0	50.0	60.0	40.0	20.0
θ_2	10.0	20.0	28.1	40.0	50.0	60.0	40.0	20.0

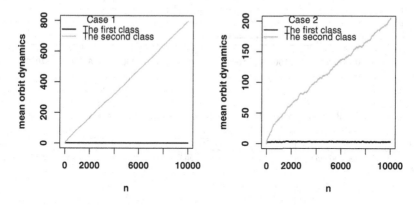

Fig. 2. Mean orbit dynamics in 'non-stable' cases 1 and 2.

Figure 2 and 3 present simulation results when the inequalities (11), (12) are violated. In case 3 the orbit rates belong to the border of the obtained stability zone and does not satisfy the system (11), (12), as both conditions are strict.

Simulation results show that in cases 1–3 the mean orbit corresponding to class-2 increases, while the mean orbit corresponding to class-1 stays bounded. Note that this effect is called *partial stability* [7]. In case 8 both orbits increase.

Figure 4 and Fig. 5 illustrate the orbits behavior when inequalities (11), (12) hold true. Note that, for the greater retrial rate, the mean orbit becomes expectedly smaller. In cases 4–6 both orbits are bounded, which indicates the stability.

The orbits behavior in case 7 (Fig. 5, the left plot) is not visually quite clear (in this case the system starts in zero initial state). To have more definite conclusion, we consider non-zero initial state and set initial values to be

$$X_m^{(1)}(1) := 100, \ X_m^{(2)}(1) := 100.$$

The results are given on Fig. 5 (the right plot) and show that the model is indeed stable.

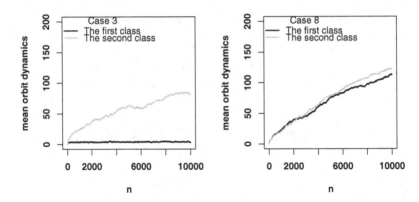

Fig. 3. Mean orbit dynamics in 'non-stable' cases 3 and 8.

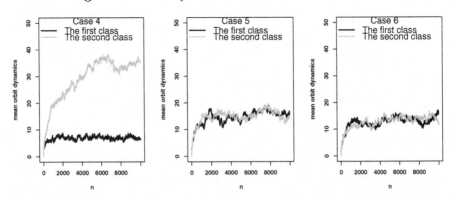

Fig. 4. Mean orbit dynamics in 'stable' cases 4–6.

Summing up, the simulation results confirm that the inequalities (5), (7), (4) give the stability criterion of the system Σ.

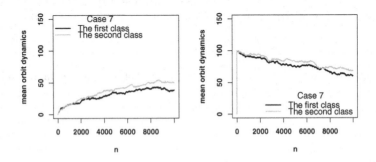

Fig. 5. Mean orbit dynamics in 'stable' case 7: zero and non-zero initial conditions.

5   Comparison of Stability Regions of Controllable and Non-controllable Systems

In this section, we discuss in brief the stability region of a controllable system and compare it with the found stability region of the corresponding original (non-controllable) system considered in Sect. 3.

The description of the main queueing system assumes that a retrial customer of an arbitrary class is accepted for service whenever the server is idle when it repeats the attempt to occupy it. It is clear that if a cost is assigned for the holding time of each class customer in the system, then the policy of accepting of any class customer will not be optimal in terms of minimising the average cost per unit of time.

Consider for example a controllable model operating according to a $c\mu$-rule, which is known to be optimal for a large class of systems with parallel queues and a single server. (The interested readers are referred e.g. to [11] and [12] for detailed description of this policy and its optimality). According to this rule, for our system the ith class customers are served with priority regardless of the arrival intensity if the factor $c_i/\tilde{\mu}_i$ takes the highest value, where c_i is a holding cost and $1/\tilde{\mu}_i$ is the total service intensity of the ith class customer. To explain, we denote by $\tilde{S}_i$ the total *holding* time which a *labeled customer* spends in the server until he leaves the system. The holding time starts when the labeled customer (being the top in the orbit queue) attempts to occupy the server and ends when the customer departs the system after service. Then $\tilde{\mu}_i = \mathsf{E}\tilde{S}_i$.

However, using a control (in particular allowing a priority), we change the dynamics of the system (e.g., the orbit-queue size) which in turn can lead to an increase of customers in the non-priority queue and as a result to the instability of the whole system. In this case, it is possible that the original (non-controllable) system will remain stable under certain parameter values, while the controllable counterpart leaves the stability region. We will now illustrate this possibility with a specific controllable model. To calculate the total service intensity in the $c\mu$-rule, we *assume* that this rule independent of the arrival flows. It is worth mentioning that this heuristic assumption nevertheless is in a remarkable agreement with the corresponding simulation results given below. Thus we consider the ith queue whose dynamics is independent of the dynamics of another queue, $i = 1, 2$. Let us introduce an auxiliary absorbing MC with transitive states $\{0, 1, 2\}$, where state 0 means that the server is idle and the labeled customer is at the head of the orbit; 1 means that the labeled customer is at the server and 2 means that the server is in a failure state and the labeled customer is at the head of the orbit as well. The absorption occurs by service completion of the customer in state 1. The mean time $\tilde{\mu}_i$ to absorption given the initial state is 0 needs to be calculated. Denote by $\tilde{\mu}_{i,j}, j \in \{0, 1, 2\}$, the conditional mean time to absorption given the initial state is j, i.e. $\tilde{\mu}_i := \tilde{\mu}_{i,0}$. According to the Markov property for the MC with an absorption we obtain a system of equations for the variables $\tilde{\mu}_{i,j}, i = 1, 2; j = 0, 1, 2$:

$$\tilde{\mu}_{i,0}\theta_i = 1 + \theta_i\tilde{\mu}_{i,1},$$
$$\tilde{\mu}_{i,1}(\alpha_i + \mu_i) = 1 + \alpha_i\tilde{\mu}_{i,2},$$
$$\tilde{\mu}_{i,2}\beta = 1 + \beta\tilde{\mu}_{i,0}.$$

After a simple algebra, we obtain the solution in the following form

$$\tilde{\mu}_i = \tilde{\mu}_{i,0} = \frac{\alpha_i + \mu_i}{\mu_i\theta_i}\left(\frac{\theta_i}{\beta}\frac{\alpha_i + \beta}{\alpha_i + \mu_i} + 1\right), \quad i = 1, 2. \tag{14}$$

Now the $c\mu$-rule prescribes the priority service of queue 1 (2) if

$$\frac{c_1}{\tilde{\mu}_1} \geq \frac{c_2}{\tilde{\mu}_2} \quad \left(\frac{c_1}{\tilde{\mu}_1} \leq \frac{c_2}{\tilde{\mu}_2}\right).$$

As a heuristic stability condition for the controllable model we use the inequality

$$\lambda_1\tilde{\mu}_1 + \lambda_2\tilde{\mu}_2 < 1, \tag{15}$$

which is a well-known stability criteria of a FIFO $M/G/1$ queueing system with two classes of customers, see e.g., Theorem 2 and Remark 4 in [13]. (This stability condition is in general heuristic since we assumed above that the queue-orbits are independent).

In Fig. 6 we combine the stability regions (11), (12) and (15) respectively of the basic (non-controllable) model and the controllable model operating under the $c\mu$-rule. The marked points (1–8) specify exactly the same eight special cases which are used for the values of the retrial rates of the basic system (see Fig. 1 and Table 1).

Using the same parameters that are used in simulation in Sect. 4, we now simulate the non-controllable and controllable systems when $c_1 = 1$ and $c_2 = 15$ and calculate the (sample) mean performance measures, such as the mean number of customers in the system

$$\bar{N} = \bar{N}_1 + \bar{N}_2,$$

and the mean cost

$$g = c_1\bar{N}_1 + c_2\bar{N}_2,$$

where $\bar{N}_i$ is the mean number of the ith class customers in the system for the fixed values of system parameters (10), and then we vary the retrial rates. The experimental results are summarized in Table 2, where we denote with a subindex $c\mu$ the values for the controllable model. In all experiments, $c_1/\tilde{\mu}_1 < c_2/\tilde{\mu}_2$, and hence the second class has a priority. We see that the controllable model, as expected, is preferable with respect to the mean cost criterion. On the other hand, the mean number of the customers in the basic (non-controllable) system is much less than that in the controllable system. In Cases 5 and 6 both systems are evidently stable. The Cases 3, 4 and 7 correspond to stability of the basic system and instability of the controllable counterpart (confirming that the basic system has a 'wider' stability zone). Finally, in Cases 1, 2 and 8 the mean performance

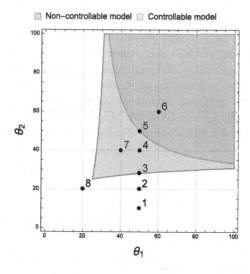

Fig. 6. Stability regions for non-controllable and controllable systems

Table 2. Mean performance values for the main and controllable system

Mean values	Cases							
	1	2	3	4	5	6	7	8
$\bar{N}$	391.75	131.58	46.68	36.01	55.92	25.58	31.60	112.05
$\bar{N}_{c\mu}$	712.55	283.75	127.57	103.83	71.01	43.82	156.72	462.32
g	5852.21	1936.38	651.44	458.61	334.34	195.09	233.76	827.41
$g_{c\mu}$	740.99	304.91	146.50	122.14	88.96	61.43	175.11	482.85

characteristics take big values in both systems, and it indicates the instability of both systems.

Thus, a controllable systems may be 'less' stable (with a smaller stability region) under the same load as in the corresponding non-controllable model, and this effect must be taken into account when designing the systems accompanied with a controller or decision maker.

6 Conclusion

We study the stability of a two-class retrial system with constant retrial rates and with unreliable server. A distinctive feature of this system is that each new customer, before to occupy server (if it is free) joins the corresponding FIFO orbit-queue. Based on results from [10] we derive the stability criterion of the two-dimensional MC describing the dynamics of this system. This result is illustrated by a numerical study and simulation a system with two equivalent classes.

Another aspect of the paper deals with an examination of the corresponding controllable system. The designers assume often that this system remains stable under the same (or almost the same) range of the parameters which guarantee stability of the corresponding non-controllable system. However, the obtained in this research results indicate that it is not quite true, and this fact must be taken into account to integrate a control component into the system.

In a future research we aim to obtain, in an explicit form, the stability conditions for more general configuration of the system and for general service time distributions.

References

1. Artalejo, J.R., Gomez-Corral, A.: Retrial Queueing Systems. Springer, Heidelberg (2008). https://doi.org/10.1007/978-3-540-78725-9
2. Bramson, M.: Stability of queueing networks. Probab. Surv. **5**, 165–345 (2008)
3. Cohen, J.W.: Analysis of Random Walks. IOS Press, Amsterdam (1992)
4. Gross, D., Shortle, J.F., Thompson, J.M., Harris, C.M.: Fundamentals of Queueing Theory, 4th edn. Wiley, Hoboken (2008)
5. Dimitriou, I.: A queueing model with two classes of retrial customers and paired services. Ann. Oper. Res. **238**, 123–143 (2016). https://doi.org/10.1007/s10479-015-2059-2
6. Dimitriou, I.: A two-class queueing system with constant retrial policy and general class dependent service times. Eur. J. Oper. Res. **270**, 1063–1073 (2018)
7. Avrachenkov, K., Morozov, E., Nekrasova, R.: Stability analysis of two-class retrial systems with constant retrial rates and general service times. ArXiv, abs/2110.09840 (2021)
8. Jain, M., Sanga, S.S.: Unreliable single server double orbit retrial queue with balking. Proc. Natl. Acad. Sci. India Sect. A Phys. Sci. **91**, 257–268 (2021). https://doi.org/10.1007/s40010-020-00725-6
9. Sanga, S.S., Jain, M.: FM/FM/1 double orbit retrial queue with customers' joining strategy: a parametric nonlinear programing approach. Appl. Math. Comput. **362**, 124542 (2019)
10. Fayolle, G., Malyshev, V., Menshikov, M.: Topics in the Constructive Theory of Countable Markov Chains, 1st edn. Cambridge University Press, Cambridge (1995)
11. Buyukkoc, C., Varaiya, P., Walrand, J.: The $c\mu$ rule revisited. Adv. Appl. Probab. **17**(1), 237–238 (1985)
12. Cox, D.R., Smith, W.L.: Queues. Methuen, London (1961)
13. Morozov, E., Rogozin, S., Nguyen, H.Q., Phung-Duc, T.: Modified Erlang loss system for cognitive wireless networks. Mathematics **10**, 2101 (2022). https://doi.org/10.3390/math10122101

The Queueing System
with Threshold-Based Direct and Inverse
General Renovation Mechanism

Viana C. C. Hilquias[1] , I. S. Zaryadov[1,2(✉)] , S. I. Matyushenko[1] ,
and T. A. Milovanova[1]

[1] Department of Applied Probability and Informatics, Peoples' Friendship University
of Russia (RUDN University), Miklukho-Maklaya str. 6, Moscow 117198, Russia
{1042195028,zaryadov-is,milovanova-ta}@rudn.ru
[2] Institute of Informatics Problems, FRC CSC RAS, IPI FRC CSC RAS,
44-2 Vavilova Str., Moscow 119333, Russia

Abstract. The paper considers a single-server queuing system with a
single-threshold mechanism for probabilistic dropping of applications
accepted into the system (general renovation). On the one hand, unlike
the previously considered systems with general renovation, this system
introduces a certain threshold value in the queue as a control parameter
of the renovation mechanism, which not only determines the moment
when the probabilistic dropping of applications accepted into the system
is enabled, but also sets a safe area in the queue from which applications accepted into the system cannot be reset. A general renovation
is a probabilistic reset of an arbitrary number of applications from the
queue outside the safe zone, which occurs at the end of the application
service on the device. For this system, the main probabilistic-time characteristics are obtained for two types of general renovation: the direct
general renovation—appllctions are dropped from the queue in the order
of arrival, and the inverse general renovation—appllctions are dropped
from the queue starting from the last one.

Keywords: Queue management · Renovation mechanism ·
Threshold · Time-probabilistic characteristic · GPSS

1 Introduction

It should be noted that all active queue management (AQM) algorithms
such as Random Early Detection (RED) [1–12], Blue and stochastic fair Blue

This paper has been supported by the RUDN University Strategic Academic Leadership
Program (recipient Zaryadov I.S.—mathematical model development, recipient Viana
C. C. Hilquias—simulation model development). Also the publication has been funded
by Russian Foundation for Basic Research (RFBR) according to the research project
No. 20-07-00804 (I. S. Zaryadov, T. A. Milovanova—numerical analysis based on the
obtained analytical results).

V. M. Vishnevskiy et al. (Eds.): DCCN 2022, LNCS 13766, pp. 309–323, 2022.
https://doi.org/10.1007/978-3-031-23207-7_24

(SFB) [13], Adaptive virtual queue algorithm [11,12], Explicit Congestion Notification (ECN) [2], or controlled delay (CoDel) [14,15]) are usually based on a rule that allow to drop incoming packets from the buffer (queue) based on some control parameters and thresholds if the level of buffer increases.

So the problem of congestion avoidance is still actual [16,17,19,22–25], and that is why new AQM algorithms are investigated [12,18,19,21–25]. The performance analysis of the effectiveness of the proposed new algorithms is mainly carried out using simulation modeling (for example, [21,26,27]), and not analytically. It is worth noting a small number of works (see, for example, [28–30,36,37,45]) in which the bridges between the available use-case results and analytic results are presented.

Unlike RED type AQMs [1–10] (when the possible packets dropping occurs at the moments of arrival and the control parameter or parameters somehow depend on the length of the queue) the idea for the the renovation-base AQM is inverse: the decision about a possible dropping is synchronized with the times of service completions (see [31–42,45]). Here we elaborate further on the mechanism of renovation and by analogy with most AQM algorithms, a control parameter was introduced into the renovation mechanism [39,42]—a certain threshold value in the queue, upon overcoming which the renovation mechanism is activated (and, depending on the model, this threshold value could also limit the area in the drive from which received applications were not dropped).

This work continues the study of systems with renovation and general renovation, first considered in [31–34]. In [39], it was proposed to use renovation and general renovation as active queue management mechanisms (the development of modern versions of which is still an urgent task [17,19]), and in [30,37] the comparison was made of the main characteristics for models with renovation and models with traditional active queue management algorithms (the classic random early detection (RED) [1]).

The key difference of the presented work from the previous ones [39,42–44] is that threshold control of the general renovation mechanism is introduced here. It is worth noting that the threshold models with renovation have already been considered by the authors, but for the case when a customer ending its service on the device (at the moment immediately before leaving the device and the system) can drop all customers from the buffer (full renovation or renovation) [42–44], and not a fixed number of applications with some given probability.

Structurally, the work consists of the following sections: Sect. 2 presents the main results for the queuing system under the first setting, in the Sect. 3 the stationary distribution of the number of applications in the system is presented, the Sect. 4 is devoted to probabilistic characteristics of the system, the Sect. 5 presents time characteristics for the direct general renovation, when applictions are dropped from the queue in the order of arrival, the Sect. 6 presents time characteristics for the inverse general renovation, when applictions are dropped from the queue starting from the last one, and the last Sect. 7 summarizes the results.

2 System Description

The idea of the general renovation mechanism is the following: at the moment of the end of its service, the packet on the server may either just leave the system with some non-zero probability p, or may drop with some renovation probability $q(i), i > 0$ exactly i applications from the queue (if there were more then i applications) or may empty the queue (if there were i or less applications)) (so, the queues with renovation are similar to queues with negative customers [46,47]) or disasters [48,49]), or to queues with unreliable servers [50–52]).

We will consider the $G|M|1|\infty$ queuing system with the general renovation mechanism, defined in [31,32], and with hysteretic control of renovation probability $q(i)$ where the threshold value Q_1 not only sets a certain boundary value in the queue, upon overcoming which the renovation mechanism is enabled (as in [42–44]), but also limits some area in the queue (buffer) from which received applications could not be dropped.

For this system, the main probabilistic-time characteristics are obtained for two types of general renovation: the direct general renovation—applictions are dropped from the queue in the order of arrival, and the inverse general renovation—applictions are dropped from the queue starting from the last one (Fig. 1).

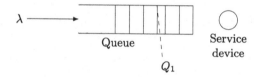

Fig. 1. Diagram of the system with threshold and renovation

The system receives a recurrent flow of applications ($A(x)$—the time distribution function between the receipt of applications), the service on the device is subject to the exponential law with the parameter μ.

The study (as in other works on this topic) will be carried out using the embedded Markov chain at the moment of receipt.

Transition probabilities $p_{i,j}, i \geq 0, j \geq 0$ (the probability that at the time of receipt the application will find j other applications in the system if the previous received application found i applications in the system) of the embedded Markov chain will be further either denoted, or calculate by using probabilities $A_{i-j+1}^{(k)}, i \geq 0, j \geq 0, k = 1, 2, 3, 4$—probabilities that between successive arrivals of applications to the system, exactly $i+j+1$ applications will leave the system (will be serviced and/or will be dropped).

$$p_{i,i+1} = A_0 = \int_0^\infty e^{-\mu x} dA(x) = \alpha(\mu). \tag{1}$$

For the system under consideration, the options are:

- **a)** $0 \leq i \leq Q_1$ (at the time of receipt of the application, the threshold value Q_1 has not been overcome).
- **b)** $i > Q_1$ (taking into account the received application, the threshold value Q_1 is overcome): **b.1)** $j > Q_1 + 1$ (by the time of a new application arrival, the threshold value will be overcome); **b.2)** $j = Q_1 + 1$ (by the time of a new application arrival, the threshold will be reached, but not overcome); **b.3)** $j = Q_1$; **b.4)** $0 \leq j < Q_1$ (by the time of a new application arrival, the threshold is not overcome).

We introduce auxiliary probabilities $\pi_k(l)$—the probability that k served applications will drop exactly l other applications from the queue ($l \geq 0, k \geq 0$).

$$\pi_1(l) = \begin{cases} q(l), & l > 1, \\ q(0) = p, & l = 0, \end{cases} \qquad \pi_k(l) = \sum_{i=0}^{l} \pi_1(i)\pi_{k-1}(l-i), \quad k > 1, l \geq 0. \quad (2)$$

and $\tilde{\pi}(l)$—the probability that k serviced applications will drop at least l applications from the queue outside the safe zone.

$$\tilde{\pi}_1(l) = \sum_{i=l}^{\infty} q(l), \qquad \tilde{\pi}_k(l) = \sum_{i=0}^{l-1} \pi_k(i)\tilde{\pi}_1(l-1), \quad k > 1, l \geq 0. \quad (3)$$

a) $0 \leq i \leq Q_1$—the renovation mechanism is not enabled, leaving the system is only possible by service.

$$p_{i,j} = A_{i+1-j}^{(1)} = \int_0^{\infty} \frac{(\mu x)^{i+1-j}}{(i+1-j)!} e^{-\mu x} dA(x), \qquad p_{i,0} = 1 - \sum_{j=1}^{i+1} A_{i+1-j}^{(1)}. \quad (4)$$

b) $i > Q_1$—the renovation mechanism is enabled at the current moment of arrival. Subcases:

b.1) $Q_1 + 1 < j < i+1$—the renovation mechanism will remain enabled by the time of a new application arrival.

$$p_{i,j} = A_{i+1-j}^{(2)} = \int_0^{\infty} \sum_{k=1}^{i+1-j} \pi_k(i+1-j-k) \frac{(\mu x)^k}{k!} e^{-\mu x} dA(x), \quad (5)$$

b.2) $j = Q_1 + 1$—by the time of a new request arrival, the renovation mechanism will be disabled.

$$p_{i,Q_1+1} = A_{i-Q_1}^{(2)} = \int_0^{\infty} \sum_{k=1}^{i-Q_1} \pi_k(i - Q_1 - k) \cdot \frac{(\mu x)^k}{k!} e^{-\mu x} dA(x). \quad (6)$$

b.3) $j = Q_1$—by the time of a new request arrival, the renovation mechanism will be disabled.

$$p_{i,Q_1} = A_{i+1-Q_1}^{(3)} = \int_0^\infty \sum_{k=1}^{i-Q_1} \pi_k(i - Q_1 - k) \cdot \frac{(\mu x)^{k+1}}{(k+1)!} e^{-\mu x} dA(x)$$

$$+ \int_0^\infty \sum_{k=1}^{i-Q_1} \widetilde{\pi}_k(i - Q_1 - k) \cdot \frac{(\mu x)^k}{k!} dA(x), j \neq i+1.$$

(7)

b.4) $0 < j < Q_1$—by the time of a new request arrival, the renovation mechanism will be disabled.

$$p_{i,j} = A_{i+1-j}^{(4)} = \int_0^\infty \int_0^x A_{i-1-Q_1}^{(3)}(y) dy A_{Q_1-j}^{(1)}(x - y) dA(x),$$

(8)

where

$$A_{i-1-Q_1}^{(3)}(y) = \sum_{k=1}^{i-Q_1} \pi_k(i - Q_1 - k) \cdot \frac{(\mu y)^{k-1}}{(k-1)!} e^{-\mu y} + \sum_{k=1}^{i-Q-1} \widetilde{\pi}_k(i - Q_1 - k) \cdot \frac{(\mu y)^k}{k!} e^{-\mu y},$$

$$A_{Q_1-j}^{(1)}(x - y) = \frac{(\mu(x - y))^{Q_1-j}}{(Q_1 - j)!} e^{\mu(x-y)},$$

$$p_{i,0} = 1 - \sum_{j=1}^{i+1} p_{i,j} = 1 - \left(\sum_{j=1}^{Q_1-1} A_{i+1-j}^{(4)} + A_{i+1-Q_1}^{(3)} \sum_{j=Q_1+1}^{i+1} A_{i-Q_1}^{(2)} \right).$$

(9)

3 Stationary Distribution of the Number of Applications in the System over the Embedded Markov Chain

Let $\overrightarrow{\pi}$ be the stationary distribution of the number of applications in the system over the embedded Markov chain (just before the receipt of a new application in the system).

System of equilibrium equations in matrix form is

$$\overrightarrow{\pi} = \overrightarrow{\pi} P$$

and in scalar form is

$$\begin{cases} \pi_0 = \sum\limits_{i=0}^\infty A_i^{(5)} \cdot pi_i, \\ \pi_k = \sum\limits_{i=k-1}^{Q_1-1} \pi_i \cdot A_{i+1-k}^{(1)} + \pi_{Q_1} \cdot A_{i+1-Q_1} + \sum\limits_{i=Q_1+1}^\infty \pi_i \cdot A_{i+1-Q_1}^{(3)}, \\ \pi_{Q_1} = \pi_{Q_1-1} A_0 + \pi_{Q_1} \cdot A_1^{(3)} + \sum\limits_{i=Q_1}^\infty \pi_i A_{i+1-Q_1}^{(3)}, \\ \pi_k = \sum\limits_{i=k-1}^\infty \pi_i \cdot A_{i+1-k}^{(2)}, k > Q_1. \end{cases}$$

(10)

As in [32–34, 42] we assume that π_k can be representable as

$$\pi_k = g^{k-Q_1-1}\pi_{Q_1+1}, \quad (k > Q_1), \tag{11}$$

where g—some constant defined by the Eq. 12.

$$g = \alpha(\mu(1 - g\widehat{\pi}(g))), 0 < g < 1, \tag{12}$$

where $\widehat{\pi}(g) = \sum\limits_{i=0}^{\infty} g^i q(i).$

4   Probabilistic Characteristics of the System

We will assume that applications are serviced and dropped in the order of arrival (the following characteristics will be the same for other variants of service discipline and renovation, see [33, 34]). Let $p^{(serv)}$ be the probability that the application received in the system will be serviced and let $p^{(loss)}$ be the probability that the application received in the system will be dropped.

$$p^{(serv)} + p^{(loss)} = 1.$$

To calculate them, we introduce auxiliary probabilities $p_i^{(serv)}$ and $p_i^{(loss)}$— the probability that the application, that found in the system at the time of its arrival $i, (i \geq 0)$ other applications, will be served or dropped.

$$p^{(serv)} = \sum_{i=0}^{\infty} p_i^{(serv)} \pi_i, \quad p^{(loss)} = \sum_{i=0}^{\infty} p_i^{(loss)} \pi_i.$$

1. If, at the time of receipt of the application under consideration, the threshold value Q_1 has not been overcome (i.e. the renovation mechanism is not enabled), then the application will enter into the safe zone $(0 \leq i \leq Q_1)$, so

$$p_i^{(serv)} = 1, \quad p_i^{(loss)} = 0.$$

2. If $i \geq Q_1 + 1$, then at the time of receipt of the application, the safe zone is completely filled and the incoming application can be dropped in the future.

$$p_i^{(serv)} = \sum_{k=1}^{i-Q_1-1} \sum_{l=0}^{i-Q_1-k} \pi_k(l), \quad p_i^{(loss)} = \sum_{k=1}^{i-Q_1-1} \widetilde{\pi}_k(i - Q_1 - k).$$

As result we obtain:

$$p^{(serv)} = 1 - \pi_{Q_1+1} \cdot \frac{1 - \widehat{\pi}(g)}{(1-g)(1-g\widehat{\pi}(g))}, \quad p^{(loss)} = \pi_{Q_1+1} \cdot \frac{(1 - \widehat{\pi}(g))}{(1-g)(1-g\widehat{\pi}(g))}. \tag{13}$$

5   Time Characteristics: FCFS Service Discipline and the Direct General Renovation

We will assume that applications are serviced and dropped in the order of arrival.

Let $W^{(serv)}(x)$ be the distribution function of waiting time for the start of service by the application received in the system, $W^{(loss)}(x)$—be the distribution function of the time spent in the queue by the dropped application.

$W_i^{(serv)}(x)$ and $W_i^{(loss)}(x)$—conditional distribution functions of the time spent in the queue by the application, that found at the moment of its arrival $i(i \geq 1)$ other applications in the system, and which will bw servwed or droppe respectively.

$$W^{(serv)}(x) = \frac{1}{p^{(serv)}} \sum_{i=0}^{\infty} W_i^{(serv)}(x) \cdot \pi_i, \quad W^{(loss)}(x) = \frac{1}{p^{(loss)}} \sum_{i=0}^{\infty} W_i^{(loss)}(x) \cdot \pi_i.$$
$$(14)$$

We also will use the Laplace-Stieltjes transforms $w^{(serv)}(s)$, $w^{(loss)}(s)$, $w_i^{(serv)}(s)$ and $w_i^{(loss)}(s)$ of functions $W^{(serv)}(x)$, $W^{(loss)}(x)$, $W_i^{(serv)}(x)$ and $W_i^{(loss)}(x)$ respectively.

5.1   Serviced Task

a) $i = 0$ (the system is empty at the time of receipt of the application under consideration)

$$W_0^{(serv)}(x) = 1, x \geq 0.$$

b) $0 < i \leq Q_1$ (the system is not empty, but there is at least one free space in the safe zone, the renovation mechanism is not enabled)

$$W_i^{(serv)}(x) = H_i(x) = \int_0^{\infty} \frac{\mu^i x^{i-1}}{(i-1)!} e^{-\mu x} dx, \quad x \geq 0, i \geq 1.$$

$$w_i^{(serv)}(s) = \left(\frac{\mu}{\mu+s}\right)^i, \quad i \geq 1.$$

c) $i \geq Q_1 + 1$ (at the time of receipt of the application under consideration, the safe zone is filled and the renovation mechanism is enabled).

$$W_{Q_1+i}^{(serv)}(x) = \sum_{j=0}^{i} H_{Q_1+j}(x) \cdot \pi_j(i-j),$$

$$w_{Q_1+i}^{(serv)}(s) = \sum_{j=0}^{i} \left(\frac{\mu}{\mu+s}\right)^{Q_1+i} \cdot \pi_j(i-j).$$

Then

$$w^{(serv)}(s) = \frac{1}{p^{(serv)}} \sum_{i=0}^{\infty} w_i^{(serv)}(s) \cdot \pi_i$$

$$= \frac{1}{p^{(serv)}} \left(\pi_0 + \sum_{i=0}^{Q_1} \left(\frac{\mu}{\mu+s} \right)^i + \pi_{Q_1+1} \left(\frac{\mu}{\mu+s} \right)^{Q_1+1} \frac{\widehat{\pi}(g)(\mu+s)}{\mu+s-\mu g\widehat{\pi}(g)} \right).$$

$$(15)$$

and the average waiting time for the start of service is:

$$w^{(serv)} = \frac{1}{p^{(serv)}} \left(\frac{1}{\mu} \sum_{i=1}^{Q_1} i\pi_i + \frac{\pi_{Q_1+1} \cdot \widehat{\pi}(g) \cdot (Q_1 + 1 - Q_1 g\widehat{\pi}(g))}{\mu(1 - g\widehat{\pi}(g))^2} \right). \quad (16)$$

5.2   Dropped Application

If the system is empty or there is at least one place in the safe zone in the queue, then the application entering the system cannot be dropped:

$$W_i^{(loss)}(x) = 0, \quad 0 \le i \le Q_1.$$

If the safe zone is completely filled (so the general renovation mechanism is enabled) at the time of arrival of the considered application, then and only then the application under consideration may be dropped:

$$W_{Q_1+1}^{(loss)}(x) = \widetilde{\pi}_1(1) \cdot H_1(x),$$

$$W_i^{(loss)}(x) = \widetilde{\pi}_1 \cdot (i - Q_1) \cdot H_1(x)$$

$$+ \sum_{k=2}^{i-Q_1} H_k(x) \sum_{j=0}^{i-Q_1-k} \pi_{k-1}(j) \cdot \widetilde{\pi}_1(i - Q_1 + 1 - k - j) \quad , i > Q_1 + 1.$$

In terms of the Laplace-Stieltjes transform we get:

$$w_i^{(loss)}(s) = 0, \quad 0 \le i \le Q_1, \quad w_{Q_1+1}^{(loss)}(s) = \widetilde{\pi}_1(1)\frac{\mu}{\mu+s},$$

$$w_i^{(loss)}(s) = \widetilde{\pi}_1(i - Q_1) \cdot \frac{\mu}{\mu+s}$$

$$+ \sum_{k=2}^{i-Q_1} \left(\frac{\mu}{\mu+s} \right)^k \sum_{j=0}^{i-Q_1-k} \pi_{k-1}(j) \cdot \widetilde{\pi}_1(i - Q_1 + 1 - k - j), \quad i > Q_1 + 1.$$

The final Laplace-Stieltjes transform for the distribution function of the time spent in the queue by the dropped application:

$$w^{(loss)}(s) = \frac{1}{p^{(loss)}} \sum_{i=0}^{\infty} w_i^{(loss)}(s)\pi_i = \frac{\mu\pi_{Q_1+1}}{p^{(loss)}} \cdot \frac{1 - \widehat{\pi}(g)}{1 - g} \cdot \frac{\mu+s}{\mu+s-\mu g\widehat{\pi}(g)}. \quad (17)$$

The average time spent in the queue by a dropped application

$$w^{(loss)} = -\left(\omega^{(loss)}(s)\right)'_{s=0} = \frac{\pi_{Q_1+1}}{p^{(loss)}} \cdot \frac{1 - \widehat{\pi}(g)}{1 - g} \cdot \frac{1}{\mu(1 - g\widehat{\pi}(g))^2}. \tag{18}$$

5.3 Time Characteristics of an Arbitrary Application

Let $W(x)$ be the distribution function of the time spent in the queue by an arbitrary application.
Then:

$$W(x) = p^{(serv)} \cdot W^{(serv)}(x) + p^{(loss)} \cdot W^{(loss)}(x). \tag{19}$$

In terms of Laplace-Stieltjes transform

$$\omega(s) = p^{(serv)} \cdot \omega^{(serv)}(s) + p^{(loss)} \cdot \omega^{(loss)}(s) = \sum_{i=0}^{Q_1} \pi_i \left(\frac{\mu}{\mu + s}\right)^i$$

$$+ \pi_{Q_1+1} \cdot \frac{\mu + s}{\mu + s - \mu g\widehat{\pi}(g)} \left(\left(\frac{\mu}{\mu + s}\right)^{Q_1+1} \widehat{\pi}(g) + \frac{\mu}{\mu + s} \cdot \frac{1 - \widehat{\pi}(g)}{1 - g}\right). \tag{20}$$

Let w be the average time spent in the queue by an arbitrary application.

$$w = \frac{1}{\mu}\left(\sum_{i=1}^{Q_1} i\pi_i + \pi_{Q_1+1} \cdot \frac{1 + Q_1\widehat{\pi}(g)(1 - g)}{(1 - g)(1 - g\widehat{\pi}(g))}\right). \tag{21}$$

6 Time Characteristics: LCFS Service Discipline and the Inverse General Renovation

We will assume that applications are served and dropped from the last one and will use the notation from the previous section.

6.1 Serviced Task

a) $i = 0$ (the system is empty at the time of receipt of the application under consideration)

$$W_0^{(serv)}(x) = 1, x \geq 0.$$

b) $0 < i \leq Q_1$ (the system is not empty, but there is at least one free space in the safe zone, the renovation mechanism is not enabled).
We will consider two subcases:
b.1) The first one, when $i + j \leq Q_1$ (the number of applications in the queue does not exceed the threshold Q_1).

$$w_{i,j}^{(serv)}(x) = \frac{\mu^{j+1}x^j}{j!}e^{-\mu x}\bar{A}(x) + \int_0^x \sum_{k=0}^j \frac{(\mu y)^k}{k!}e^{-\mu y}dA(y)w_{i,j-k+1}^{(serv)}(x - y).$$

$$w_{i,j}^{(serv)}(s) = \frac{\mu^{j+1}(-1)^j}{j!}\bar{\alpha}^{(j)}(\mu + s) + \sum_{k=0}^{j}\frac{(-\mu)^k}{k!}\alpha^{(k)}(\mu + s)w_{i,j-k+1}^{(serv)}(s).$$

b.2) The second subcase, when $i + j > Q_1$ (the threshold Q_1 is overcome and the renovation mechanism is enabled).

$$w_{i,j}^{(serv)}(x) = \sum_{k=1}^{j+i-Q_1}\frac{\mu^{k+Q_1-i+1}x^{k+Q_1-i}}{(k+Q_1-i)!}e^{-\mu x}\tilde{\pi}_k(i+j-Q_1-k)\bar{A}(x)$$

$$+ \int_0^x \sum_{k=0}^{i+j-Q_1}\frac{(\mu y)^k}{k!}e^{-\mu y}\sum_{l=0}^{i+j-Q_1-k}\pi_k(l)dA(y)w_{i,j-k-l+1}^{(serv)}(x-y)$$

$$+ \int_0^x \sum_{k=1}^{i+j-Q_1}\tilde{\pi}_k(i+j-Q_1-k)\sum_{l=0}^{Q_1-i}\frac{(\mu y)^{k+l}}{(k+l)!}e^{-\mu y}dA(y)w_{i,Q_1-i-l+1}^{(serv)}(x-y).$$

$$w_{i,j}^{(serv)}(s) = \sum_{k=1}^{j+i-Q_1}\frac{\mu^{k+Q_1-i+1}(-1)^{k+Q_1-i}}{(k+Q_1-i)!}\tilde{\pi}_k(i+j-Q_1-k)\bar{\alpha}^{(k+Q_1-i)}(\mu+s)$$

$$+ \sum_{k=0}^{i+j-Q_1}\frac{(-\mu)^k}{k!}\sum_{l=0}^{i+j-Q_1-k}\pi_k(l)\alpha^{(k)}(\mu+s)w_{i,j-k-l+1}^{(serv)}(s)$$

$$+ \sum_{k=1}^{i+j-Q_1}\tilde{\pi}_k(i+j-Q_1-k)\sum_{l=0}^{Q_1-i}\frac{(-\mu)^{k+l}}{(k+l)!}\alpha^{(k)}(\mu+s)w_{i,Q_1-i-l+1}^{(serv)}(s).$$

c) $i \geq Q_1 + 1$ (the safe zone is filled and the renovation mechanism is enabled).

$$w_{i,j}^{(serv)}(x) = \sum_{k=1}^{j+1}\frac{\mu^k x^{k-1}}{(k-1)!}\pi_k(j-k+1)\bar{A}(x)+$$

$$+ \int_0^x \sum_{k=0}^{j}\frac{(\mu y)^k}{k!}e^{-\mu y}\sum_{l=0}^{j-k}\pi_k(l)dA(y)w_{i,j-k-l+1}^{(serv)}(x-y),$$

$$w_{i,j}^{(serv)}(s) = \sum_{k=1}^{j+1}\frac{\mu^k(-1)^{k-1}}{(k-1)!}\pi_k(j-k+1)\bar{\alpha}^{(k-1)}(\mu+s)$$

$$+ \sum_{k=0}^{j}\frac{(-\mu)^k}{k!}\sum_{l=0}^{j-k}\pi_k(l)\alpha^{(k)}(\mu+s)w_{i,j-k-l+1}^{(serv)}(s).$$

Then

$$w^{(serv)}(s) = \frac{1}{p^{(serv)}}\sum_{i=0}^{\infty}w_{i,0}^{(serv)}(s)\pi_i. \tag{22}$$

and our future task is to obtain the final expression for $w^{(serv)}(s)$ by using the same methods as for a dropped application [33,34].

6.2   Dropped Application

If the system is empty or there is at least one place in the safe zone in the queue, then the application entering the system cannot be dropped:

$$W_i^{(loss)}(x) = 0, \quad 0 \le i \le Q_1.$$

If the safe zone is completely filled (so the general renovation mechanism is enabled) at the time of arrival of the considered application, then and only then the application under consideration may be dropped:

$$w_{i,j}^{(loss)}(x) = \sum_{k=1}^{j+1} \frac{\mu^k x^{k-1}}{(k-1)!} \tilde{\pi}_k(j-k+2)\bar{A}(x)$$

$$+ \int_0^x \sum_{k=0}^j \frac{(\mu y)^k}{k!} e^{-\mu y} \sum_{l=0}^{j-k} \pi_k(l) dA(y) w_{i,j-k-l+1}^{(loss)}(x-y).$$

In terms of the Laplace-Stieltjes transform we get:

$$\omega_i^{(loss)}(s) = \sum_{k=1}^{j+1} \frac{\mu^k (-1)^{k-1}}{(k-1)!} \tilde{\pi}_k(j-k+2)\bar{a}^{(k-1)}(\mu+s)$$

$$+ \sum_{k=0}^j \sum_{l=0}^{j-k} \frac{(-\mu)^k}{k!} \pi_k(l) a^{(k)}(\mu+s)\omega_{i,j-k-l+1}^{(loss)}(s).$$

The final Laplace-Stieltjes transform for the distribution function of the time spent in the queue by the dropped application:

$$\omega^{(loss)}(s) = \frac{1}{p^{(loss)}} \sum_{i=0}^\infty \omega_{i,0}^{(loss)}(s)\pi_i = \frac{\pi_{Q_1+1}}{(1-g)p^{(loss)}} \cdot \frac{\mu(1-\hat{\pi}(z(s)))}{\mu+s-\mu g\hat{\pi}(g)}, \quad (23)$$

where $z(s) = \alpha(\mu+s-\mu z\hat{\pi}(z))$. If $s=0$, then $z=g$ defined by Eq. 12. This Laplace-Stieltjes transform for the distribution function of the time spent in the queue by the dropped application has the same form as in [33, 34].

7   Conclusion

In this paper, we considered a single-line queuing system with an infinite capacity queue, with a threshold and renovation mechanism. Analytical expressions of the distribution of the number of applications in the system were found for this system, expressions for calculating time and probabilistic characteristics were obtained.

The results obtained for the considered system coincide with the results presented in the paper [34], if threshold value $Q_1 = 0$.

In the future, it is planned to study the time characteristics of the system for the following service and renovation options:

– service in the order of arrival (first come, first serve basis), applications are dropped in reverse order (starting from the last one);
– service in reverse order (last come, first serve basis), applications are dropped in direct order (from the first one).

Also we are planning to study a model of general renovation with two threshold values that determine both the renovation (drop) probability and the safe zone in the queue.

References

1. Floyd, S., Jacobson, V.: Random early detection gateways for congestion avoidance. IEEE/ACM Trans. Netw. **4**(1), 397–413 (1993). https://doi.org/10.1109/90.251892
2. Ramakrishnan, K., Floyd, S., Black, D.: The addition of explicit congestion notification (ECN) to IP. RFC 3168. Internet Engineering Task Force. https://tools.ietf.org/html/rfc3168. Accessed 27 May 2021
3. Floyd, S., Gummadi, R., Shenker, S.: Adaptive RED: an algorithm for increasing the robustness of RED's active queue management (2001). http://www.icir.org/floyd/papers/adaptiveRed.pdf
4. Class-based weighted fair queueing and weighted random early detection. http://www.cisco.com/c/en/us/td/docs/ios/12_0s/feature/guide/fswfq26.html. Accessed 27 June 2022
5. Floyd, S., Gummadi, R., Shenker, S.: Adaptive RED: an algorithm for increasing the robustness of RED's active queue management (2001). http://www.icir.org/floyd/papers/adaptiveRed.pdf. Accessed 19 May 2022
6. Changwang, Z., Jianping, Y., Zhiping, C., Weifeng, C.: RRED: robust RED algorithm to counter low-rate denial-of-service attacks. IEEE Commun. Lett. **14**(5), 489–491 (2010). https://doi.org/10.1109/LCOMM.2010.05.091407
7. Ott, T.J., Lakshman, T.V., Wong, L.H.: SRED: stabilized RED. In: Proceedings IEEE INFOCOM 1999, New York, NY, USA, vol. 3, pp. 1346–1355. IEEE (1999). https://doi.org/10.1109/INFCOM.1999.752153
8. Lin, D., Morris, R.: Dynamics of random early detection. Comput. Commun. Rev. **27**(4), 127–137 (1997)
9. Anjum, F.M., Tassiulas, L.: Balanced RED: an algorithm to achieve fairness in the internet. Technical research report (1999). http://www.dtic.mil/dtic/tr/fulltext/u2/a439654.pdf
10. Aweya, J., Ouellette, M., Montuno, D.Y.: A control theoretic approach to active queue management. Comput. Netw. **36**, 203–235 (2001)
11. Chrysostomoua, C., Pitsillidesa, A., Rossidesa, L., Polycarpoub, M., Sekercioglu, A.: Congestion control in differentiated services networks using fuzzy-RED. Control. Eng. Pract. **11**, 1153–1170 (2003)
12. Feng, W.-C.: Improving internet congestion control and queue management algorithms. http://thefengs.com/wuchang/umich_diss.html. Accessed 29 May 2021
13. Feng, W., Kandlur, D.D., Saha, D., Shin, K.G.: BLUE: a new class of active queue management algorithms. UM CSE-TR-387-99 (1999). https://www.cse.umich.edu/techreports/cse/99/CSE-TR-387-99.pdf
14. Nichols, K., Jacobson, V., McGregor, A., Iyengar, J.: Controlled delay active queue management. RFC 8289. Internet Engineering Task Force. https://tools.ietf.org/html/rfc8289. Accessed 29 Aug 2019

15. Hoeiland-Joergensen, T., McKenney, P., Taht, D., Gettys, J., Dumazet, E.: The flow queue codel packet scheduler and active queue management algorithm. Internet Engineering Task Force (2018). https://tools.ietf.org/html/rfc8290

16. IETF working group on active queue management and packet scheduling (AQM): description of the working group. http://tools.ietf.org/wg/aqm/charters. Accessed 29 June 2022

17. Baker, F., Fairhurst, G.: IETF recommendations regarding active queue management. RFC 7567. Internet Engineering Task Force. https://tools.ietf.org/html/rfc7567. Accessed 27 May 2022

18. McKenney, P.E.: Stochastic fairness queueing. In: Proceedings of IEEE International Conference on Computer Communications, San Francisco, CA, USA, vol. 2, pp. 733–740. IEEE (1990). https://doi.org/10.1109/INFCOM.1990.91316

19. Adams, R.: Active queue management: a survey. IEEE Commun. Surv. Tutor. **15**(3), 1425–1476 (2013)

20. Paul, A.K., Kawakami, H., Tachibana, A., Hasegawa, T.: An AQM based congestion control for eNB RLC in 4G/LTE network. In: 2016 IEEE Canadian Conference on Electrical and Computer Engineering (CCECE), Vancouver, BC, Canada, pp. 1–5. IEEE (2016). https://doi.org/10.1109/CCECE.2016.7726792

21. Menth, M., Veith, S.: Active queue management based on congestion policing (CP-AQM). In: German, R., Hielscher, K.-S., Krieger, U.R. (eds.) MMB 2018. LNCS, vol. 10740, pp. 173–187. Springer, Cham (2018). https://doi.org/10.1007/978-3-319-74947-1_12

22. Irazabal, M., Lopez-Aguilera, E., Demirkol, I.: Active queue management as quality of service enabler for 5G networks. In: European Conference on Networks and Communications (EuCNC), Valencia, Spain, pp. 421–426. IEEE (2019). https://doi.org/10.1109/EuCNC.2019.8802027

23. Pesántez-Romero, I.S., Pulla-Lojano, G.E., Guerrero-Vásquez, L.F., Coronel-González, E.J., Ordoñez-Ordóñez, J.O., Martinez-Ledesma, J.E.: Performance evaluation of hybrid queuing algorithms for QoS provision based on DiffServ architecture. In: Yang, X.-S., Sherratt, S., Dey, N., Joshi, A. (eds.) Proceedings of Sixth International Congress on Information and Communication Technology. LNNS, vol. 216, pp. 333–345. Springer, Singapore (2022). https://doi.org/10.1007/978-981-16-1781-2_31

24. George, J., Santhosh, R.: Congestion control mechanism for unresponsive flows in internet through active queue management system (AQM). In: Shakya, S., Bestak, R., Palanisamy, R., Kamel, K.A. (eds.) Mobile Computing and Sustainable Informatics. LNDECT, vol. 68, pp. 765–777. Springer, Singapore (2022). https://doi.org/10.1007/978-981-16-1866-6_58

25. Singha, S., Jana, B., Jana, S., Mandal, N.K.: A novel congestion control algorithm using buffer occupancy RED. In: Das, A.K., Nayak, J., Naik, B., Dutta, S., Pelusi, D. (eds.) Computational Intelligence in Pattern Recognition. AISC, vol. 1349, pp. 519–528. Springer, Singapore (2022). https://doi.org/10.1007/978-981-16-2543-5_44

26. Apreutesey, A.M.Y., Korolkova, A.V., Kulyabov, D.S.: Hybrid modelling of the RED algorithm in the Julia language. In: 7th International Young Scientists Conference on Information Technology, Telecommunications and Control Systems, ITTCS 2020, Journal of Physics: Conference Series, vol. 1694 (2020). https://doi.org/10.1088/1742-6596/1694/1/012025

27. Apreutesey, A.M.Y., Korolkova, A.V., Kulyabov, D.S.: Computer simulation of the stochastic RED algorithm. In: CEUR Workshop Proceedings, vol. 1694, pp. 45–53 (2021). http://ceur-ws.org/Vol-2946/paper-04.pdf

28. Chydzinski, A., Chrost, L.: Analysis of AQM queues with queue size based packet dropping. Int. J. Appl. Math. Comput. Sci. **21**(3), 567–577 (2011). https://doi.org/10.2478/v10006-011-0045-7

29. Chydzinski, A.: On the transient queue with the dropping function. Entropy **22**, 825 (2020). https://doi.org/10.3390/e22080825

30. Konovalov, M.G., Razumchik, R.V.: Numerical analysis of improved access restriction algorithms in a $GI|G|1|N$ system. J. Commun. Technol. Electron. **63**(6), 616–625 (2018). https://doi.org/10.1134/S1064226918060141

31. Kreinin, A.: Queueing systems with renovation. J. Appl. Math. Stochast. Anal. **10**(4), 431–443 (1997). https://doi.org/10.1155/S1048953397000464

32. Bocharov, P.P., Zaryadov, I.S.: Probability distribution in queueing systems with renovation. Bulletin of Peoples' Friendship University of Russia. Mathematics Information Sciences Physics, no. 1–2, pp. 15–25 (2007). (in Russian)

33. Zaryadov, I.S.: The $GI/M/n/\infty$ queuing system with generalized renovatione. Autom. Remote. Control. **71**(4), 663–671 (2010). https://doi.org/10.1134/S0005117910040077

34. Zaryadov, I.S., Pechinkin, A.V.: Stationary time characteristics of the $GI/M/n/\infty$ system with some variants of the generalized renovation discipline. Autom. Remote. Control. **70**(12), 2085–2097 (2009). https://doi.org/10.1134/S0005117909120157

35. Zaryadov, I., Razumchik, R., Milovanova, T.: Stationary waiting time distribution in G—M—n—r with random renovation policy. In: Vishnevskiy, V.M., Samouylov, K.E., Kozyrev, D.V. (eds.) DCCN 2016. CCIS, vol. 678, pp. 349–360. Springer, Cham (2016). https://doi.org/10.1007/978-3-319-51917-3_31

36. Konovalov, M., Razumchik, R.: Queueing systems with renovation vs. queues with RED (2017). Supplementary Material. https://arxiv.org/abs/1709.01477

37. Konovalov, M., Razumchik, R.: Comparison of two active queue management schemes through the M/D/1/N queue. Informatika i ee Primeneniya (Inform. Appl.) **12**(4), 9–15 (2018). https://doi.org/10.14357/19922264180402

38. Bogdanova, E.V., Zaryadov, I.S., Milovanova, T.A., Korolkova, A.V., Kulyabov, D.S.: Characteristics of lost and served packets for retrial queueing system with general renovation and recurrent input flow. In: Vishnevskiy, V.M., Kozyrev, D.V. (eds.) DCCN 2018. CCIS, vol. 919, pp. 327–340. Springer, Cham (2018). https://doi.org/10.1007/978-3-319-99447-5_28

39. Hilquias, V.C.C., et al.: The general renovation as the active queue management mechanism. Some aspects and results. In: Vishnevskiy, V.M., Samouylov, K.E., Kozyrev, D.V. (eds.) DCCN 2019. CCIS, vol. 1141, pp. 488–502. Springer, Cham (2019). https://doi.org/10.1007/978-3-030-36625-4_39

40. Meykhanadzhyan, L.A., Zaryadov, I.S., Milovanova, T.A.: Stationary characteristics of the two-node Markovian tandem queueing system with general renovation. Sistemy i Sredstva Informatiki [Syst. Means Inform.] **30**(3), 14–31 (2020)

41. Gorbunova, A., Lebedev, A.: Queueing system with two input flows, preemptive priority, and stochastic dropping. Autom. Remote. Control. **81**, 2230–2243 (2020). https://doi.org/10.1134/S0005117920120073

42. Hilquias, V.C.C., Zaryadov, I.S., Milovanova, T.A.: Two types of single-server queueing systems with threshold-based renovation mechanism. In: Vishnevskiy, V.M., Samouylov, K.E., Kozyrev, D.V. (eds.) DCCN 2021. LNCS, vol. 13144, pp. 196–210. Springer, Cham (2021). https://doi.org/10.1007/978-3-030-92507-9_17

43. Hilquias, V.C.C., Zaryadov, I.S.: Single-server queuing systems with exponential service times and threshold-based renovation. In: International Congress on Ultra Modern Telecommunications and Control Systems and Workshops 2021, pp. 91–97 (2021). https://doi.org/10.1109/ICUMT54235.2021.9631585

44. Hilquias, V.C.C., Zaryadov, I.S.: Comparison of two single-server queueing systems with exponential service times and threshold-based renovation. In: Workshop on Information Technology and Scientific Computing in the Framework of the ITTMM 2021, CEUR Workshop Proceedings, vol. 2946, pp. 54–63 (2021). http://ceur-ws.org/Vol-2946/paper-05.pdf
45. Konovalov, M., Razumchik, R.: Finite capacity single-server queue with Poisson input, general service and delayed renovation. Eur. J. Oper. Res. https://doi.org/10.1016/j.ejor.2022.05.047
46. Gelenbe, E.: Product-form queueing networks with negative and positive customers. J. Appl. Probab. **28**(3), 656–663 (1991). https://doi.org/10.2307/3214499
47. Pechinkin, A.V., Razumchik, R.V.: The stationary distribution of the waiting time in a queueing system with negative customers and a bunker for superseded customers in the case of the LAST-LIFO-LIFO discipline. J. Commun. Technol. Electron. **57**(12), 1331–1339 (2012). https://doi.org/10.1134/S1064226912120054
48. Semenova, O.V.: Multithreshold control of the $BMAP/G/1$ queuing system with MAP flow of Markovian disasters. Autom. Remote. Control. **68**(1), 95–108 (2007). https://doi.org/10.1134/S0005117907010092
49. Dudin, A.N., Klimenok, V.I., Vishnevsky, V.M.: Mathematical methods to study classical queuing systems. In: The Theory of Queuing Systems with Correlated Flows, pp. 1–61. Springer, Cham (2020). https://doi.org/10.1007/978-3-030-32072-0_1
50. Krishnamoorthy, A., Pramod, P.K., Chakravarthy, S.R.: Queues with interruptions: a survey. TOP **22**(1), 290–320 (2014). https://doi.org/10.1007/s11750-012-0256-6
51. Rykov, V., Kozyrev, D.: On the reliability function of a double redundant system with general repair time distribution. Appl. Stochast. Models Bus. Ind. **35**, 191–197 (2019). https://doi.org/10.1002/ASMB.2368
52. Nguyen, D.P., Kozyrev, D.: Reliability analysis of a multirotor flight module of a high-altitude telecommunications platform operating in a random environment. In: 2020 International Conference Engineering and Telecommunication (En&T), pp. 1–5 (2020). https://doi.org/10.1109/EnT50437.2020.9431312

Effective Algorithm of Estimation the Performance Measures of Group of Servers with Dependence of Call Repetition on the Type of Call Blocking

Mikhail S. Stepanov$^{1(\boxtimes)}$ ⓘ, Sergey N. Stepanov^{1,2} ⓘ, and Fedor S. Kroshin^{1} ⓘ

^{1} Department of Communication Networks and Commutation Systems,
Moscow Technical University of Communication and Informatics,
8A, Aviamotornaya str., Moscow 111024, Russia
{m.s.stepanov,s.n.stepanov,f.s.kroshin}@mtuci.ru
^{2} Kotel'nikov Institute of Radio Engineering and Electronics of RAS,
Mokhovaya 11-7, Moscow 125009, Russia

Abstract. A mathematical model of a fully accessible communication system has been constructed and investigated, taking into account the repetition of a call after the preliminary service stage and due to the busy service devices. The probability of call retry depends on the stage of the connection establishment at which the denial of service was received. The model also assumes that the user's stay in the call repetition state is limited by a random variable that has an exponential distribution, which determines the aging time of the transmitted information. It is shown that the exact values of the stationary characteristics of the constructed model can be calculated by an efficient recursive algorithm developed for the model where the user repeats the call only because the service devices are busy. The constructed model and methods of its analysis can be used for evaluation of the positive effect that the use of the preliminary service like chatbot creates when planning the required number of call center operators. The model also makes it possible to estimate the share of incoming requests that need to be redirected to other call centers in an overload situation. Numerical examples are given to illustrate the practical use of the results obtained.

Keywords: Repeated calls · System of state equations · Recursive algorithm · Service resource planning

1 Introduction

An important feature of the formation of flows of requests for the provision of communication services is the presence of repeated attempts to connect after receiving a denial of service. Streams of repeated requests can initiate an avalanche-like increase in traffic in individual network segments, sharply degrading the quality of its work. Under these conditions, the use of traditional planning

V. M. Vishnevskiy et al. (Eds.): DCCN 2022, LNCS 13766, pp. 324–337, 2022.
https://doi.org/10.1007/978-3-031-23207-7_25

techniques based on models with losses or waiting leads to significant errors. To eliminate the noted shortcomings, a special class of models is introduced, in which the joint service of both primary and repeated calls is studied (see survey of obtained results in [1–6]). It is especially important to take into account the impact of repeated calls in various kinds of reference and information services. These are call and contact centers, emergency services, situational centers, etc. [7–10]. The theoretical analysis of such models is not an easy task due to the appearance of a dependence between successive call arrivals. Nevertheless, for some simple models, it was possible to build efficient algorithms for calculating characteristics. The first results were published in [11,12]. Generalization of these results to models with a more detailed account of the specifics of the formation of retrial flows will be the subject of analysis in this paper.

Among models with taking into account the customer behavior after getting refusal exist one model that has efficient recursive algorithm for calculation of stationary characteristics [6,11–13]. This is a Markov model of a full availability group of servers with Poisson flow of primary calls and possibility of call repetition only because all servers are busy. Further we refer to this model as a basic model with repeated calls. For this model a lot of other results are obtained. Among them are asymptotic formulae for calculation performance measures in case of large load [14,15], a variety of approximate algorithms [6] and some other results [6].

The introducing into the basic model of other important from the practical point of view reasons of the call losses and subsequent repetition, for example, losses on preliminary stages of servicing, the dependence of the probability of repeating a call on the connection establishment stage at which the denial of service was received, aging of transmitted information, inner blocking, priorities of incoming flows etc. seriously complicate a model. Stationary characteristics of such models can be calculated only by solution of the system of state equations with help of iterative or matrix-geometric algorithms [16–21]. Both approaches are time consuming and not applicable for values of input parameters interesting for practice applications.

The aim of the paper is to show that generalized model with retrials with taking into account some of the above mentioned reasons of call repetition can be numerically analyzed with help of the basic model where call can be repeated only because all servers are busy. The closest queueing system to the model considered in this paper was analyzed in [22]. The model describes the functioning of a group of servers with a Poisson flow of incoming primary calls and the possibility of blocking before and after takeover a free server. Also it is supposed that, a call can leave the process of getting connection with server because of the aging of transmitted information. After each loss, there is some probability that a customer repeats the call. The call retry probability depends on the type of call (primary or retry), but does not depend on the stage of connection establishment at which the refusal was received.

The absence of dependence of the retrial probability on the connection establishment phase at which the failure was received limits the scope of the results obtained in [22]. In particular, the results of [22] cannot be used to estimate the proportion of incoming requests that need to be redirected to other call centers in an overload situation. It will be shown in the paper that the noted limitation is not essential. This important feature of the behavior of the subscriber after getting refusal in servicing can be built into the model considered in [22] and at the same time it will be possible to estimate the performance measures of the generalized in this way model by using the corresponding characteristics of the basic model. It means that all results obtained for the basic model [6,11–13] can be applied for generalized model.

A short outline of this paper is as follows. In Sect. 2, the mathematical description of the basic and generalized models with retrials are presented. In Sect. 3 the number of probabilistic and algebraic transforms of constructed model are presented that allows to perform effective estimation of performance measures of the studied model. Section 4 contains the numerical results that show the practical applications of the analyzed model. Section 5 concludes the paper.

2 Model Description

The analyzed model may have a different technical interpretation. Its choice depends on the definition of the resource used to service incoming requests. Let's assume that we are talking about the process of forming and servicing the flow of requests to the call center. Denote by v the total number of operators. We will assume that the arrival of primary claims obeys the Poisson law with intensity λ. If all v operators are busy, then the customer is refused and with probability H_1 for the first connection attempt and with probability H_2 for the second and all subsequent connection attempts repeats the request after a random time that has an exponential distribution with the parameter μ, and with additional probabilities $1 - H_1$ and $1 - H_2$, respectively, give up trying to get service at the call center. Without loss of generality, we can suppose that service time of a primary or repeated request does not depend on its type and has an exponential distribution with the parameter equals to one. The total intensity of repeated requests is determined by a random number of customers repeating the call. The constructed model describes the functioning of the basic model (see previous section), taking into account repeated calls [6,11–13]. Its performance measures can be calculated by efficient recursive algorithm [6,11–13].

Let us consider a generalization of the basic model based on the possibility of preliminary servicing an incoming request. This stage corresponds to the interaction of the customer and a robotic operator like chatbot or other facilities with similar functions. It is assumed that all incoming requests go through this stage of service. For the primary attempt, after preliminary servicing, with probability $(1 - b_p)$, the service stage begins at the operator, with probability $b_p(1 - H_{1,p})$, the customer receives the required information and completes the service, and with probability $b_p H_{1,p}$ the customer repeats the request after a random time that has an exponential distribution with the parameter μ. For a retry, the same steps are performed with different probabilities. After preliminary servicing with probability $(1 - b_r)$, the service stage begins at the operator, with probability $b_r(1 - H_{2,b})$ the customer receives the required information and completes the service, and with probability $b_r H_{2,b}$ the customer tries to connect again.

We also assume that with the beginning of the process of repeating the request, the process of aging of the transmitted information begins. If, after a random time that has an exponential distribution with the σ parameter, the customer does not get service, then he stops trying to establish a connection. The functional model of the studied service system is presented in Fig. 1.

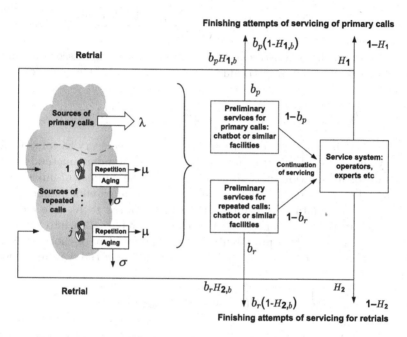

Fig. 1. The generalized model with retrials and preliminary servicing of incoming requests.

Let $j(t)$ denote the number of customers repeating the call at time t, and let $i(t)$ denote the number of busy servers at time t. The functioning of the model is described by a random process $r(t) = \big(j(t), i(t)\big)$ defined on the state space $S = \big\{(j, i), j = 0, 1, \ldots, i = 0, 1, \ldots, v\big\}$. Conditions for the existence of a stationary regime are written out in the standard way for $r(t)$ [23].

Denote by $p(j, i)$ the probabilities of stationary states. They have an interpretation of the proportion of time the model is in the (j, i) state and can be used to determine the main performance measures. These include: $p(v)$—share of busy time for all operators; $J(v)$—average number of customers repeating a call and being in a busy state of all operators; I—average number of customers busy; J—average number of customers repeating a call; τ—share of repeated calls in the total flow of incoming requests; M—average number of repeated applications per one primary; π_c—share of lost initial and repeated requests. The formal definitions of the introduced characteristics in terms of the values of $p(j,i)$ have the form:

$$\pi_p = \sum_{j=0}^{\infty} p(j,v); \quad J(v) = \sum_{j=0}^{\infty} p(j,v)j; \quad I = \sum_{j=0}^{\infty}\sum_{i=0}^{v} p(j,i)\,i; \qquad (1)$$

$$J = \sum_{j=0}^{\infty}\sum_{i=0}^{v} p(j,i)\,j; \quad \tau = \frac{J\mu}{\lambda + J\mu}; \quad M = \frac{J\mu}{\lambda};$$

$$\pi_c = \frac{\lambda(b_p + (1 - b_p)p(v)) + \mu(Jb_r + (1 - b_r)J(v))}{\lambda + J\mu}.$$

All introduced characteristics and many others can be expressed through the values of auxiliary variables $p(i)$ and $J(i)$, $i = 0, 1, \ldots, v$ which are defined as follows:

$$p(i) = \sum_{j=0}^{\infty} p(j,i), \quad J(i) = \sum_{j=0}^{\infty} p(j,i)j, \quad i = 0, 1, \ldots, v.$$

To evaluate the listed performance measures, it is necessary to compose and solve a system of equilibrium equations. To use the standard methods of linear algebra for its solution, it is necessary to limit the number of unknowns in the system of equilibrium equations. Assume that the maximum number of customers who can simultaneously repeat the call is limited to N. The customer in the initial attempt to connect to the operator, having fallen into the (N, v) state, refuses service with a probability of one. The problem of choosing N will be discussed later. The system of equilibrium equations of the truncated model has the form:

$$p(j,i)\Big\{\lambda(1 - b_p)I(i < v) + \lambda b_p H_{1,b} I(j < N, i < v) \qquad (2)$$

$$+ \lambda(1 - (b_p(1 - H_{1,b}) + (1 - b_p)(1 - H_1)))I(j < N, i = v)$$

$$+ j(\sigma + \mu(1 - b_r H_{2,b}))I(i < v)$$

$$+ j(\sigma + \mu(1 - b_r H_{2,b} - (1 - b_r)H_2))I(i = v) + iI(i > 0)\Big\}$$

$$= p(j, i - 1)\lambda(1 - b_p)I(i > 0)$$

$$+ p(j - 1, i)\lambda I(j > 0)\big[b_p H_{1,b} + (1 - b_p)H_1 I(i = v)\big]$$

$$+ p(j + 1, i - 1)(j + 1)\mu(1 - b_r)I(j < N, i > 0)$$

$$+ p(j + 1, i)(j + 1)I(j < N)\big[\sigma + \mu b_r(1 - H_{2,b}) + \mu(1 - b_r)(1 - H_2)I(i = v)\big]$$

$$+ p(j, i + 1)(i + 1)I(i < v).$$

The indicator function $I(\cdot)$ takes the value one if the condition formulated in parentheses is satisfied and zero otherwise. The values of $p(j,i)$ satisfy the normalization condition. Multiplying the Eq. (2) containing the probability $p(j,i)$

on the left side, successively by i and j and summing the resulting expressions over all $j = 0, \ldots, N$ and $i = 0, \ldots, v$, we obtain the expressions connecting the intensities of incoming and served requests:

$$\lambda + J\mu = \lambda(b_p + (1 - b_p)p(v)) + \mu(Jb_r + (1 - b_r)J(v)) + I; \qquad (3)$$

$$J(\mu + \sigma) = \lambda b_p H_{1,b} + \lambda(1 - b_p)p(v)H_1 + J\mu b_r H_{2,b} + J(v)\mu(1 - b_r)H_2 - \gamma, \quad (4)$$

$$\gamma = \lambda(1 - b_p)H_1 p(N, v) + \lambda b_p H_{1,b} \sum_{i=0}^{v} p(N, i).$$

The value of γ can be used to estimate the error introduced by truncating the original infinite state space [13].

Using the (2) it can be proved that the following relations are true

$$p(i)\lambda(1 - b_p) + J(i)\mu(1 - b_r) = p(i + 1)(i + 1), \qquad i = 0, 1, \ldots, v - 1. \quad (5)$$

The matrix (2) does not have any special properties, therefore, to solve (2), it remains to use iterative or matrix-geometric algorithms [17–20]. Both approaches are time consuming and not applicable for values of input parameters interesting for practice applications. In the following sections it will be shown that the exact values of the stationary characteristics of the constructed model can be calculated by an efficient recursive algorithm developed for the model where the user repeats the call only because the service devices are busy. Early we call this model as a basic. This model is a particular case of the constructed model and can be obtained for the following choice of input parameters: $b_p = b_r = H_{1,b} = H_{2,b} = \sigma = 0$.

3 Model Simplifications

Consider a sequence of probabilistic and algebraic transformations of the constructed model that allow to perform effective estimation of it's performance measures. Further to simplify the notations we will use at each transformation step the same symbols to denote the changing of value of input parameters and characteristics, differing them by a superscript. The value of the digit used in the superscript for performance measures means the number of simplification. For input parameters, the digit denotes the number, of changing by parameter its value. A parameter without a superscript has the same value as in the generalized model.

At the first step by using the basic property of exponentially distributed random variables we remodel the process of call repetition and the process of aging the transmitted information. We suppose that after being refused servicing in first attempt, a customer with probabilities $H_{1,b}^{(1)} = \frac{H_{1,b}\mu}{\mu+\sigma}$ or $H_1^{(1)} = \frac{H_1\mu}{\mu+\sigma}$ depending of the phase of servicing repeats a call after random time having exponential distribution with parameter $\mu + \sigma$ and with additional probabilities after the same time a customer leaves the system. If customer refused servicing in repeated attempt, a customer with probabilities $H_{2,b}^{(1)} = \frac{H_{2,b}\mu}{\mu+\sigma}$ or $H_2^{(2)} = \frac{H_2\mu}{\mu+\sigma}$ depending of the phase of servicing repeats a call after random time having exponential distribution with parameter $\mu + \sigma$ and with additional probabilities after the same time a customer leaves the system. If we take out of consideration for repeating customer the time interval before leaving the system we obtain particular case of the generalized model only without aging of the transmitted information. The process of server occupation remains the same as in the studied model. We see that model obtained at the first simplification step is a particular case of the studied model if we choose the following values of its input parameters:

$$H_{1,b}^{(1)} = \frac{H_{1,b}\mu}{\mu+\sigma}; \quad H_{2,b}^{(1)} = \frac{H_{2,b}\mu}{\mu+\sigma}; \tag{6}$$

$$H_1^{(1)} = \frac{H_1\mu}{\mu+\sigma}; \quad H_2^{(1)} = \frac{H_2\mu}{\mu+\sigma}; \mu^{(1)} = \mu + \sigma; \quad \sigma^{(1)} = 0.$$

Other parameters do not change their values.

The functioning of simplified model is described by two-dimensional Markov process $r(t) = (j(t), i(t))$, where $j(t)$ is the number of repeating customers and $i(t)$ is the number of busy servers at time t defined on finite state space S, $(j, i) \in S$, $j = 0, 1, \ldots, N$, $i = 0, 1, \ldots, v$. Let us denote by $P^{(1)}(j, i)$ the probability of stationary state (j, i) of $r(t)$. The values of $P^{(1)}(j, i)$ are related by system of state equations

$$p^{(1)}(j, i)\Big\{ \lambda(1 - b_p)I(i < v) + \lambda b_p H_{1,b}^{(1)}I(j < N, i < v) \tag{7}$$

$$+ \lambda(1 - (b_p(1 - H_{1,b}^{(1)}) + (1 - b_p)(1 - H_1^{(1)})))I(j < N, i = v)$$

$$+ j\mu^{(1)}(1 - b_r H_{2,b}^{(1)})I(i < v)$$

$$+ j\mu^{(1)}(1 - b_r H_{2,b}^{(1)} - (1 - b_r)H_2^{(1)})I(i = v) + iI(i > 0)\Big\}$$

$$= p^{(1)}(j, i - 1)\lambda(1 - b_p)I(i > 0)$$

$$+ p^{(1)}(j - 1, i)\lambda I(j > 0)(b_p H_{1,b}^{(1)} + (1 - b_p)H_1^{(1)}I(i = v))$$

$$+ p^{(1)}(j + 1, i - 1)(j + 1)\mu^{(1)}(1 - b_r)I(j < N, i > 0)$$

$$+ p^{(1)}(j + 1, i)(j + 1)\mu^{(1)}I(j < N)$$

$$\times \big[b_r(1 - H_{2,b}^{(1)}) + (1 - b_r)(1 - H_2^{(1)})I(i = v)\big]$$

$$+ p^{(1)}(j, i + 1)(i + 1)I(i < v).$$

The values of $p^{(1)}(j, i)$ satisfy the normalization condition. Using the (7) it can be proved that the following relations are true

$$p^{(1)}(i)\lambda(1 - b_p) + J^{(1)}(i)\mu^{(1)}(1 - b_r) = p^{(1)}(i + 1)(i + 1), \quad i = 0, 1, \ldots, v - 1. \tag{8}$$

It follows from (5) and (8) that the following relations are true

$$p(i) = p^{(1)}(i), \quad J(i) = J^{(1)}(i)\frac{\mu+\sigma}{\mu}, \quad i = 0, 1, \ldots, v. \tag{9}$$

Relations (9) allow us to find the performance measures of studied model if we can calculate the performance measures of the simplified model. Let us continue the process of simplification.

The next simplifying step is to single out a separate group in the total mass of customers repeating the call, the change in the number of which is functionally independent of the number of busy servers. Let's give this group an index one. The remaining users who repeat the call form the second group. In the first group, the number of repeating customers increases by one with probability $H_1^{(1)}b_p$ each time when primary call appears in the first simplified model. Repeating customer of this group leaves the group with probability $1-H_2^{(1)}b_r$ after a random time having exponential distribution with parameter $\mu^{(1)}$ and with additional probability continue to stay in the first group. The considered model describes in more detail the process of generating a random number of customers repeating a call, but its main performance measures are the same as in the first simplified model.

Let us denote by $j_1(t)$ and $j_2(t)$ the number of repeating customers at time t in the first and second groups respectively and by $i(t)$ we denote the number of busy servers at time t. The dynamic of changing the model states is described by a three-dimensional Markov process of the type $r(t) = (j_1(t), j_2(t), i(t))$ with finite number of states S, $(j_1, j_2, i) \in S$, $j_1 = 0, 1, \ldots, N_1$ $j_2 = 0, 1, \ldots, N_2$ $i = 0, 1, \ldots, v$. This will be the second simplified model. Let us denote by $P^{(2)}(j_1, j_2, i)$ the probability of stationary state (j_1, j_2, i) of $r(t)$.

Acting in a standard way, it is not difficult to write down a system of equilibrium equations that relates the quantities $P^{(2)}(j_1, j_2, i)$ and calculate the probabilistic characteristics of the second simplified model. It is clear that this is a more time-consuming way of estimating the characteristics compared to solution (7). However, the constructed model with a detailed description of the number of subscribers repeating the call can significantly simplify the solution of this problem. Let's us suppose that the following relations are true

$$P^{(2)}(j_1, j_2, i) = P^{(2)}(0, j_2, i)\frac{D^{j_1}}{j_1!}, \quad (j_1, j_2, i) \in S, \tag{10}$$

where

$$D = \frac{\lambda H_1^{(1)}b_p}{\mu^{(1)}\left(1 - H_2^{(1)}b_r\right)}$$

and $N_1 = N_2 = \infty$.

After substitution of (10) into system of equilibrium equations that relates the quantities $P^{(2)}(j_1,j_2,i)$ with changing $P^{(2)}(0,j_2,i)$ into $P^{(3)}(j,i)$ we obtain the following system of linear equations

$$P^{(3)}(j,v)\left(\lambda^{(1)}H_1^{(2)}I(j<N)+j(\mu^{(2)}(1-H_2^{(1)})+\sigma^{(2)})+v\right) \quad (11)$$

$$= P^{(3)}(j,v-1)\lambda^{(1)}+P^{(3)}(j-1,v)\lambda^{(1)}H_1^{(2)}I(j>0)$$

$$+P^{(3)}(j+1,v-1)(j+1)\mu^{(2)}I(j<N)$$

$$+P^{(3)}(j+1,v)(j+1)(\mu^{(2)}(1-H_2^{(1)})+\sigma^{(2)})I(j<N),$$

$$j=0,1,\ldots,N; \quad i=v;$$

$$P^{(3)}(j,i)\left(\lambda^{(1)}+j(\mu^{(2)}+\sigma^{(2)})+i\right)=P^{(3)}(j,i-1)\lambda^{(1)}I(i>0) \quad (12)$$

$$+P^{(3)}(j+1,i-1)(j+1)\mu^{(2)}I(j<N,i>0)$$

$$+P^{(3)}(j+1,i)(j+1)\sigma^{(2)}I(j<N)+P^{(3)}(j,i+1)(i+1)I(i<v),$$

$$j=0,1,\ldots,N; \quad i=0,1,\ldots v-1,$$

where

$$\lambda^{(1)}=\lambda\big((1-b_p)+(1-b_r)A\big), \quad (13)$$

$$A=\frac{H_{1,b}^{(1)}b_p}{1-H_{2,b}^{(1)}b_r},$$

$$\mu^{(2)}=\mu^{(1)}(1-b_r),$$

$$\sigma^{(2)}=\mu^{(1)}(1-H_{2,b}^{(1)})b_r,$$

$$H_1^{(2)}=\frac{(1-b_p)H_1^{(1)}+(1-b_r)H_2^{(1)}A}{(1-b_p)+(1-b_r)A}.$$

It is easy to see that the system of linear equations (10) is a system of equations of statistical equilibrium of a particular case of the model under study, obtained with the following choice of input parameters $b_p=b_r=H_{1,b}=H_{2,b}=0$. This special case is the same as the basic model, except for the presence of aging of the transmitted information. This possibility can be removed from the model if we use the transformations used in the construction of the first simplified model at the beginning of this section. The performed transformations make it possible to use the basic model to evaluate the characteristics of the constructed generalized model.

Let us denote the parameters and characteristics of the corresponding basic model with the same symbols that were used for the generalized model, only with the index e at the top. Thus, in the basic model, λ^e the intensity of the flow of primary requests; μ^e parameter for exponential distribution of time between successive repeated calls; H_1^e and H_2^e, respectively, the call retry probabilities for the primary request and retrials. In the same way, we denote the characteristics of servicing requests introduced by the equalities (1). These are: $p^e(v)$; $J^e(v)$; I^e J^e; τ^e; M^e and π_c^e.

By changing the model of forming the flow of repeated calls and the procedure of aging of the transmitted information, it can be shown that the characteristics of the generalized model can be calculated using the characteristics of the basic model, if we select its parameters from the relations:

$$\lambda^e = \lambda((1 - b_p) + (1 - b_r)A); \quad A = \frac{\mu b_p H_{1,b}}{\sigma + \mu(1 - b_r H_{2,b})}; \tag{14}$$

$$\mu^e = \sigma + \mu(1 - b_r H_{2,b});$$

$$H_1^e = \frac{((1 - b_p)H_1 + (1 - b_r)H_2 A)\mu(1 - b_r)}{((1 - b_p) + (1 - b_r)A)(\sigma + \mu(1 - b_r H_{2,b}))};$$

$$H_2^e = \frac{H_2 \mu(1 - b_r)}{\sigma + \mu(1 - b_r H_{2,b})}.$$

The values of the characteristics of generalized model are determined from the expressions:

$$p(v) = p^e(v); \quad I = I^e; \tag{15}$$

$$J = \frac{\lambda b_p H_{1,b}}{\sigma + \mu(1 - b_r H_{2,b})} + \frac{\sigma + \mu(1 - b_r H_{2,b})}{\mu(1 - b_r)} J^e;$$

$$J(v) = \frac{\lambda b_p H_{1,b}}{\sigma + \mu(1 - b_r H_{2,b})} p^e(v) + \frac{\sigma + \mu(1 - b_r H_{2,b})}{\mu(1 - b_r)} J^e(v).$$

The remaining characteristics τ, M and π_c are calculated using expressions (15) and definitions of (1). The calculation process is straightforward and carried out by efficient recursive algorithm based on the sequential decomposition of the system of state equations of the basic model into N systems of linear equations with a tridiagonal matrix [6, 13].

4   Numerical Assessment

The constructed model can be used to solve various problems encountered in practical applications. Let's evaluate the positive effect that the use of the chatbot service creates when planning the required number of call center operators. To solve the formulated problem, it is necessary to change the previously accepted interpretation of the input parameters of the generalized model. In the new terms, b_p denotes the proportion of primary requests whose service was limited to contact with the chatbot service. Of these, $b_p(1 - H_{1,p})$ part was satisfied with the service, and $b_p H_{1,p}$ part was not satisfied, which led to a repetition of the request. Similarly, b_r denotes the proportion of repeat requests whose service was limited to contact with the chatbot service. Of these, $b_r(1 - H_{2,b})$ part was satisfied with the service, and $b_r H_{2,b}$ part was not satisfied, which led to a repetition of the request. For used interpretation of parameters, the definition of π_c the share of lost requests changes. Now it looks like:

$$\pi_c = \frac{\lambda(b_p H_{1,b} + (1 - b_p)p(v)) + \mu(J b_r H_{2,b} + (1 - b_r)J(v))}{\lambda + J\mu}.$$

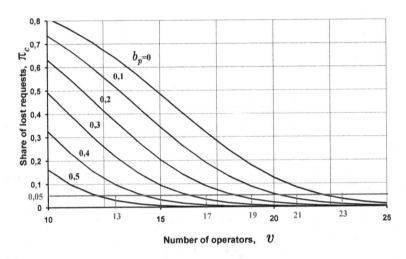

Fig. 2. The results of estimating the number of operators for different modes of operation of the chatbot service.

In Fig. 2 for parameter values: $v = 10$; $\lambda = 15$; $N = 40$; $\mu = 10$; $b_r = 0,1$; $H_{1,b} = 0$; $H_{2,b} = 0,9$; $H_1 = 0,7$; $H_2 = 0,95$; $\sigma = 0,1$ the results of estimating the number of operators to ensure the loss of requests at the level 0,05 for different proportions of requests b_p served on the chatbot service are given. The above data show that the use of the chatbot service can significantly reduce the required number of operators. The constructed model and algorithms for estimating its characteristics make it possible to numerically evaluate the positive effect of applying pre-servicing of incoming requests on the number of operators planning.

The constructed model and algorithms for estimating its probabilistic characteristics can also be used to eliminate the negative consequences of overloading information centers. The simplest and most effective solution is to redirect part of incoming requests to other centers that have the same functionality for servicing them. The value of portion of redirected calls can be found with help of constructed model.

Let us consider a numerical example that illustrated the solution of formulated problem. Model input parameters have the following values: $v = 30$; $\lambda = 35$; $b_r = 0,2$; $H_{1,b} = 0,05$; $H_{2,b} = 0,7$; $H_1 = 0,8$; $H_2 = 0,9$; $\sigma = 0,5$; $N = 40$.

Figure 3 illustrates the dependence of π_c on the portion of redirected primary calls $b_p(1 - H_{1,b})$ that varies from 0 (no redirection) to 0,57 for different values of intensity of repetition μ that takes values 1; 5; 10; 20. The content of the figure shows that uploading some of the initial requests to other information services allows to meet the requirements for the quality of their service in an overloaded information center. The required portion of redirected requests increases with increasing of intensity of call repetition.

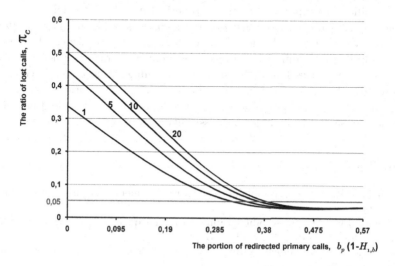

Fig. 3. Dependence of π_c on the portion of redirected primary calls $b_p(1 - H_{1,b})$ for different values of μ.

5 Conclusion

In this paper the mathematical model of a fully accessible communication system has been constructed and investigated, taking into account the repetition of a call after the preliminary service stage and due to the busy service devices. The model also assumes that the user's stay in the call repetition state is limited by a random variable that has an exponential distribution, which determines the aging time of the transmitted information. It is assumed that the arrival of primary claims obeys the Poisson law. The total intensity of repeated requests is determined by a random number of subscribers repeating the call. The probabilities of call repetition depend on the reason of failure and the number of attempt.

The constructed model generalizes the well-known basic model of a fully accessible system of service devices with repeated calls [6, 11–13], by taking into account the stage of preliminary servicing of incoming requests and aging of the transmitted information and the model considered in [22] by taking into account the dependence of call retry probability on the stage of connection establishment at which the refusal was received. The paper shows that despite the refinement of the process of servicing of incoming requests and the formation of repeated calls, the service characteristics of the generalized model can be calculated by an efficient recursive algorithm developed for the basic model [13]. This approach is based on the truncation of used state space and the sequential decomposition of the system of state equations of the basic model into N systems of linear equations with a tridiagonal matrix [13]. The error caused by truncation is found and algorithm of constructing the borders of truncated state space providing the given relative error of performance measures estimation is developed [13]. This results significantly expands the area of practical use of the model.

The constructed model and methods of its analysis can be used for evaluation of the positive effect that the use of the preliminary service like chatbot creates when planning the required number of call center operators. The model also makes it possible to estimate the share of incoming requests that need to be redirected to other call centers in an overload situation. Numerical examples are given to illustrate the practical use of the results obtained. Proposed model can be further developed to include the possibility of waiting for blocked requests.

References

1. Artalejo, J.R.: Accessible bibliography on retrial queues. Math. Comput. Model. **30**, 1–6 (1999)
2. Artalejo, J.R.: Accessible bibliography on retrial queues: progress in 2000–2009. Math. Comput. Model. **51**, 1071–1081 (2010)
3. Stepanov, S.N.: Numerical Methods for Analyzing Models with Retrials. Nauka, Moscow (1983). (in Russian)
4. Falin, G.I., Templeton, J.G.C.: Retrial Queues. Chapman and Hall, London (1997)
5. Artalejo, R., Gomez-Corral, A.: Retrial Queueing Systems: A Computational Approach. Springer, Berlin (2008). https://doi.org/10.1007/978-3-540-78725-9
6. Stepanov, S.N.: Teletraffic Theory: Concepts, Models, Applications. Goriachay Linia-Telecom, Moscow (2015). (in Russian)
7. Koole, G.M.: A Deep Dive into Call Center Workforce Management. MG Books, Amsterdam (2020)
8. Aksin, O.Z., Armony, M., Mehrotra, V.: The modern call-center: a multidisciplinary perspective on operations management research. Prod. Oper. Manag. **16**, 665–688 (2007)
9. Gans, N., Koole, G.M., Mandelbaum, A.: Telephone call centers: tutorial, review, and research prospects. Manuf. Serv. Oper. Manag. **5**, 79–141 (2003)
10. Koole, G., Li, S.: A practice-oriented overview of call center workforce planning. arXiv (2021). arXiv:2101.10122
11. Ionin, G.L., Sedol, S.S.: Telephone systems with repeated calls. In: Proceedings of the 6th International Teletraffic Congress, Munich, Germany (1970)
12. Kornyshev, Y.N.: A single-line system with repeated orders and preliminary servicing. Eng. Cybern. **15**, 63–68 (1977)
13. Stepanov, S.N.: Markov models with retrials: the calculation of stationary performance measures based on the concept of truncation. Math. Comput. Model. **30**, 207–228 (1999)
14. Stepanov, S.N.: Generalized model with retrials in case of extreme load. Queueing Syst. **27**, 131–151 (1998)

15. Stepanov, S.N., Stepanov, M.S., Shishkin, M.O.: Performance measures of emergency services in case of overload. In: Vishnevskiy, V.M., Samouylov, K.E., Kozyrev, D.V. (eds.) DCCN 2020. LNCS, vol. 12563, pp. 436–449. Springer, Cham (2020). https://doi.org/10.1007/978-3-030-66471-8_33

16. Ding, S., Koole, G., van der Mei, R.D.: On the estimation of the true demand in call centers with redials and reconnects. Eur. J. Oper. Res. **246**, 250–262 (2015)

17. Stepanov, S.N., Stepanov, M.S.: Construction and analysis of a generalized contact center model. Autom. Remote. Control. **75**(11), 1936–1947 (2014). https://doi.org/10.1134/S0005117914110046

18. Stepanov, S.N., Stepanov, M.S.: Algorithms for estimating throughput characteristics in a generalized call center model. Autom. Remote. Control. **77**(7), 1195–1207 (2016). https://doi.org/10.1134/S0005117916070067

19. Kim, C.S., Klimenok, V.I., Dudin, A.N.: Priority tandem queueing system with retrials and reservation of channels as a model of call center. Comput. Ind. Eng. **96**, 61–71 (2016)

20. Dudin, S., Dudina, O.: Retrial multi-server queuing system with PHF service time distribution as a model of a channel with unreliable transmission of information. Appl. Math. Model. **65**, 676–695 (2019)

21. Anisimov, V., Artalejo, J.: Approximation of multiserver retrial queues by means of generalized truncated models. TOP Off. J. Span. Soc. Stat. Oper. Res. **10**, 51–66 (2002). https://doi.org/10.1007/BF02578940

22. Stepanov, S., Stepanov, M.: Estimation of the performance measures of a group of servers taking into account blocking and call repetition before and after server occupation. Mathematics 9(21), 2811:1–2811:24 (2021). https://doi.org/10.3390/math9212811

23. Deul, N.: Stationary conditions for multi-server queueing systems with repeated calls. J. Inf. Process. Cybern. **16**, 607–613 (1980)

Transient Behaviour of Finite-Source Single-Line Queueing Systems with Jumps of Network Traffic

V. M. Vishnevsky[1], K. A. Vytovtov[1]([✉]), E. A. Barabanova[1],
G. K. Vytovtov[2], and A. V. Dvorkovich[3]

[1] V. A. Trapeznikov Institute of Control Sciences of RAS, Moscow, Russia
vytovtov_konstan@mail.ru
[2] Astrakhan State Technical University, Astrakhan, Russia
[3] Moscow Institute of Physics and Technology (National Research University),
Zhukovsky, Russia

Abstract. In this paper, the accurate analytical method for studying the transient behavior of finite-source single-line queueing systems with jumps of network traffic is presented. It based on the method of the probability translation matrix. The mathematical model allows us to analyze the transient mode of Local Area Network when a switch has been turned on or an arrival and service rate has been changed. We've obtained the analytical expressions for the system state probabilities including the expression for the loss probability and the transient time. The numerical calculations for the five-state queuing system $M/M/1/n$ are presented too. The calculation results for the considered particular cases fully correspond to the well-known analytical and numerical results for the system in the stationary mode.

Keywords: Local area network · Queueing system · Transient time

1 Introduction

Finite-source queueing systems are often used for description modern telecommunication systems. Some of practical applications of finite-source queueing systems are presented in papers [1–17]. The considered examples of such systems include telecommunication and computer ones. For example, in [1] the so-called Engset model has been used for describing circuit switch networks. In the work the steady state expressions of blocking probability and the number of ongoing calls have been found. Another example of Finite-Source queueing systems is considered in [2] where the multi-server heterogeneous queueing systems are investigated. For this type of system the performance characteristics and state

The work was carried out at the Moscow Institute of Physics and Technology with the financial support of the Ministry of Science and Higher Education of the Russian Federation under the project No. 075-11-2021-047.

probabilities in stationary mode are calculated. The problems of performance and reliability of telecommunication networks are investigated in [3, 4]. The paper [3] concerns with a single unreliable server queueing system with a finite number of heterogeneous sources of calls. Here it is found the analytical expressions of stationary probabilities of the system and shown the effect of the breakdown of the server on the mean response times of the requests and the mean number of calls staying at the system. Another mathematical methods as asymptotic methods and stochastic simulation are used in [4] to determine stationary performance measures of the system, such as mean number of calls staying at the server, mean queue lengths, average waiting, etc. A retrial queueing system of type M/M/1/N with a non-reliable server and impatient customers is considered in paper [5]. By the use of a self-developed simulation program the authors have obtained the graphical dependencies of steady state distribution of the customers, the mean response time of a successfully served customer, the probability of abandonment of a customer and other graphical results. Despite the large number of works devoted to the study of the queueing system stationary characteristics investigating only stationary mode of queueing system sometimes does not allow to describe telecommunication system functioning. It is because performance characteristics in transient mode can substantially differ from analogously ones in stationary mode [8–11]. For example, the authors of the paper [9] have shown that there is significant disparity between the transient and steady state values of single-server queue with batch arrivals under heavy load conditions. It should be noted that the mathematical methods of transient behavior investigation are more complex than stationary mode investigation ones. So, for example, in the paper [10] transient probabilities of queueing systems with catastrophes and impatient customers are presented by using continued fractions and generating function techniques. The accurate analytical method for the analysis of queueing system transient behavior and inhomogeneous continuous Markov processes with piecewise constant transition intensities has been proposed in [11–13]. Most other methods are numerical and their examples have been considered in introductions of [11, 12]. Recently, there are more and more applications of queuing systems transient behaviour results in telecommunications. In addition to installation and reboot [11], a transient mode occurs in the network with sharp load fluctuations [12]. Such fluctuations of traffic may be the results of long-term network attacks and in some cases the system may not go to stationary mode at all. So, investigating transient behaviour of finite-source single-line queueing systems with jumps of network traffic is the aim of this work and relevant problem for modern telecommunications. The paper is organized in the following manner. In Sect. 2 the statement of the problem is presented. Section 3 describes the detailed mathematical model. Numerical results are presented in Sect. 4.

2   The Statement of the Problem

A single-line queuing system with limited buffer $M/M/1/n$ is considered in this paper. Requests are arrived into this system from the finite number n of sources. Additionally, the source cannot generate requests until its previous request is serviced. For example, such the system is the model of a local area network that consists of n work stations served by a single threaded web server. The investigated system is described by the graph presented in the Fig. 1. In this figure: S_0 is the state when the system is idle, S_1 is the state when requests arrive to the web-server from one work station, S_i is the state when information packets from i work stations are in the system, and S_n is the state when requests from n work stations have arrived to the web-server. In the last state, no one request can enter the system, so this state is information loss one. The arrive $\lambda(t)$ or service $\mu(t)$ rates of information flows are piecewise-constant functions of t which are distributed exponentially within each of the intervals (Fig. 2). Such changes of the parameters $\lambda(t)$ and $\mu(t)$ can be caused by influence of an anomalous traffic. This fact can lead to the occurrence of transient processes in the system. As a result the probabilities of states, including the probability of losses, are not equal to ones in the stationary mode. Therefore, the study of the transient mode, taking into account traffic jumps is the important practical problem to predict the performance metrics of a local area network.

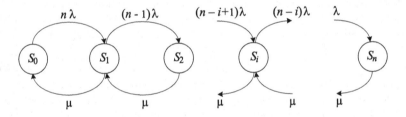

Fig. 1. The queuing system state graph.

The main aim of this paper is investigating transient modes of the studied system by using the analytical method developed in [11–13, 15].

3   The Mathematical Method

First of all, in according to the method [13] it is necessary to construct the so-call translation matrix for each interval with the constant parameters. For this, let us to write the system of Kolmogorov equations for j-th interval with the constant parameters in accordance to the graph (Fig. 1) as

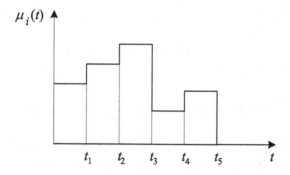

Fig. 2. The service rate time dependence graph

$$\frac{dP_{0j}(t)}{dt} = -n\lambda_j P_{0j}(t) + \mu_j P_{1j}(t)$$

$$\frac{dP_{1j}(t)}{dt} = n\lambda_j P_{0j}(t) - [(n-1)\lambda_j + \mu_j]P_{1j}(t) + \mu_j P_{2j}(t)$$

....

$$\frac{dP_{ij}(t)}{dt} = (n-i+1)\lambda_j P_{i-1,j}(t) - [(n-1)\lambda_j + \mu_j]P_{i,j}(t) + \mu_j P_{i+1,j}(t)$$ (1)

....

$$\frac{dP_{nj}(t)}{dt} = \lambda_j P_{n-1,j}(t) + \mu_j P_{nj}(t)$$

Here $p_{ij}(t)$ is the probability of i-th state of the system on the j-th interval. In other words, $p_{ij}(t)$ is the probability that i sources are currently transmitting information. Obviously, that the solution of the system (1) with the constant coefficients can be written in the form of the $n \times n$ probability translation matrix $\mathbf{M}$ [13]. This matrix relates the state probabilities at the moment t to those probability at $t = t_0$:

$$\mathbf{P}(t) = \mathbf{M} \cdot \mathbf{P}(t_0)$$ (2)

Here $\mathbf{P}(t) = \{p_{ij}(t)\}^T, i = \overline{1,n}, j = \overline{1,L}$ is the column-vector, T is the transpose operator, L is the number of the intervals. The matrix $\mathbf{M}$ in (2) can be found as the exponential of the coefficient matrix of the system (1). However, such an algorithm is very unwieldy and it is inapplicable in analysis of the systems with the piecewise-constant coefficients of (1). Therefore, we use the analytical method presented in [11,13,15]. The matrix in (2) allows us to describe the system behaviour at any time. Moreover, the proposed approach allows us to calculate such an important characteristic as the time of transient modes. It has been shown in the papers [12,13] that the transient time can be determined as $\tau_{tr} = (3 \div 5)\tau$, where $\tau = \max\{\tau_i, i = \overline{1,n}\} = \max 1/|\gamma_i|, i = \overline{1,n}$ is so-call time constant, γ_i is the eigenvalue of the coefficient matrix in (2). For service rates which have jumps at the moments $t_1, t_2, ..., t_{(L-1)}$ (Fig. 2), the probability translation matrix $\mathbf{M}_L$ describes the system behaviour within the L interval [11–13] as

$$\mathbf{M}_L = \mathbf{M}_L(t - t_{L-1}) \prod_{i=L-1}^{1} \mathbf{M}_i(\Delta t_i) \qquad (3)$$

Thus, the expression (3) allows us to construct the probability translation matrix in the accurate analytical form at any t for the piesewice-constant functions $\lambda_{(t)}$ and $\mu(t)$.

4   Numerical Calculation

In this section we consider the five-state queuing system $M/M/1/n$ described by the graph shown in Fig. 1. This system describes the example of small local area network (LAN) consists of four work stations. The probabilities of system states can be found by using the following Kolmogorov equations:

$$\frac{dP_0(t)}{dt} = -4\lambda P_0(t) + \mu P_1(t)$$

$$\frac{dP_1(t)}{dt} = 4\lambda(t)P_0(t) - (\mu(t) + 3\lambda(t))P_1(t) + \mu(t)P_2(t)$$

$$\frac{dP_2(t)}{dt} = 3\lambda(t)P_1(t) - (\mu(t) + 2\lambda(t))P_2(t) + \mu(t)P_3(t) \qquad (4)$$

$$\frac{dP_3(t)}{dt} = 2\lambda(t)P_2(t) - (\mu(t) + \lambda(t))P_3(t) + \mu(t)P_4(t)$$

$$\frac{dP_4(t)}{dt} = \lambda(t)P_3(t) - \mu(t)P_4(t)$$

First of all the system with the constant λ and μ is studied. The LAN is based on Ethernet technology with the packet size L = 1500 bytes and the information transfer rate $G = 1Gbit/s$. The service rate μ calculated as $\mu = G/L$ is $83.3 \cdot 10^3 packets/s$.

Here we investigate the three cases that is interesting for practical applications: 1) $\lambda < \mu$; 2) $\lambda = \mu$; 3) $\lambda > \mu$.

The first case is $\lambda < \mu$. In this case the characteristic equation of the system (1) for $\lambda = 16.7 \cdot 10^3 packets/s$ and $\mu = 83.3 \cdot 10^3 packets/s$ can be written as

$$-\gamma^5 - 4992\gamma^4 - 8.447 \cdot 10^{20}\gamma^3 - 5.624 \cdot 10^{15}\gamma^2 - 1.2 \cdot 10^{20}\gamma = 0, \qquad (5)$$

Here $\lambda = 16.7 \cdot 10^3 packets/s$ corresponds to input throughput $25Mbit/s$. The solution of (7) are $\gamma_0 = 0$; $\gamma_1 = -220996$; $\gamma_2 = -41277.5$; $\gamma_3 = -89644.2$;

$\gamma_4 = -147282$. The dependencies of the system state probabilities are presented in Fig. 3. The dashed line corresponds to the probability that the system is idle. The dashed-dotted line corresponds to the probability that one of four work stations is transmitting data. The solid line represents the probability that two of four computers are occupied in the system. The dotted line corresponds to the probability that three of four work stations are busy. The dotted-solid line corresponds to the probability that all four work stations are busy (loss probability). As it is seen from the Fig. 3. The steady-state loss probability is 0.012. This value coincides with the value calculated by the well-known formula for steady state probabilities [14]:

$$\pi_i = \pi_0 \rho^i \frac{m!}{(m-i)!}, 1 \le i \le m, \tag{6}$$

where

$$\pi_0 = \left(\sum_{i=0}^{4} \rho^i \frac{m!}{(m-i)!} \right)^{-1} 1 \le i \le m, \tag{7}$$

i is the number of system state, m is the number state of information loss, $\rho = \lambda/\mu$ is the system load. The transient time calculated by [13] is $\tau_{tr} = (3 \div 5)(\frac{1}{|\gamma_i|}) = (3 \div 5)\frac{1}{|41277.5|} = 1.2 \cdot 10^{-4}$ seconds.

Suppose that the service rate of the web-server has decreased to $16.7 \cdot 10^3$ packets/s. In this case $\lambda = \mu$ and the characteristic equation of system (1) has the form

$$-\gamma^5 - 232400\gamma^4 - 1.791 \cdot 10^{10}\gamma^3 - 5.306 \cdot 10^{14}\gamma^2 - 4.936 \cdot 10^{18}\gamma = 0, \tag{8}$$

And the solutions of (10) are $\gamma_0 = 0$; $\gamma_1 = -108635$; $\gamma_2 = -17330$; $\gamma_3 = -38711.3$; $\gamma_4 = -67723.7$. Therefore the transient time is $\tau_{tr} = (3 \div 5)(1/|17330|) = 2.9 \cdot 10^{-4}$ s. The steady state loss probability calculated by using the proposed method and the expressions (8) and (9) is increased to 0.37. This fact confirms the correctness of the translation matrix method using for such a class of the problems (Fig. 4).

Now let us suppose that the server service rate decreased even more due to the network errors. In this case $\lambda > \mu$. For example, it is assumed that $\lambda = 16.7 \cdot 10^3 packets/s$ and $\mu = 10.42 \cdot 10^3 packets/s$. The characteristic equation of system (5) is

$$-\gamma^5 - 207680\gamma^4 - 1.445 \cdot 10^{10}\gamma^3 - 3.92 \cdot 10^{14}\gamma^2 - 3.412 \cdot 10^{18}\gamma = 0 \tag{9}$$

and $\gamma_0 = 0$; $\gamma_1 = -95341.1$; $\gamma_2 = -16782.1$; $\gamma_3 = -35511.1$; $\gamma_4 = -60072.8$. The transient time is $\tau_{tr} = (3 \div 5) = (3 \div 5)(1/|16782.1|) = 3 \cdot 10^{-4}$ s. And the loss probability increased in comparison with the previous cases to 0.54.

Let us consider the transient behaviour of the system under the influence of service rate jumps. The first jump of service rate from $\mu_0 = 83.3 \cdot 10^3 packets/s$ to $\mu_1 = 10.42 \cdot 10^3 packets/s$ occurs at $t = 0.0003$ s (Fig. 6). This jump occurs

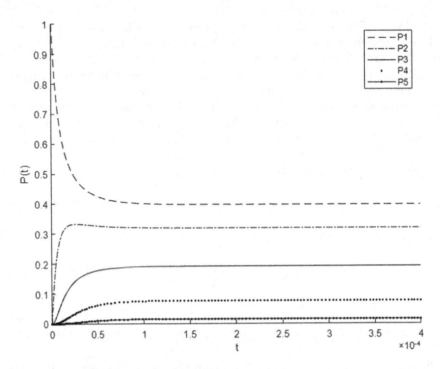

Fig. 3. Dependencies of the system state probabilities for $\lambda < \mu$

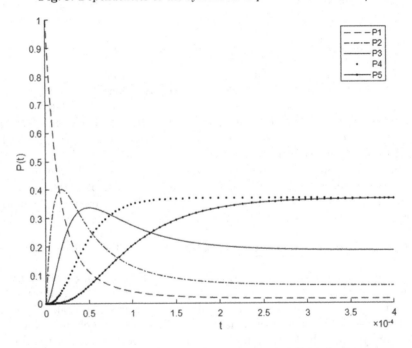

Fig. 4. Dependencies of the system state probabilities for $\lambda = \mu$

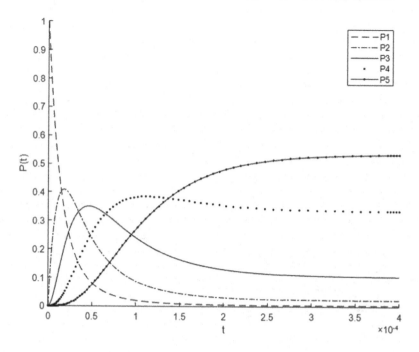

Fig. 5. Dependencies of the system state probabilities for $\lambda > \mu$

abruptly and it may be the result of anomalous traffic impact. Here the arrive rate to be constant $\lambda = 16.7 \cdot 10^3 packets/s$. In this case the jump of the service rate occurs in the stationary state of the system. As it is seen in Fig. 6 the probabilities of states and transient time after the jump does not depend on the moment of the jump and the values of arrive and service rates before the jump (Fig. 5).

In the second case the first jump of the service rate occurs from $\mu_0 = 83.3 \cdot 10^3 packets/s$ to $\mu_1 = 10.42 \cdot 10^3 packets/s$ at $t = 0.0001$ s. And the second jump from $\mu_1 = 10.42 \cdot 10^3 packets/s$ to $\mu_2 = 16.7 \cdot 10^3 packets/s$ occurs at the time $t = 0.0004$ s. The arrive rate is $\lambda = 16.7 \cdot 10^3 packets/s$. In this case the jumps are in the transient modes of the system. As it is seen in Fig. 7 the transient time in this case is the sum of the transient time before the jumps and after the jumps. Also, the probabilities of states after the jumps depend only on the values of arrive and service rates in the moments of jump occurrences and they do not depend on the probabilities of states before the jumps.

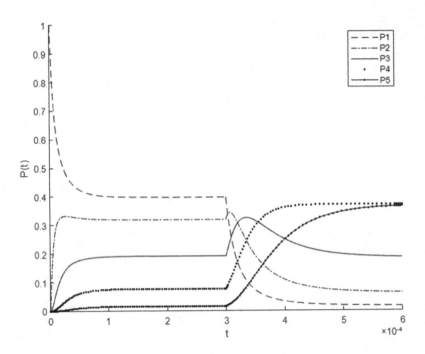

Fig. 6. Dependencies of the system state probabilities after the jump of $\mu(\mu_0 > \mu_1)$

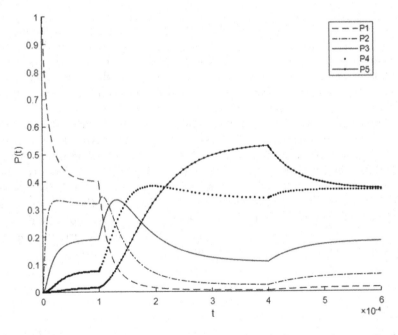

Fig. 7. Dependencies of the system state probabilities after two jumps of $\mu(\mu_0 > \mu_1, \mu_1 < \mu_2, \mu_2 = \lambda)$

5   Conclusion

In this paper, transient behavior of finite-source single-line queueing systems with jumps of network traffic is studied. Such systems describe widely used modern local networks under influence of an anomalous traffic. Obviously, studding a transient mode in the considered type of networks is the important theoretical and practical problem. The analytical method of calculation of transient state probabilities for constant and jump functions of arrival and service rates are proposed. The proposed method allows us to evaluate network performance metrics at any time, to assess their compliance with the Quality of Service, to predict network behavior, and to determine the necessity of its modernization. Moreover, the proposed approach allows us solving the network design problem since this approach is accurate analytical. The numerical calculation results of the five states system for the cases without jumps and with one and two jumps are presented. The corresponding graphs of the system states transient probabilities time dependences are presented. The probabilities numerical values in stationary made completely match with the steady state probabilities values obtained with using early presented methods. Also, important performance measures such as loss probability and transient time are calculated.

References

1. Bonald, T.: Insensitive traffic models for communication networks. Discrete Event Dyn. Syst. **17**, 405–417 (2007). https://doi.org/10.1007/s10626-007-0012-5
2. Efrosinin, D., Stepanova, N., Sztrik, J.: Algorithmic analysis of finite-source multi-server heterogeneous queueing systems. Mathematics **9**, 2624 (2021)
3. Sztrik, J., Kim, C.S.: Tool supported performability investigations of heterogeneous finite-source retrial queues. Annales Univ. Sci. Budapest. Sect. Comp. **32**, 201–220 (2010)
4. Sztrik, J., Kim, C.S.: Markov-modulated finite-source queueing models in evaluation of computer and communication systems. Math. Comput. Model. **38**(7–8), 961–968 (2003)
5. Tóth, A., Sztrik, J.: Simulation of finite-source retrial queuing systems with collisions, non-reliable server and impatient customers in the orbit. In: Proceedings of the 11th International Conference on Applied Informatics, pp. 408–419. CEUR Workshop Proceedings Eger, Hungary, (2020)
6. Sztrik, J.: Finite - source queueing systems and their applications. Formal Meth. Comput. **1**, 7–10 (2001)
7. Nazarov, A., Sztrik, J., Kvatch, A.: Recent results in finite-source retrial queues with collisions. In: Queueing Theory 1: Advanced Trends, pp. 213–259 (2020)
8. Rubino., G.: Transient analysis of Markovian queueing systems: a survey with focus on closed forms and uniformization. In: Queueing Theory 2: Advanced Trends, pp. 269–307 (2020)
9. Oduol, V.: Transient analysis of a single-server queue with batch arrivals using modeling and functions akin to the modified bessel functions. Int. J. Appl. Sci. Eng. Technol. **5**(1), 34–39 (2009)

10. Kumar, B., Lakshmi, S.R.A., Anbarasu, S.: Transient and steady-state analysis of queueing systems with catastrophes and impatient customers. Int. J. Math. Oper. Res. **6**(5), 523–549 (2014)
11. Parygin, D.S., Finogeev, A.G., Kamaev, V.A., Finogeev, A.A., Gnedkova, E.P., Tyukov, A.P.: A convergent model for distributed processing of Big Sensor Data in urban engineering networks. J. Phys Conf. Ser. **803**, 012112 (2017)
12. Vytovtov, K.A., Barabanova, E.A.: An analytical method for the analysis of inhomogeneous continuous Markov processes with piecewise constant transition intensities. Autom. Remote. Control. **82**(12), 2111–2123 (2021). https://doi.org/10.1134/S0005117921120043
13. Vishnevsky, V., Vytovtov, K., Barabanova, E., Semenova, O.: Transient behavior of the $MAP/M/1/N$ queuing system. Mathematics **9**(20), 2559 (2021)
14. Vytovtov, K., Barabanova, E., Vishnevsky, V.: The analytical method of transient behavior of the $M|M|1|n$ queuing system for piece-wise constant information flows. Lect. Notes Comput. Sci. **13144**, 167–181 (2021). https://doi.org/10.1007/978-3-030-92507-9_15
15. Dudin, A.N., Klimenok, V.I., Vishnevsky, V.M.: Methods to study queuing systems with correlated arrivals. In: The Theory of Queuing Systems with Correlated Flows, pp. 63–146. Springer, Cham (2020). https://doi.org/10.1007/978-3-030-32072-0_2
16. Hnatushenko, V.V., Vytovtov, G.K.: Analysis of the queueing systems at jumping variable information flow intensity. Appl. Probl. Math. Model. **4**(2.1), 77–83 (2021)
17. Kozyrev, D., et al.: Mobility-centric analysis of communication offloading for heterogeneous Internet of Things devices. Wirel. Commun. Mob. Comput. **2018**, 3761075 (2018)

Probability Density of the Interval Duration Between Events in the Generalized MAP with Its Incomplete Observability

Anastasia Keba[✉] and Ludmila Nezhel'skaya

National Research Tomsk State University, 36 Lenina Ave., Tomsk 634050, Russia
mir.na.mig7@mail.ru

Abstract. We consider a generalized MAP (Markovian Arrival Process) with an arbitrary number of states under conditions of its incomplete observability (in the presence of unextendable dead time of fixed duration). An explicit form of the probability density of the values of the interval duration between the moments of the events occurrence for a two-state flow is found.

Keywords: Generalized MAP · Probability density of the interval duration between events · Unextendable dead time · Hidden markov process · Embedded markov chain

1 Introduction

The intensive development of computer systems, the expansion of information communication networks gave impetus to the formation of a wide range of applications of the queuing theory apparatus, namely: the design, operation of information and computer networks, satellite communication systems, telecommunication networks, etc. This was the reason for introducing into consideration a new mathematical model of the incoming flow of events, which adequately describes the real information flows of random events.

Since in practice all flow parameters are often known partially or completely unknown, or change in time (often the changes are random), then doubly stochastic flows [1] were considered as mathematical models of real event flows in the mid-80s of the last century. These flows are characterized by double randomness: the moments when events occur are random and the intensity of the flow is a random process.

In this paper, we consider one of the types of doubly stochastic event flows, which are mathematical models of request information flows that operate in modern telecommunication networks. Doubly stochastic flows are divided into two classes:

1. The flow intensity is a continuous random process [2].

© The Author(s), under exclusive license to Springer Nature Switzerland AG 2022
V. M. Vishnevskiy et al. (Eds.): DCCN 2022, LNCS 13766, pp. 349–360, 2022.
https://doi.org/10.1007/978-3-031-23207-7_27

2. The flow intensity is a piecewise constant random process with finitely many states [3].

Note that the flows of the second-class include a generalized MAP event flow with an arbitrary (finite) number of states [4–6], which is an adequate mathematical model of real flows of random events.

In most cases, incoming flow models are considered when the events are fully observable. However, in practice, any device spends some time registering an event (message), in other words, an event received by the device generates a dead time period [7], during which other flow events that have occurred are unavailable for observation and do not cause its extension (unextendable dead time). This effect is typical for most real systems. In this regard, it can be considered that dead time period acts as a distorting factor in solving the estimation problem, since the effect of dead time entails the loss of flow events, which negatively affects the estimation of both states and flow parameters. In order to estimate the loss of events in a flow due to the effect of dead time, it is necessary to estimate the value of its duration. The period of unobservability of the flow of events can last for some fixed time [8–12], and can also be random [13–16]. It is assumed that this period of unobservable flow has a fixed duration T.

In this paper, we study a generalized MAP with an arbitrary number of states under conditions of unextendable dead time of a fixed duration. An explicit form of the probability density of the interval duration between the moments of the events occurrence of a two-state flow operating under conditions of unextendable dead time of a fixed duration is found. This article is a direct development of [4–6].

2   Statement of the Problem

We consider a generalized MAP event flow with an arbitrary number of states (hereinafter, a flow), which operates in a steady state (the flow operation time tends to infinity). The accompanying random process $\lambda(t)$ of the flow is a piecewise constant fundamentally unobservable random process with n states: $S_1, S_2, ..., S_n$. It is assumed that for $\lambda(t) = \lambda_i$ the i-th state (S_i), $i = \overline{1,n}$, of the process $\lambda(t)$ (the flow) takes place. In this case, $\lambda_1 > \lambda_2 > ... > \lambda_n > 0$. The observables are the time moments of the events occurrence of the flow $t_1, t_2, ...$.

The distribution function of a random variable - the duration of the stay of the process $\lambda(t)$ in the state S_i - is exponential: $F_i(t) = 1 - e^{-\lambda_i t}$, $t \geq 0$, $i = \overline{1,n}$. If the process $\lambda(t)$ is in the S_i state at time t, then on the half-interval $[t, t + \Delta t)$, where Δt is a sufficiently small value, at time the end of the state S_i, $i = \overline{1,n}$, the state is played.

1. Flow event occurs, and the process $\lambda(t)$ transits from the state S_i to the state S_j with probability $P_1(\lambda_j|\lambda_i)$, $i,j = \overline{1,n}$.
2. Flow event don't occurs, and process $\lambda(t)$ transits from the state S_i to the state S_j with probability $P_0(\lambda_j|\lambda_i)$, $i,j = \overline{1,n}$.

Note that for the introduced probabilities we have

$$\sum_{j=1}^{n} P_0(\lambda_j|\lambda_i) + \sum_{j=1}^{n} P_1(\lambda_j|\lambda_i) = 1, \ i = \overline{1,n}.$$

Remark 1. The generalization to a MAP event flow with an arbitrary number of states consists in introducing the probability $P_0(\lambda_i|\lambda_i) \neq 0$, $i = \overline{1,n}$, of the process $\lambda(t)$ switching from the state S_i to the state S_i without a flow event occurrence.

Statement 1. For a generalized MAP with n states, the process $\lambda(t)$ is a hidden Markov process.

Proof. The behavior of the random process $\lambda(t)$ after an arbitrary time t^0 in a probabilistic sense does not depend on the prehistory, that is, on how the process behaved before the time t^0, but depends only on the value of the process at time t^0.

Let at time t^0 the value of the process be $\lambda(t^0) = \lambda_i$. Since the exponential distribution has the property of no aftereffect, the remaining part of the stay duration of the process $\lambda(t)$ in the state S_i after the time t^0 does not depend on how long it has already been in this state S_i . The moment of transition of the process $\lambda(t)$ to the state S_i or the state S_j after t^0 is determined by the moment of the end of the stay duration of the process $\lambda(t)$ in the state S_i, $i,j = \overline{1,n}$. *Statement 1 is proven.*

The block matrix of infinitesimal characteristics of the process $\lambda(t)$ has the form $\mathbf{D} = \|\mathbf{D}_0|\mathbf{D}_1\|$, where

$$\mathbf{D}_0 = \left\| \begin{array}{cccc} -\lambda_1(1 - P_0(\lambda_1|\lambda_1)) & \lambda_1 P_0(\lambda_2|\lambda_1) & \dots & \lambda_1 P_0(\lambda_n|\lambda_1) \\ \lambda_2 P_0(\lambda_1|\lambda_2) & -\lambda_2(1 - P_0(\lambda_2|\lambda_2)) & \dots & \lambda_2 P_0(\lambda_n|\lambda_2) \\ \lambda_3 P_0(\lambda_1|\lambda_3) & \lambda_3 P_0(\lambda_2|\lambda_3) & \dots & \lambda_3 P_0(\lambda_n|\lambda_3) \\ \dots & \dots & \dots & \dots \\ \lambda_n P_0(\lambda_1|\lambda_n) & \lambda_n P_0(\lambda_2|\lambda_n) & \dots & -\lambda_n(1 - P_0(\lambda_n|\lambda_n)) \end{array} \right\|_{n \times n}$$

,

$$\mathbf{D}_1 = \left\| \begin{array}{cccc} \lambda_1 P_1(\lambda_1|\lambda_1) & \lambda_1 P_1(\lambda_2|\lambda_1) & \dots & \lambda_1 P_1(\lambda_n|\lambda_1) \\ \lambda_2 P_1(\lambda_1|\lambda_2) & \lambda_2 P_1(\lambda_2|\lambda_2) & \dots & \lambda_2 P_1(\lambda_n|\lambda_2) \\ \lambda_3 P_1(\lambda_1|\lambda_3) & \lambda_3 P_1(\lambda_2|\lambda_3) & \dots & \lambda_3 P_1(\lambda_n|\lambda_3) \\ \dots & \dots & \dots & \dots \\ \lambda_n P_1(\lambda_1|\lambda_n) & \lambda_n P_1(\lambda_2|\lambda_n) & \dots & \lambda_n P_1(\lambda_n|\lambda_n) \end{array} \right\|_{n \times n}$$

The elements $d_{ii}^{(0)}$, $i = \overline{1,n}$, of the matrix $\mathbf{D}_0$ are the output intensities of the process $\lambda(t)$ from their states, taken with the opposite sign; the elements $d_{ij}^{(0)}$ are the intensities of transitions from the state S_i to the state S_j without the occurrence of an event, $i,j = \overline{1,n}$. The elements $d_{ij}^{(1)}$ of the matrix $\mathbf{D}_1$ are the intensities of transitions of the process $\lambda(t)$ from the the state S_i to the state S_j with the occurrence of the event, $i, j = \overline{1, n}$.

Each registered event of the flow creates an unobservable period of fixed duration T (dead time), during which other events of the original flow are not observed (lost); in addition, events occurring during the dead time period do not cause it to be extended (unextendable dead time). The first thread event after the dead time period again creates a dead time period of duration T, and so on.

As an illustration, Fig. 1 shows one realization of the process $\lambda(t)$ and the observed flow, where λ_i is the value of the process $\lambda(t)$ in the state S_i, $i = \overline{1, n}$; shading indicates periods of dead time; events of the generalized MAP that are inaccessible to observation are marked with black circles; $t_1, t_2, \dots$ - the event occurrence times in the observed flow.

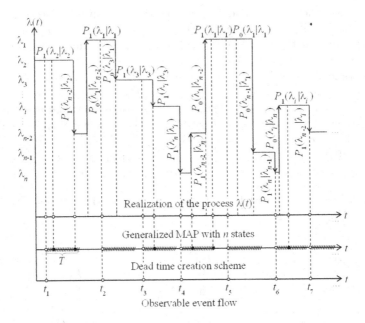

Fig. 1. Realization of Generalized MAP with an arbitrary number of states under conditions of unextendable dead time of duration T

Let $\tau_k = t_{k+1} - t_k$, $\tau_k \geq 0$, be the duration of the k-th interval between neighboring events of the observed flow t_k and t_{k+1}, $k = 1, 2, \dots$. For the probability density of the values τ_k, due to the operation of the flow in the stationary mode, $p_T(\tau_k) = p_T(\tau)$, $\tau \geq 0$, is true for any $k \geq 1$, which allows, without loss of generality, to put the moment of event occurrence t_k is equal to zero or, what is the same, the moment of event occurrence is $\tau = 0$.

The problem is to find an explicit form of the probability density distribution of the interval duration between the moments of neighboring events occurrence in the generalized two-state MAP with unextendable dead time. In what follows, we assume the number of flow states $n = 2$.

3   Derivation of the Probability Density of the Interval Duration Between Neighboring Events in the Generalized MAP Under Conditions of Partial Observability

Let us consider the interval $(0, \tau)$ of duration $\tau = T + t$ between neighboring events of the observed flow, where T is the value of the duration of the dead time, t is the value of the duration of the interval between the end of the dead time and the onset of next event, $t > 0$.

Statement 2. The sequence $t_1, t_2, ...$ forms a Markov chain $\{\lambda(t_k)\}$, $k = 1, 2, ...,$ embedded at times of occurrence of events.

Proof. Consider the behavior of the $\lambda(t)$ process.

1. Flow events can occur only at the moment of the end of the state S_i, $i = \overline{1, n}$, of the process $\lambda(t)$.
2. The stay duration of the process $\lambda(t)$ in one or another state S_i, $i = \overline{1, n}$, is distributed according to an exponential law with the property of no aftereffect.

Then the behavior of the process $\lambda(t)$ after the moment of time t_k, $k = 1, 2, ...,$ of the event occurrence in the observed flow does not depend on the prehistory, but depends only on the value of the process $\lambda(t)$ at time t_k, i.e. from $\lambda(t_k)$. Thus, the sequence $\{\lambda(t_k)\}$ is a Markov chain embedded in the time axis. *Statement 2 is proved.*

By Statement 2, the density $p_T(\tau)$ is defined as

$$
p_T(\tau) = \begin{cases} 0, & 0 \leq \tau < T, \\ \sum_{i=1}^{2} \pi_i(0\,T) \sum_{j=1}^{2} q_{ij}(T) \sum_{k=1}^{2} \widetilde{p}_{jk}(\tau - T), & \tau \geq T, \end{cases} \tag{1}
$$

where $q_{ij}(T)$ is the transition probability that during the dead time of duration T the process $\lambda(\tau)$ will go from the state S_i at time $\tau = 0$ to the state S_j at time $\tau = T$, $i, j = 1, 2$; $\pi_i(0|T)$ is the conditional stationary probability that the process $\lambda(\tau)$ at time $\tau = 0$ is in the state S_i, $i = 1, 2$, with the condition that at a given moment of time an event of the observed flow has occurred, giving rise to the dead time period of duration T; $\widetilde{p}_{jk}(t)$ is the density corresponding to the conditional probability $p_{jk}(t)$ that there are no flow events on the interval $(0, t)$ and at time t is the value of the process $\lambda(t) = \lambda_k$ provided that at time $t = 0$ the value of the process is $\lambda(t) = \lambda_j$, $j, k = 1, 2$.
  The following lemmas are valid for the quantities introduced.

Lemma 1. *The transition probabilities* $q_{ij}(T)$, $i, j = 1, 2$, *in the generalized MAP with dead time are determined by the formulas*

$$q_{11}(T) = \pi_1 + \pi_2 e^{-aT}, \quad q_{12}(T) = \pi_2(1 - e^{-aT}),$$
$$q_{21}(T) = \pi_1(1 - e^{-aT}), \quad q_{22}(T) = \pi_2 + \pi_1 e^{-aT},$$
$$a = \lambda_1(P_0(\lambda_2|\lambda_1) + P_1(\lambda_2|\lambda_1)) + \lambda_2(P_0(\lambda_1|\lambda_2) + P_1(\lambda_1|\lambda_2)), \qquad (2)$$

$$\sum_{j=1}^{2} q_{ij}(T) = 1, \quad i = 1, 2,$$

where the final a priori probabilities of the states S_1 *and* S_2 *for* $n = 2$ *are respectively [4]*

$$\pi_1 = \frac{\lambda_2(P_0(\lambda_1|\lambda_2) + P_1(\lambda_1|\lambda_2))}{a}, \quad \pi_2 = \frac{\lambda_1(P_0(\lambda_2|\lambda_1) + P_1(\lambda_2|\lambda_1))}{a}, \qquad (3)$$

$$\pi_1 + \pi_2 = 1.$$

Proof. Let the process $\lambda(\tau)$ at time $\tau + \Delta\tau$ take the value $\lambda(\tau + \Delta\tau) = \lambda_j$, i.e. is in the state S_j, $j = 1, 2$. Let us describe possible situations on the interval $(0, \tau + \Delta\tau)$.

1. The process $\lambda(\tau)$ at time $\tau = 0$ takes the value $\lambda(0) = \lambda_i$, i.e. is in the state S_i, $i = 1, 2$, and on the interval $(0, \tau)$ transits from the state S_i to the state S_j, $i \neq j$, with the probability $q_{ij}(\tau)$, $i, j = 1, 2$, $i \neq j$, and on the half-interval $[\tau, \tau + \Delta\tau)$ the state S_j of the process $\lambda(\tau)$ does not end with probability $e^{-\lambda_j \Delta\tau}$; the probability of this situation is equal to $e^{-\lambda_j \Delta\tau} q_{ij}(\tau)$, $i, j = 1, 2$, $i \neq j$.

2. The process $\lambda(\tau)$ at time $\tau = 0$ takes the value $\lambda(0) = \lambda_i$, i.e. is in the state S_i, $i = 1, 2$, and on the interval $(0, \tau)$ transits from the state S_i to the state S_j, $i \neq j$, with the probability $q_{ij}(\tau)$, $i, j = 1, 2$, $i \neq j$, and on the half-interval $[\tau, \tau + \Delta\tau)$ the state S_j of the process $\lambda(\tau)$ ends with probability $1 - e^{-\lambda_j \Delta\tau}$, and the process $\lambda(\tau)$ transits from the state S_j to the state S_j without an event occurring, the probability of this is $P_0(\lambda_j|\lambda_j)$; the probability of this situation is written as $P_0(\lambda_j|\lambda_j)(1 - e^{-\lambda_j \Delta\tau}) q_{ij}(\tau)$, $i, j = 1, 2$, $i \neq j$.

3. The process $\lambda(\tau)$ at time $\tau = 0$ takes the value $\lambda(0) = \lambda_i$, i.e. is in the state S_i, $i = 1, 2$, and on the interval $(0, \tau)$ transits from the state S_i to the state S_j, $i \neq j$, with the probability $q_{ij}(\tau)$, $i, j = 1, 2$, $i \neq j$, and on the half-interval $[\tau, \tau + \Delta\tau)$ the state S_j of the process $\lambda(\tau)$ ends with probability $1 - e^{-\lambda_j \Delta\tau}$, and the process $\lambda(\tau)$ transits from the state S_j to the state S_j with the occurrence of an event, the probability of this is $P_1(\lambda_j|\lambda_j)$; the probability of this situation is written as $P_1(\lambda_j|\lambda_j)(1 - e^{-\lambda_j \Delta\tau}) q_{ij}(\tau)$, $i, j = 1, 2$, $i \neq j$.

4. The process $\lambda(\tau)$ at time $\tau = 0$ takes the value $\lambda(0) = \lambda_i$, i.e. is in the state S_i, $i = 1, 2$, and on the interval $(0, \tau)$ transits from the state S_i to the state S_i with the probability $q_{ii}(\tau)$, $i = 1, 2$, and on the half-interval $[\tau, \tau + \Delta\tau)$ the state S_j of the process $\lambda(\tau)$ ends with probability $1 - e^{-\lambda_i \Delta\tau}$, and the process $\lambda(\tau)$ transits from the state S_i to the state S_j without occurrence of an event, the probability of this is $P_0(\lambda_i|\lambda_j)$; the probability of this situation is written as $P_0(\lambda_i|\lambda_j)(1 - e^{-\lambda_i \Delta\tau}) q_{ii}(\tau)$, $i, j = 1, 2$, $i \neq j$.

5. The process $\lambda(\tau)$ at time $\tau = 0$ takes the value $\lambda(0) = \lambda_i$, i.e. is in the state S_i, $i = 1, 2$, and on the interval $(0, \tau)$ transits from the state S_i to the state S_i with the probability $q_{ii}(\tau)$, $i = 1, 2$, and on the half-interval $[\tau, \tau + \Delta\tau)$ the state S_j of the process $\lambda(\tau)$ ends with probability $1 - e^{-\lambda_i \Delta\tau}$, and the process $\lambda(\tau)$ transits from the state S_i to the state S_j with the occurrence of the event, the probability of this is $P_1(\lambda_i|\lambda_j)$; the probability of this situation is written as $P_1(\lambda_i|\lambda_j)(1 - e^{-\lambda_i \Delta\tau})q_{ii}(\tau)$, $i, j = 1, 2$, $i \neq j$.

Taking into account all possible transitions of the process $\lambda(\tau)$ on the interval $(0, \tau + \Delta\tau)$ and performing the necessary transformations, we obtain the following systems of differential equations for the introduced transition probabilities $q_{ij}(\tau)$, $0 \leq \tau \leq T$, $i, j = 1, 2$, with initial conditions:

$$q'_{i1}(\tau) = -\lambda_1(P_0(\lambda_2|\lambda_1) + P_1(\lambda_2|\lambda_1))q_{i1}(\tau) + \lambda_2(P_0(\lambda_1|\lambda_2) + P_1(\lambda_1|\lambda_2))q_{i2}(\tau),$$
$$q'_{i2}(\tau) = \lambda_1(P_0(\lambda_2|\lambda_1) + P_1(\lambda_2|\lambda_1))q_{i1}(\tau) - \lambda_2(P_0(\lambda_1|\lambda_2) + P_1(\lambda_1|\lambda_2))q_{i2}(\tau),$$
$$q_{ii}(0) = 1, \ q_{ij}(0) = 0, \ i, j = 1, 2, \ i \neq j.$$

Solving systems of equations by the elimination method or by the Euler method, determining the unknown constants from the initial conditions and replacing the time τ in the found solution by the time T, we arrive at (2). *Lemma 1 is proved.*

Lemma 2. *The probability densities $\widetilde{p}_{jk}(\tau)$, $j, k = 1, 2$, in the generalized MAP are determined by the formulas*

$$\widetilde{p}_{j1}(\tau) = \lambda_1 P_1(\lambda_1|\lambda_1)p_{j1}(\tau) + \lambda_2 P_1(\lambda_1|\lambda_2)p_{j2}(\tau),$$
$$\widetilde{p}_{j2}(\tau) = \lambda_1 P_1(\lambda_2|\lambda_1)p_{j1}(\tau) + \lambda_2 P_1(\lambda_2|\lambda_2)p_{j2}(\tau), \ j = 1, 2, \tag{4}$$

where the conditional probabilities $p_{jk}(\tau)$, $j, k = 1, 2$, have the form [17]

$$p_{11}(\tau) = \frac{[\lambda_2(1 - P_0(\lambda_2|\lambda_2)) - z_1]e^{-z_1\tau} - [\lambda_2(1 - P_0(\lambda_2|\lambda_2)) - z_2]e^{-z_2\tau}}{z_2 - z_1},$$

$$p_{12}(\tau) = \frac{\lambda_1 P_0(\lambda_2|\lambda_1)}{z_2 - z_1}\{e^{-z_1\tau} - e^{-z_2\tau}\},$$

$$p_{22}(\tau) = \frac{[\lambda_1(1 - P_0(\lambda_1|\lambda_1)) - z_1]e^{-z_1\tau} - [\lambda_1(1 - P_0(\lambda_1|\lambda_1)) - z_2]e^{-z_2\tau}}{z_2 - z_1},$$

$$p_{21}(\tau) = \frac{\lambda_2 P_0(\lambda_1|\lambda_2)}{z_2 - z_1}\{e^{-z_1\tau} - e^{-z_2\tau}\},$$

$$z_{1,2} = \frac{\lambda_1(1 - P_0(\lambda_1|\lambda_1)) + \lambda_2(1 - P_0(\lambda_2|\lambda_2)) \mp \sqrt{D}}{2},$$

$$D = [\lambda_1(1 - P_0(\lambda_1|\lambda_1)) - \lambda_2(1 - P_0(\lambda_2|\lambda_2))]^2 + 4\lambda_1\lambda_2 P_0(\lambda_2|\lambda_1)P_0(\lambda_1|\lambda_2),$$

$$0 < z_1 < z_2.$$

Proof. Let us determine the joint probabilities $p_{ij}(\tau)\lambda_j P_1(\lambda_d|\lambda_j)\Delta\tau + o(\Delta\tau)$ of the fact that, without the occurrence of an event on the interval $(0, \tau)$, the process $\lambda(\tau)$ on this interval transits from the state S_i to the state S_j, $i, j = 1, 2$, on the half-interval $[\tau, \tau + \Delta\tau)$ the state S_j of the process $\lambda(\tau)$ ends, and at the time of the occurrence of the event, the process $\lambda(\tau)$ transits from the state S_j to the state S_d, $d, j = 1, 2$:

$$
\begin{aligned}
& p_{11}(\tau)\lambda_1 P_1(\lambda_1|\lambda_1)\Delta\tau + o(\Delta\tau), \; p_{11}(\tau)\lambda_1 P_1(\lambda_2|\lambda_1)\Delta\tau + o(\Delta\tau), \\
& p_{12}(\tau)\lambda_2 P_1(\lambda_1|\lambda_2)\Delta\tau + o(\Delta\tau), \; p_{12}(\tau)\lambda_2 P_1(\lambda_2|\lambda_2)\Delta\tau + o(\Delta\tau), \\
& p_{21}(\tau)\lambda_1 P_1(\lambda_1|\lambda_1)\Delta\tau + o(\Delta\tau), \; p_{21}(\tau)\lambda_1 P_1(\lambda_2|\lambda_1)\Delta\tau + o(\Delta\tau), \\
& p_{22}(\tau)\lambda_2 P_1(\lambda_1|\lambda_2)\Delta\tau + o(\Delta\tau), \; p_{22}(\tau)\lambda_2 P_1(\lambda_2|\lambda_2)\Delta\tau + o(\Delta\tau).
\end{aligned}
$$

Note that each of the above joint probabilities can be represented as

$$
p_{ij}(\tau)\lambda_j P_1(\lambda_d|\lambda_j)\Delta\tau + o(\Delta\tau) = \int_{\tau}^{\tau+\Delta\tau} \widetilde{p}_{ij}^{(d)}(u)\,du = \widetilde{p}_{ij}^{(d)}(\tau)\Delta\tau + o(\Delta\tau),
$$

$$
i, j = 1, 2, \; d = 1, 2,
$$

where $\widetilde{p}_{ij}^{(d)}(\tau)$ is the corresponding probability $p_{ij}(\tau)\lambda_j P_1(\lambda_d|\lambda_j)$ probability density, $i, j = 1, 2$, $d = 1, 2$.

Let us write the last equality in the form

$$
p_{ij}(\tau)\lambda_j P_1(\lambda_d|\lambda_j) + \frac{o(\Delta\tau)}{\Delta\tau} = \widetilde{p}_{ij}^{(d)}(\tau) + \frac{o(\Delta\tau)}{\Delta\tau}, \; i, j = 1, 2, \; d = 1, 2.
$$

Letting $\Delta\tau \to 0$ here, we find

$$
\widetilde{p}_{ij}^{(d)}(\tau) = p_{ij}(\tau)\lambda_j P_1(\lambda_d|\lambda_j), \; i, j = 1, 2, \; d = 1, 2.
$$

Then the probability density $\widetilde{p}_{ij}(\tau)$ of the fact that without the occurrence of an event on the interval $(0, \tau)$ and the occurrence of an event at time τ, the process $\lambda(\tau)$ transits from the state S_i to the state S_j, $i, j = 1, 2$, according to the above equality will be written in the form (4). *Lemma 2 is proved.*

Let us introduce into consideration the transition probabilities $\pi_{ij}(T)$ – the probability that during the time that passes from the moment $\tau = 0$ of the occurrence of the flow event to the moment of the next event in the observed flow, the process $\lambda(\tau)$ will transits from the state S_i (at the moment $\tau = 0$) to the state S_j (the moment of the next event in the observed flow), $i, j = 1, 2$.

Lemma 3. *In the generalized MAP with dead time, the transition probabilities* $\pi_{ij}(T)$, $i, j = 1, 2$, *have the form*

$$\pi_{11}(T) = p_{11} - \pi_2(p_{11} - p_{21})(1 - e^{-aT}),$$
$$\pi_{12}(T) = p_{12} + \pi_2(p_{22} - p_{12})(1 - e^{-aT}),$$
$$\pi_{21}(T) = p_{21} + \pi_1(p_{11} - p_{21})(1 - e^{-aT}),$$
$$\pi_{22}(T) = p_{22} - \pi_1(p_{22} - p_{12})(1 - e^{-aT}), \tag{5}$$

$$\sum_{j=1}^{2} \pi_{ij}(T) = 1, \ i = 1, 2,$$

where π_1, π_2 *are defined in* (3), *value* a *in* (2), *transition probabilities* p_{ij}, $i, j = 1, 2$ *are given by the formulas [17]*

$$p_{1i} = \frac{P_1(\lambda_i|\lambda_1)(1 - P_0(\lambda_2|\lambda_2)) + P_0(\lambda_2|\lambda_1)P_1(\lambda_i|\lambda_2)}{(1 - P_0(\lambda_1|\lambda_1))(1 - P_0(\lambda_2|\lambda_2)) - P_0(\lambda_1|\lambda_2)P_0(\lambda_2|\lambda_1)},$$
$$p_{2i} = \frac{P_1(\lambda_i|\lambda_2)(1 - P_0(\lambda_1|\lambda_1)) + P_0(\lambda_1|\lambda_2)P_1(\lambda_i|\lambda_1)}{(1 - P_0(\lambda_1|\lambda_1))(1 - P_0(\lambda_2|\lambda_2)) - P_0(\lambda_1|\lambda_2)P_0(\lambda_2|\lambda_1)}, \tag{6}$$

$$\sum_{j=1}^{2} p_{ij} = 1, \ i = 1, 2.$$

Proof. Since the process $\lambda(\tau)$ (Statement 1) is Markovian, the probabilities $q_{ij}(T)$ and p_{ij} defined in (2) and (6) respectively, allow us to write the equations for the probabilities $\pi_{ij}(T)$, $i, j = 1, 2$, in the form

$$\pi_{ij}(T) = q_{i1}(T)p_{1j} + q_{i2}(T)p_{2j}, \ i, j = 1, 2.$$

Substituting the expressions (2) and (6) (for the corresponding i, j) into the last formula, we arrive at (5). *Lemma 3 is proved.*

Lemma 4. *In the generalized MAP with dead time, the conditional stationary probabilities* $\pi_i(0|T)$, $i = 1, 2$, *are defined by the expressions*

$$\pi_1(0\,T) = \frac{p_{21} + \pi_1(p_{11} - p_{21})(1 - e^{-aT})}{p_{12} + p_{21} + (p_{22} - p_{12})(1 - e^{-aT})},$$
$$\pi_2(0\,T) = \frac{p_{12} + \pi_2(p_{22} - p_{12})(1 - e^{-aT})}{p_{12} + p_{21} + (p_{22} - p_{12})(1 - e^{-aT})}, \tag{7}$$

where π_1, π_2 *are defined in* (3), *value* a *in* (2), *transition probabilities* p_{ij}, $i, j = 1, 2 - $ *in* (6)

Proof. For an embedded Markov chain, the equations for final probabilities

$$\pi_1(0\,T) = \pi_1(0\,T)\pi_{11}(T) + \pi_2(0\,T)\pi_{21}(T),$$
$$\pi_2(0\,T) = \pi_1(0\,T)\pi_{12}(T) + \pi_2(0\,T)\pi_{22}(T),$$

$$\pi_1(0\,T) + \pi_2(0\,T) = 1.$$

Substituting (5) into the written system of equations, we find (7). *Lemma 4 is proved.*

Theorem 1. *The probability density of the interval duration between neighboring events in a generalized two-state MAP operating under dead time conditions has the form*

$$p_T(\tau) = \begin{cases} 0, & 0 \le \tau < T, \\ \gamma(T)z_1 e^{-z_1(\tau-T)} + (1 - \gamma(T))z_2 e^{-z_2(\tau-T)}, & \tau \ge T, \end{cases}$$

$$\gamma(T) = \frac{1}{z_2 - z_1}(z_2 - \lambda_1\pi_1(T)[P_1(\lambda_2|\lambda_1) + P_1(\lambda_2|\lambda_2)] \tag{8}$$
$$-\lambda_2\pi_2(T)[P_1(\lambda_1|\lambda_2) + P_1(\lambda_2|\lambda_2)]),$$
$$\pi_1(T) = \pi_1 - [\pi_1 - \pi_1(0\,T)]e^{-aT}, \quad \pi_2(T) = \pi_2 - [\pi_2 - \pi_2(0\,T)]e^{-aT},$$

where π_i, $i = 1, 2$, are defined in (3), value a in (2), $\pi_i(0|T)$, $i = 1, 2$ – in (7).

Proof. Substituting the expressions (2), (4) and (7) into the formula (1), after the necessary transformations we get (8). *The theorem 1 is proved.*

The resulting formula (8) allows you to find the average number of lost events of the original flow per unit time due to the distoring factor (unextendable dead time):

$$\Delta_T = \frac{1}{M(\tau)} - \frac{1}{M_T(\tau)},$$

$$M(\tau) = \int_0^\infty \tau p(\tau)d\tau,$$

$$p(\tau) = \gamma(0)z_1 e^{-z_1\tau} + (1 - \gamma(0))z_2 e^{-z_2\tau},$$

$$\gamma(0) = \frac{1}{z_2 - z_1}(z_2 - \lambda_1\pi_1(0)[P_1(\lambda_2|\lambda_1) + P_1(\lambda_2|\lambda_2)] - \lambda_2\pi_2(0)[P_1(\lambda_1|\lambda_2) + P_1(\lambda_2|\lambda_2)]),$$

$$\pi_1(0) = \frac{p_{21}}{p_{12} + p_{21}}, \quad \pi_2(0) = \frac{p_{12}}{p_{12} + p_{21}},$$

where p_{ij}, $i, j = 1, 2$, are defined in (6);

$$M_T(\tau) = \int_T^\infty \tau p_T(\tau)d\tau,$$

where $p_T(\tau)$ is defined in (8).
Making the necessary calculations, we find

$$M(\tau) = \frac{\gamma(0)}{z_1} + \frac{1 - \gamma(0)}{z_2}, \quad M_T(\tau) = T + \frac{\gamma(T)}{z_1} + \frac{1 - \gamma(T)}{z_2}.$$

4   Conclusion

The model of the generalized MAP considered in the article with unextendable dead time belongs to the class of doubly stochastic flows, which are generally correlated flows. In this regard, the study of such flows, which are mathematical models of real information flows of messages (requests), is the modern content of the tasks of designing, implementing, operating information and computer networks [18–22], satellite communication systems, telecommunication networks, etc.

The explicit form (8) of the probability density $p_T(\tau)$, $\tau \geq 0$, the interval duration between neighboring events of the observed flow is obtained in the article. The result obtained can be used to solve the problem of estimating the unknown parameters of a generalized MAP with two states, in particular, to estimate the duration of the unobservability period T.

References

1. Grandell, J.: Doubly Stochastic Poisson Processes. SpringerVerlag, Berlin, Berlin-Heidelberg (1976)
2. Cox, D.R., Miller, H.D.: The Theory of Stochastic Processes, pp. 203–251. Wiley, New York (1965)
3. Neuts, M.F.: A versatile Markov point process. J. Appl. Probab. **16**, 764–779 (1979)
4. Nezhel'skaya, L.A., Keba, A.V.: Optimal state estimation of a generalized MAP event flow with an arbitrary number of states observability. Autom. Remote Control. **82**(5), 798–811 (2021)
5. Keba, A.V., Nezhel'skaya, L.A.: Optimal state estimation of generalized MAP with an arbitrary number of states under conditions of unextendable dead time. Tomsk. State Univ. J. Control. Comput. Sci. **56**, 68–80 (2021)
6. Keba, A.V., Nezhel'skaya, L.A.: Statistical experiments on a simulation model of a generalized MAP with an arbitrary number of states in conditions of unextendable dead time. In: Proceedings of the Thirteenth International Conference on New information technologies in the study of complex structures, pp. 85–86. TSU, Tomsk (2020)
7. Apanasovich, V.V., Kolyada, A.A., Chernyavsky, A.F.: Statistical analysis of random flows in a physical experiment. University, Minsk (1988)
8. Nezhel'skaya, L.: Probability density function for modulated MAP event flows with unextendable dead time. In: Dudin, A., Nazarov, A., Yakupov, R. (eds.) ITMM 2015. CCIS, vol. 564, pp. 141–151. Springer, Cham (2015). https://doi.org/10.1007/978-3-319-25861-4_12
9. Gortsev, A.M., Nezhel'skaya, L.A., Solov'ev, A.A.: Optimal state estimation in MAP-event flows with unextendable dead time. Autom. Remote Control. **73**(8), 1316–1326 (2012)
10. Nezhel'skaya, L.A.: Estimation of the unextendable dead time period in a flow of physical events by the method of maximum likelihood. Russ. Phys. J. **59**(5), 651–662 (2016)
11. Nezhelskaya, L.A., Sidorova, E.F.: Estimation of the unextendable dead time duration in correlated synchronous generalized flow of the second order Tomsk. State Univ. J. Control. Comput. Sci. **48**, 21–30 (2019)

12. Kalyagin, A.A., Nezhelskaya, L.A.: Comparison of MP- and MM-estimates of the duration of the dead time in generalized semi-synchronous event flow. Tomsk. State Univ. J. Control. Comput. Sci. **3**(32), 23–32 (2015)

13. Vasilyeva, L.A.: Estimation of the parameters of a doubly stochastic flow of events in the presence of dead time. Tomsk. State Univ. J. **S1**(1), 9–13 (2002)

14. Gortsev, A.M., Vetkina, A.V.: MM-estimation of the parameter of the uniform distribution of the duration of unextendable random dead time in the semi-synchronous events flow in the special case. Tomsk. State Univ. J. Control. Comput. Sci. **58**, 58–70 (2022)

15. Gortsev, A.M., Zavgorodnyaya, M.E.: Estimation of the parameter of unextendable dead time random duration in the Poisson flow of events. Tomsk. State Univ. J. Control. Comput. Sci. **40**, 32–40 (2017)

16. Nezhel'skaya, L.A., Pershina, A.A.: Estimate of the parameter of unextendable random dead time in a recurrent generalized asynchronous flow of physical events. Russ. Phys. J. **63**(1), 99–104 (2020). https://doi.org/10.1007/s11182-020-02007-5

17. Keba, A.V., Nezhel'skaya, L.A.: Estimation of the probability density parameters of the interval duration between events in generalized MAP with two states. Tomsk. State Univ. J. Control. Comput. Sci. **57**, 62–73 (2021)

18. Vishnevsky, V.M., Dudin, A.N., Klimenok, V.I.: Stochastic Systems with Correlated Flows. Theory and Applications in Telecommunication Networks. Technosphere, Moscow (2018)

19. Dudin, A.N., Kim, C.S., Dudin, S.A.: Optimal control by a node of wireless sensor network with quality of transmission depending on the amount of harvested energy. In: Proceedings of the Twenty Second International Scientific Conference on Distributed Computer and Communication Networks: Control, Computation, Communications (DCCN-2019), pp. 29–36. ICS RAS, Moscow (2019)

20. Dudin, A.N., Nazarov, A.A.: The MMAP/M/R/0 queueing system with reservation of servers operating in a random environment. Probl. Inf. Transm. **51**(3), 289–298 (2015)

21. Chakravarthy, S.R.: The Markovian arrival process: a review and future work. Adv. Probab. Theory Stochast. Processes **1**, 21–49 (2001)

22. Dudin, A.N., Dudin, S.A., Dudina, O.S.: Study of a system with a phase-type service time distribution and using the processor division discipline when the limit of the total intensity of request servicing is exceeded. In: Proceedings of the Twenty Second International Scientific Conference on Distributed Computer and Communication Networks: Control, Computation, Communications (DCCN-2019), pp. 75–82. ICS RAS, Moscow (2019)

Inventory Control with Returns and Controlled Markov Queueing Systems

S. Granin[1], V. Laptin[2], and A. Mandel[3(✉)]

[1] Moscow Institute of Physics and Technology, Moscow, Russia
[2] M.V. Lomonosov Moscow State University, Lenin's Mountings 1, Moscow, Russia
straqker@bk.ru
[3] V.A. Trapeznikov Institute of Control Sciences RAS, Profsoyuznaya 65, Moscow, Russia
almandel@yandex.ru

Abstract. A new model of inventory control with returns is considered, when it is possible for consumers to return (under certain conditions) the products they have purchased and a similar model of multi-channel controlled queueing system with switching of service channels. An optimal inventory control strategy in such a system turns out to be four-level. Then we investigate a model of a multilinear queueing system (QS) with channel switching under uncertainty and study the analogy of this model with the model of inventory control with returns.

Keywords: Inventory control with returns · Optimal inventory control strategies · Multi-channel queueing systems · Switching of service channels

1 Introduction

A fundamentally new model of inventory control with returns is considered, which differs from the classical models of the theory of inventory control [1–3]. In this case, it is assumed that the consumer can not only purchase the goods stored in the warehouse or in the store, but also return them to the seller (possibly on terms different from the ones of the purchase). In the same way, it is assumed that the warehouse itself can not only submit an order for replenishment, but also return previously received consignments of goods to the same supplier.

A few words about the main motivation for exploring this new inventory control model. It was due to the analogy identified between this formulation of the inventory control problem and management problems for one class of queueing systems (QS) first described in [4]. In this paper, new deeper features of this analogy are revealed. Let us refer to a remarkable review by V. Rykov on controlled QS [5]. The threshold nature of control processes was proved for the first time in [5]. In a QS discussed by the authors the control strategy consisted of enabling or disabling redundant service channels [6,7]. Based on this analogy, for a completely applied class of problems in the theory of controlled queueing systems, real algorithms for optimal channel switching in queueing systems were constructed.

© The Author(s), under exclusive license to Springer Nature Switzerland AG 2022
V. M. Vishnevskiy et al. (Eds.): DCCN 2022, LNCS 13766, pp. 361–370, 2022.
https://doi.org/10.1007/978-3-031-23207-7_28

2   Parametric Strategies in the Inventory Control Theory

Parametrization of strategies for inventory control is extremely important in the context of supply chain management, since in this case it is possible to take into account the optimization aspects of solving local problems of inventory control by entering the calculated values of strategy parameters (such as order size, reorder point, etc.) in the macro-model of the supply chain. Similar studies are described, for example, in [8]–[12]. This is why the parametrization of optimal inventory control strategies, first described by S. Karlin [13] remains relevant and significant for each model under study.

Besides classical two-level (R, r)-strategies[1] of inventory control [1–3], some other parametric strategies are popular in the theory and practice of inventory control (logistics of inventories), due to the fact that the value T of the minimum interval between two supplies is quite frequently introduced as one more additional parameter. It is this additional parameter that could be introduced in the new model of inventory control with returns, which we will not do in our paper, since it is assumed that starting from Sect. 4, this value T is specified.

Nevertheless, to clarify the possibilities of this form of strategy parameterization too, let us refer to [14], which considers a system using the (mT, R, r)-strategy of inventory control, where R and r mean the same as in footnote 1 of this page.

Now, in [14] the task is to find an algorithm that could become a tool for specialists who manage industrial production with a large fleet of different machinery. The manager's concern is to select a strategy for replenishing spare parts in order to avoid downtime caused by machine breakdowns, while minimizing the cost of storing these parts. A Bayesian model of spare parts consumption is presented, together with a sophisticated inventory control strategy of type (mT, R, r). The novelty of this study lies in the fact that the spare parts consumption process is described by a set of generic scenarios ("consumption scenarios"), which are defined by a system of performance indicators that use the states of the variables in the Bayesian model. Once the above-mentioned indicators are specified, a Bayesian network is defined, which allows, by means of Bayesian simulation, to calculate for a given planning period the desired combinations of parameters such as: length of the recovery period, order size, type of spare parts (new or used), downtime risk (and related losses), order costs and storage costs.

The problem of keeping production machines up and running is typical of any branch of production. The task of managing spare parts inventories in a facility is complicated by their high cost and expensive storage, as well as by market volatility (which translates into fluctuations in spare part prices, availability and delivery times).

The authors offer two wordings of the causal model: (1) determine the optimal period of forced replenishments (with constant frequency) under uncertainty caused by fluctuating delivery times and fluctuating demand for parts and (2)

---

[1] In case of two-level R is the level of replenishment and r is so called the reorder point.

determine cost-minimizing replenishment periods for constant stock levels: fully stocked and safe.

For the end user (i.e., the specialist authorized to make purchasing decisions), these scenarios are the building blocks (for each type of spare part) that help to make purchasing decisions. It is possible to calculate these blocks if you have information about the inventory level at the start of each period, the cost of spare parts (each type), and the indication of the type of delivery (deferred or forced). The final inventory management strategy will consist of such blocks (which is the subject of a separate study).

The model (T, R, r) allows the use of forced restocking, while a modified model (mT, R, r) with a varying control period, allows minimizing the cost of procurement. This approach was verified by a model built in BayesiaLab.

3 Inventory Control Model with Returns

A multistep model of inventory control is considered during the planning period $T = (0, N\tau)$, where N is a sufficiently large natural number, of one type of product with backlogging of the outstanding demand (so-called backorders). The demand at each of the steps of the process is described by a model of independent in the aggregate, identically distributed random variables $\{z(n), n = 1, 2, ..., N\}$ with the distribution function $F(z)$. We will also assume that the delivery lag time is equal to 0. In this case, it is customary to associate the state of the inventory management system not with the stock on hand, but with a so-called inventory position. An inventory position (with a delivery time equal to 0) is defined [1] as the stock on hand minus the backordered demand.

Such a model describes well the systems for managing the stocks of mass consumption products, the demand for which is not (or is not very susceptible) to seasonal changes. This type of products include staple foods (baked goods, meat, dairy products, etc.), medicines for chronic diseases, and many, many others.

Unlike the classical multi-step inventory control model, we will assume that the domain of the demand distribution function at one step $F(z)$ is the entire real axis: $-\infty < z < \infty$. Let also the mathematical expectation of this distribution be positive. The possibility of negative values of demand will be interpreted as the return by the consumer of the products purchased at the warehouse. We will also assume that when the goods are returned to the consumer, the entire amount that he paid when purchasing it is returned. The warehouse also gets the opportunity to return the goods to its supplier, paying a fixed amount A_2 for this, but the price c_2, at which the money is returned by the supplier, is less than the price of its purchase from the supplier: $c_2 < c_1$.

Let the criterion for the optimal functioning of the warehouse be the minimum of the total average costs during the planning period $[0, T]$, where $T = N\tau$, and τ is the duration of one-step between the moments of making order decisions. Let $C_n^*(x)$ denote the minimum possible value of average costs for the inventory control system, which has n steps left until the end of the planning period $[0, T]$

and which at the beginning of this interval has an inventory position x. Then we can write the following discrete dynamic programming equations:

$$C_0^*(x) \equiv 0; C_n^*(x) = \min_u \{(A_1 + c_1 u) \times \mathbf{1}(u) + (A_2 + c_2 u) \times \mathbf{1}(-u) + g(x+u)$$

$$+ \alpha \int_{-\infty}^{\infty} C_{n-1}^*(x+u-\xi) dF(\xi)\}, n = 1, 2, \ldots, N,$$

(1)

where $g(y) = h \int_{-\infty}^{\max\{y,0\}} (y-z) dF(z) + d \int_{\max\{y,0\}}^{+\infty} (z-y) dF(z)$, α is discount coefficient, $0 \leq \alpha \leq 1$, and $\mathbf{1}(u)$ is the unit jump function (Heaviside function). Equations (1) can be rewritten as follows

$$C_0^*(x) \equiv 0; \quad C_n^*(x) = -c_1 x + \min \begin{cases} A_1 + \min_{y>x} G_n(y), \\ \min\left\{G_n(x), \tilde{G}_n(x)\right\}, \\ A_2 + (c_1 + c_2)x + \min_{y<x} \tilde{G}_n(y), \end{cases}$$

$$n = 1, 2, , \ldots, N,$$

(2)

where the function $G_n(y)$ is given by the formula

$$G_n(y) = c_1 y + h \int_{-\infty}^{\max\{y,0\}} (y-z) dF(z) + d \int_{\max\{y,0\}}^{+\infty} (z-y) dF(z)$$

$$+ \alpha \int_{-\infty}^{+\infty} C_{n-1}^*(y-z) dF(z)$$

(3)

and the function $\tilde{G}_n(y)$ by the formula

$$\tilde{G}_n(y) = -c_2 y + h \int_{-\infty}^{\max\{y,0\}} (y-z) dF(z) + d \int_{\max\{y,0\}}^{+\infty} (z-y) dF(z)$$

$$+ \alpha \int_{-\infty}^{+\infty} C_{n-1}^*(y-\xi) dF(\xi)$$

(4)

In the framework of the economics of calculating costs, the function $\min_{y>x} G_n(y)$ describes the minimum variable part of the costs when making a decision to place an order (excluding the fixed part A_1 and the current inventory position x and including the case of failure to submit an order when $u = 0$), and the function $\min_{y<x} \tilde{G}_n(y)$ is equal to variable part of the minimum cost when deciding whether to return the goods to the supplier (excluding the fixed part A_2 and the current inventory position x). Note also that, due to the form of the functions $G_n(y)$ and $\tilde{G}_n(y)$, the point S_n of the absolute minimum of the function $\tilde{G}_n(y)$ is located to the right of the point R_n of the absolute minimum of the function $G_n(y)$ [2].

---

[2] This statement will be proved in the next section.

In the next Fig. 1 shows the hypothetical form and relative position of the functions $G_n(y)$ and $\tilde{G}_n(y)$.

For the functions $G_n(y)$ and $\tilde{G}_n(y)$ shown in Fig. 1 the optimal rule of inventory control is given by the formula:

$$u_n^*(x) = \begin{cases} R_n - x, & \text{if } x \leq r_n, \\ 0, & \text{if } r_n < x < s_n, \\ x - S_n, & \text{if } x \geq s_n. \end{cases} \quad (5)$$

So, if the functions $G_n(y)$ and $\tilde{G}_n(y)$ for all values of n have the properties that are qualitatively characterized in Fig. 1, then the optimal inventory control strategy turns out to be a four-level (R_n, r_n, S_n, s_n)-strategy $(r_n < R_n < S_n < s_n)$. This strategy is arranged so that:

(a) If the inventory position x at the time of making decisions n steps before the end of the planning period is less than or equal to r_n, then an order is submitted that replenishes the inventory position in the warehouse to the value R_n.

(b) If the inventory position x at the time of making decisions n steps before the end of the planning period is greater than or equal to s_n, then part of the stock from the warehouse is returned to the supplier so as to bring the inventory position in the warehouse to the value S_n.

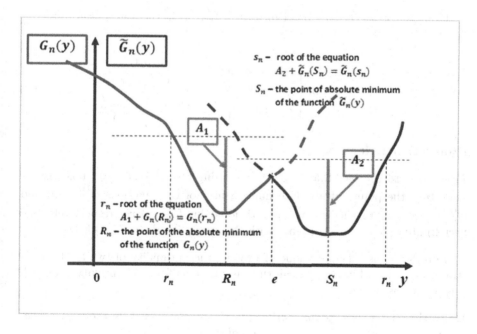

Fig. 1. Type and arrangement of functions $G_n(y)$ and $\tilde{G}_n(y)$.

4   Proof of Optimality of (R_n, r_n, S_n, s_n)-Strategies

We transform Eq. (2) by introducing the auxiliary functions $G_{order}^{(n)}(y)$ and $G_{return}^{(n)}(y)$. The function $G_{order}^{(n)}(y)$ describes the average cost for n steps before the end of the planning period at the current inventory position x and when deciding on a positive order size u (after which the inventory position becomes equal to $y = x + u$) and choosing the optimal inventory control strategy at subsequent $(n-1)^{th}$ steps without taking into account the constant component of costs when placing an order A_1 (this is the subscript "order").The function $G_{return}^{(n)}(y)$ describes the average cost for n steps before the end of the planning period when deciding on a negative order size u (after which the inventory level also becomes equal to $y = x + u$) and the choice of the optimal inventory management strategy at the subsequent $(n-1)^{th}$ steps without taking into account the constant component of costs when placing an order A_2 (this is the subscript "return"). Let

$$G_{order}^{(n)}(y) = c_1 y + g(y) + \alpha \int_{-\infty*}^{+\infty} C_{n-1}^*(y-z)dF(z), n \in \overline{1, N}, \qquad (6)$$

$$G_{return}^{(n)}(y) = -c_2 y + g(y) + \alpha \int_{-\infty*}^{+\infty} C_{n-1}^*(y-z)dF(z), n \in \overline{1, N}. \qquad (7)$$

Using functions (6)–(7), one can rewrite Eq. (2) in the following form:

$$C_0^*(x) \equiv 0; \quad C_n^*(x) = \min \begin{cases} -c_1 x + \min \begin{cases} A_1 + \min_{u>0} G_{order}^{(n)}(x+u), \\ G_{order}^{(n)}(x,0), \end{cases} \\ c_2 x + \min \begin{cases} A_2 + \min_{u<0} G_{return}^{(n)}(x+u), \\ G_{return}^{(n)}(x,0), \end{cases} \end{cases} \qquad (8)$$

Proposition 1. $R_n < S_n$.

Proof. Due to the fact that $c_1 > c_2$ and using formulas (7)–(8), it is easy to show that the points of absolute minima in y of the functions $G_{order}^{(n)}(y)$ and $G_{return}^{(n)}(y)$ lead to such values of y denoted by R_n and S_n respectively (see Sect. 2) are ordered in accordance with the inequality $R_n < S_n$. Q.E.D.

In order to use the technique of proving the optimality of two-level (R, r)-strategies proposed in [1,13] recall the basic notion of the A-convexity, on which this proof was based.

Definition 1. *An everywhere differentiable function $f(y)$ is called A-convex $(A \geq 0)$ if, for any $a > 0$, the next inequality holds*

$$A + f(x+\alpha) - f(x) - f'(x)\alpha \geq 0. \qquad (9)$$

It turns out that in order to prove the optimality of (R_n, r_n, S_n, s_n)-strategies, it is necessary to introduce a generalization of this concept.

Definition 2. *Let there exist a real number a, which has the property that it is to the right of the point of absolute minimum of the everywhere differentiable function $f(x)$ and for all real y and z such that the points y and $y + z$ are to the left of the point a, that is belong to the interval $(-\infty, a)$, the next inequality holds*

$$A + f(y + z) - f(y) - f'(y)z \geq 0, A > 0. \tag{10}$$

Then the function $f(x)$ is called A-convex on the left in the interval $(-\infty, a]$.

Definition 3. *Let there exist a real number b, which has the property that it is to the left of the point of absolute minimum of the everywhere differentiable function $f(x)$ and for all real y and z such that the points y and $y + z$ are to the right of the point b, that is belong to the interval (b, ∞), the inequality (10) holds then the function $f(x)$ is called A-convex on the right in the interval $[b, \infty)$.*

Proposition 2. *All functions $C_n^*(x)$ are A_1-convex on the left and A_2-convex on the right.*

Proof. Using formulas (6)–(7) for the case $n = 1$, it is easy to show that the functions $G_{order}^{(n)}(y)$ and $G_{return}^{(n)}(y)$ are convex, that is, 0-convex , and by the characteristic properties of the A-convexity construction [1], are A_1- and A_2-convex, respectively. Hence, as in ([1], by virtue of formula (7) for the case $n = 1$, it is established that the function $C_1^*(x)$ is simultaneously A_1-convex on the left and A_2-convex on the right.

Let us state the hypothesis of mathematical induction that for some number $n > 1$ the functions $C_n^*(x)$ are A_1-convex on the left and A_2-convex on the right. Further progress consists in establishing the fact that the $G_{order}^{(n+1)}(y)$ function is A_1-convex and the $G_{return}^{(n+1)}(y)$ function is A_2-convex. This follows from formulas (7) and (8) and the scheme of the proof, as in the case $n = 1$, repeats the scheme of the proof given in [1]. Then it will follow from Eq. (9) and Proposition 1 that the function $C_{n+1}^*(x)$ will also be A_1-convex on the left and A_1-convex on the right. In this case, the role of points a and b from Definitions 2 and 3 is played by the common point e of intersection of the graphs of the curves (see Fig. 1). Q.E.D.

As a result, the optimal inventory management strategy will be determined by the rule established by formula (6). In this case, the parameters r_n and s_n in formula (5) are solutions of the following equations (see Fig. 1):

$$A_1 + G_{order}^{(n)}(R_n) = G_{order}^{(n+1)}(r_n) \text{ for } r_n \leq R_n \tag{11}$$

$$A_2 + G_{return}^{(n)}(S_n) = G_{return}^{(n+1)}(s_n) \text{ for } s_n < S_n \tag{12}$$

It is easy to see that the proof of the above mathematical induction hypothesis methodologically repeats the proof of a similar proposition (for two-level strategies) in the classical theory of inventory control [1], adjusted for the alternativeness (orders and returns) of the objective functional.

5   Multichannel QS with Switching of Channels

As stated in [7], in the studied QS the number of working service channels can be changed at the moments of control, which are separated from each other for a fixed time (the control step). In this case, it is considered that the QS receives the simplest incoming flow, the intensity of which $\lambda(t)$ is constant throughout the step, and at the moments of control it undergoes jump-like Markov changes, taking a finite number k of values λ_i from the discrete set $\Lambda = \{\lambda_i, i \in \overline{1,k}\}$. The task is to form a strategy for switching working channels (disabling some of the working channels or putting into operation reserve channels), which minimizes the average costs of the QS for a given N-step planning period. In this case, as in [1], it is assumed that the matrix of transition probabilities of the corresponding homogeneous Markov chain $P = \|p_{ij}\|$ is given, where p_{ij} is the transition probability (at the time of control) from the intensity λ_i, $i \in \overline{1,k}$, at the previous step to the intensity λ_j, $j \in \overline{1,k}$, at the next step.

In [7] was shown that solving the problem of choosing the optimal channel switching strategy is reduced to the following system of dynamic programming equations:

$$C_1^*(\lambda_i, m) = \min_{u \geq u_i} C^{(1)}(\lambda_i, m, u), \tag{13}$$

$$C_n^*(\lambda_i, m) = \min_{u \geq u_i} (C^{(1)}(\lambda_i, m, u) + \alpha \sum_{j=1}^{l} p_{ij} C_{n-1}^*(\lambda_j, u)), n \in \overline{2, N}. \tag{14}$$

where $C_n^*(\lambda_i, m)$ is the minimum possible value of the total average costs at the last n steps of the control process, when the mathematical expectation is taken along the trajectory of the incoming flow intensity, which makes Markov jumps. The variable u in Eqs. (13)–(14) is the current (n steps before the end of the planning period) value of the control decision on the number of switched working channels. It has been proved that the optimal channel switching strategy, as in the equivalent inventory control problem with returns [4], is a four-level strategy.

Equation (14) can be re-written in the following equivalent form:

$$C_n^*(\lambda_i, m) = \min \begin{cases} -c_1 m + \min \begin{cases} A_1 + \min_{u > u_i} B_{\text{on}}^{(n)}(\lambda_i, u), \\ B_{\text{on}}^{(n)}(\lambda_i, m), \end{cases} \\ -(c_1 - c_2)m + \min \begin{cases} A_2 + \min_{u > u_i} B_{\text{off}}^{(n)}(\lambda_i, u), \\ B_{\text{off}}^{(n)}(\lambda_i, m), \end{cases} \end{cases} n \in \overline{1, N}. \tag{15}$$

where the functions $B_{\text{on}}^{(n)}(\lambda_i, u)$ and $B_{\text{off}}^{(n)}(\lambda_i, u)$ are similar to the functions $G_{order}^{(n)}(y)$ and $G_{return}^{(n)}(y)$, only instead of the "order" footnote in the inventory control problems the "on" (enable the channel) footnote is used , while the "off" (disable the channel) footnote replaces the "return" footnote. In (13)–(15) the value λ_i denotes the current intensity of the incoming flow (equivalent to the demand in inventory control problems), while the value m stands for the number of enabled channels as the QS enters the next step of the control process.

6 Summary

A fundamentally new model of inventory control with returns is considered. It is proved that the optimal inventory control strategy according to this model turns out to be a four-level (R, r, S, s)-strategy. This strategy is an essential generalization of the well-known two-level (R, r)-strategies. The analogies between the optimization problems of the inventory control theory and the theory of controlled queueing systems with switching channels are noted and new deeper features of this analogy are revealed.

References

1. Hadley, G., Whitin, T.M.: Analysis of Inventory Systems. Prentice-Hall, Inc., Englewood Clifs, New Jersey (1963)
2. Pervozvansky, A.: Mathematical Models of Inventory and Production Control. Science Publ. House, Moscow (1975). (In Russian)
3. Lototsky, V., Mandel, A.: Inventory Control Models and Methods. SciencePubl. House, Moscow (1992). (in Russian)
4. Mandel, A., Granin, S. Investigation of analogies between the problems of inventory control and the problems of the controllable queuing systems. In: Proceedings of the 11th International Conference Management of Large-Scale System Development (MLSD), pp. 1–4. IEEE (2018). https://doi.org/10.1109/MLSD.2018.855185
5. Rykov, V. Controllable queueing systems probability theory. Math. Statist. Theor. Cybern. **12**, 43–153 (1975). (In Russian)
6. Mandel, A., Laptin, V.: Myopic channel switching strategies for stationary mode: threshold calculation algorithms. In: Vishnevskiy, V.M., Kozyrev, D.V. (eds.) DCCN 2018. CCIS, vol. 919, pp. 410–420. Springer, Cham (2018). https://doi.org/10.1007/978-3-319-99447-5_35
7. Mandel, A.S., Laptin, V.A.: Channel switching threshold strategies for multi-channel controllable queuing systems. In: Vishnevskiy, V.M., Samouylov, K.E., Kozyrev, D.V. (eds.) DCCN 2020. CCIS, vol. 1337, pp. 259–270. Springer, Cham (2020). https://doi.org/10.1007/978-3-030-66242-4_21
8. Ettl, M., Feigin, G.E., Lin, G.Y., Yao, D.D.: A supply network model with base-stock control and service requirements. Oper. Res. **48**(2), 216–232 (2000)
9. Graves, S.C., Wilems, S.P.: Supply chain design safety stock placement and supply chain configuration. Handb. Oper. Res. Manag. Sci. **11**, 95–132 (2003)
10. Justus, N., Meyr, H.: Designing a planning system for suppliers of the machine building industry. In: Technical program of IFORS-2014 World Conference, Barcelona, p. 3 (2014)

11. Pishchulo, G., Richter, K., Golesorkhi, S.: Supply chain contracting under asymmetric information and partial vertical integration. In: Technical program of IFORS-2014 World Conference, Barcelona, p. 18 (2014)
12. Alegoz, M., Ozturk, Z.K.: A goal programming approach to design the supply chain network. In: Technical program of IFORS-2014 World Conference, Barcelona, p. 27 (2014)
13. Karlin, S.: Mathematical Methods and Theory in Games, Programming, and Economics. Addison-Wesley Publishing Company, New York (1959)
14. Sawik, T.: Integrated supply chain scheduling under multi-level disruptions. In: Preprint of 15th IFAC Symposium on Information Control Problems in Manufacturing (INCOM 2015, Ottawa), Ottawa (2015)

On Estimating the Average Response Time of High-Performance Computing Environments

A. V. Gorbunova^(✉) and V. M. Vishnevsky

V.A. Trapeznikov Institute of Control Sciences of Russian Academy of Sciences,
65, Profsoyuznaya Street, Moscow 117997, Russia
avgorbunova@list.ru, vishn@inbox.ru

Abstract. Parallel and distributed computing have become widespread as one of the main ways to improve the performance of computing environments when processing big data. The issue of assessing the main performance indicators of such systems remains relevant to the present time due to the complexity of such analysis. A fork-join queuing system is well suited for modeling parallel data processing. The article explores the possibility of estimating the average response time of the system using an analytical expression for the upper bound of the maximum order statistics. Two options are described, depending on the amount of knowledge available about the operation of a high-performance environment. The results of the numerical experiment confirm the efficiency of the approach for a fairly wide range of parameters, one of the main advantages of which is the ease of use.

Keywords: Parallel computing · High-performance computing environment · Data-intensive applications · Fork-join queueing system · Average response time · Order statistics

1 Introduction

The need for big data processing arises in various areas, for example, in the framework of scientific experiments, in the analysis of production processes, in engineering calculations, in traffic management, in financial risk analysis, etc. Therefore, there is a growing demand for access to the services of high-performance computing environments that use various methods of parallelization to improve performance. If earlier the relevant specialists had separate parallel computers at their disposal, then with the advent and development of cloud technologies, they have access to hundreds or more virtual machines [1].

High-performance computing environments are based on parallel structures, for modeling the functioning of which the fork-join queuing system (QS) is well

This paper has been funded by RFBR according to the research project No. 19-29-06043.

suited. Upon arrival to it, the task is divided into subtasks, after which each of the subtasks enters one of the parallel queues and waits for its service. The execution time of the entire task is determined by the longest processing time of one of its subtasks.

One of the most important indicators of system performance is its average response time. A correct assessment of this characteristic is important for service providers in connection with the need to comply with the agreement on the quality of their services. In addition, based on the estimates obtained, a strategy is built to allocate the required amount of resources for the implementation of the relevant tasks. Since maintaining the system's performance is costly, it affects the overall cost of the services provided by the provider and its competitiveness.

A considerable number of works are devoted to the study of fork-join QS, in particular [2–7] and many others. However, exact results have been obtained only for the case of two subsystems. In one of the first papers in this area, [8], an expression was presented for the average response time of a system with Poisson input and exponential service time on homogeneous servers. Approximate algorithms have been developed for more complex system configurations, i.e., with a large number of subsystems and distributions of input flow and service time different from those indicated. At the same time, the scope of these numerical methods is limited by specific distribution parameters, the value of the system load factor, the number of subsystems, as well as the complexity and duration of the computational method itself. A detailed review of articles can be found in [9], and among more recent works, for example, [10–15].

The response time of a fork-join system is the maximum of all random variables of the times the subtasks stay in the subsystems, i.e., the maximum order statistics. However, the complexity of the analysis of this quantity is explained by the dependence of the sojourn times of the subtasks. This dependence arises due to the generality of the moment of their appearance in the system. From the theory of extreme values, an analytical expression for the upper bound of the maximum order statistics is known. This expression uses data only on the mathematical expectation and dispersion of the residence time in subsystems, as well as the number of subsystems.

The article explores the possibility of using the upper bound for the initial assessment of the average response time of a mathematical model of a parallel structure of a high-performance computing environment. As a service time distribution, we consider a truncated Pareto distribution obtained on the basis of empirical data for the Google search engine [13]. For the input flow, several variants of distributions with different values of the coefficient of variation, both smaller and larger than unity, were considered.

The results of numerical experiments show that the use of the upper bound as an estimate of the average response time of a fork-join system is possible for a fairly wide range of parameters, even despite the heavy tails of the considered distributions and the complexity of analyzing such a system as a whole. The approximation error does not exceed 30%–35% and corresponds to a greater extent to low values of the coefficient of variation of the arrival flow. Given that the main strategy of service providers is to over-allocate resources—of the order of 50% or more—the savings are significant, as is the time spent on this initial assessment. Note that one of the latest approaches proposed in [14–16] to estimate the average response time is a method using machine learning algorithms. Despite its universality and good accuracy, its application still requires a lot of time compared to using only analytical expressions and can serve to further correct the obtained estimates.

The article is organized as follows: Sect. 2 presents a description of the mathematical model of the parallel structure of a high-performance computing environment, Sect. 3 presents an approach to estimating the average response time of the system, Sect. 4 describes the results of a numerical experiment, Sect. 5 discusses the effectiveness of the proposed way to conduct an initial assessment of one of the main performance indicators of the system, in conclusion some results are summarized.

2 Mathematical Model of the Parallel Structure of a High-Performance Computing Environment

As a model of a parallel structure of a high-performance computing environment, consider a fork-join queuing system. The functioning of this system is as follows (Fig. 1):

1) at the moment a request enters the system, it is instantly split into K ($K \geqslant 2$) sub-requests, each of which is sent to its own subsystem and either queues for service there or starts to be served immediately if the queue is empty;
2) after the end of service, the sub-request remains in the system until all sub-requests that initially made up one request complete their service, then the entire request is instantly assembled, and only after that it is considered fully served and can leave the system.

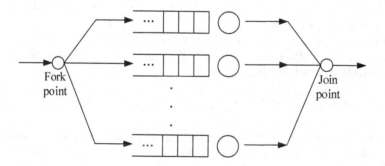

Fig. 1. The fork-join queuing system.

The response time of the system is defined as the maximum of the random residence times of its sub-requests in the corresponding subsystems

$$R_K = \max(\xi_1, ..., \xi_K).$$

For convenience, we introduce the following notation. Let $A(x)$ and $B(x)$ be the distribution functions (or densities) of time between adjacent arrivals of requests and the service time of sub-requests, respectively; a_i and b_i—i-th initial moments of the corresponding random variables ($a_1 = a$, $b_1 = b$), σ_A and σ_B—their standard deviations, and $C_A = \sigma_A/a$ and $C_B = \sigma_B/b$ are coefficients of variation, then the system load factor in the accepted notation is defined as $\rho = b/a$.

Let us describe in more detail the architecture of the fork-join QS, which we will use to simulate the functioning of the parallel structure of a high-performance computing environment. The system actually consists of K identical subsystems of type $G|G|1$. Service in subsystems occurs in the order in which subtasks arrive.

The service time on the servers follows a truncated Pareto distribution with a distribution function of the form [17]

$$B(x) = \frac{1 - (L/x)^\alpha}{1 - (L/H)^\alpha}, \quad 0 \leqslant L \leqslant x \leqslant H, \quad \alpha > 0.$$

where α is the distribution shape parameter, L is the minimum value, and H is the maximum value assumed by the service time random variable. Based on empirical data, the following values of the parameters of the truncated Pareto distribution [13, 18] were obtained:

$$\alpha = 2.0119, \quad L = 2.14, \quad H = 276.6, \tag{1}$$

i.e. the minimum service time is 2.14 ms and the maximum is 276.6 ms. The i-th moment of the truncated Pareto distribution is given by:

$$b_i = \frac{L^\alpha}{1 - (L/H)^\alpha} \frac{\alpha(L^{i-\alpha} - H^{i-\alpha})}{\alpha - i}, \quad \alpha \neq i.$$

Thus, for the given values of the parameters, the average service time is approximately 4.22 ms, the variance is 22.34, and the ratio of the root of the variance to the average service time is approximately 1.22 ($C_B \approx 1.22$).

For the distribution of the input flow, we consider several options with different values of the coefficient of variation. We start with the Erlang distribution with a density of the following form

$$A_1(x) = \frac{\lambda^k}{k!} x^{k-1} e^{-\lambda x}, \quad x \geqslant 0, \quad \lambda > 0, \quad k = 2, 3, 4, ..., \tag{2}$$

for it is known that its coefficient of variation $1/\sqrt{k}$ is always less than one.

We will not exclude the case of the Poisson input flow, which is obtained from the previous distribution for $k = 1$. We also consider a heavy-tailed distribution ($C_A > 1$), namely the Gamma distribution with density

$$A_2(x) = \frac{\lambda^k}{\Gamma(k)} x^{k-1} e^{-\lambda x}, \quad x \geqslant 0, \quad \lambda > 0, \quad k > 0, \tag{3}$$

where $\Gamma(k)$ is the gamma function.

3   On an Approach to Estimating the Average Response Time of a High-Performance Computing Environment

To estimate the average response time $E[R_K]$ of the described system, we will use an approach based on the theory of order statistics. It is known that the residence times of sub-requests in the fork-join QS are correlated (positively associated) random variables. This fact complicates the analysis of the average response time of the system as the mathematical expectation of the maximum of these random variables. If, however, we assume that the residence times of sub-requests in subsystems are independent, then it is known from the theory of order statistics for extreme values [19], that

$$E[R_K] \leqslant E[R_1] + \sqrt{Var[R_1]} \frac{K-1}{\sqrt{2K-1}}. \tag{4}$$

Thus, to estimate the upper bound of the average response time, it is necessary to find the mathematical expectation and variance (standard deviation) of the sub-request's sojourn time in the subsystem, i.e. $E[R_1]$ and $Var[R_1]$ respectively. So, for example, if the input flow and service times have an exponential distribution, then the sojourn time of a request in the system will also have an exponential distribution with a parameter equal to the difference between the rates of service and arrival of requests. However, for other variants of combinations of these distributions, it is extremely difficult or even impossible to determine the distribution of the request's sojourn time. Therefore, we analyze the possibility of estimating $E[R_1]$ and $Var[R_1]$ using the distribution parameters of the input flow and service time.

By analogy with [13], we will consider two approaches to the analysis of a high-performance computing environment. On the one hand, we can assume that we know the specific distribution of time between adjacent arrivals of tasks and, accordingly, sub-tasks into subsystems. In this case, we will deal with the mathematical model of the parallel computing environment as a "white box" model. On the other hand, if it is impossible to determine the specific distribution of the arrival flow, then modern technical means allow to evaluate its characteristics, in particular the first and second moments of time between adjacent request arrivals. [13]. In this case, we will deal with the mathematical model of the structure for parallel computing as a "black box" model.

Many analytic expressions are known for approximating the average response time of the $G|G|1$ system [20]. In addition, depending on the value of the coefficient of variation of the input flow or servicing time, the average response time of systems $M|G|1$ or $G|M|1$ can be the upper or lower bounds for the average response time of QS $G|G|1$. In this paper, we will focus on the Krämer-Langenbach-Belz formula [21]

$$E[R_1] \approx \frac{b\rho}{2(1-\rho)}(C_A^2 + C_B^2)g(\rho, C_A, C_B) + b, \tag{5}$$

where

$$g(\rho, C_A, C_B) = \begin{cases} \exp\left\{ -\frac{2(1-\rho)}{3\rho} \cdot \frac{(1-C_A^2)^2}{C_A^2+C_B^2} \right\}, & \text{if} \quad 0 \leqslant C_A \leqslant 1, \\ \exp\left\{ -(1-\rho) \cdot \frac{C_A^2-1}{C_A^2+4C_B^2} \right\}, & \text{if} \quad C_A > 1. \end{cases} \tag{6}$$

Much fewer approximations are known for calculating the variance of the sojourn time of a request in the QS $G|G|1$. In [22] a formula was proposed for approximating the variance of the sojourn time of a request in the system, derived on the basis of the isomorphism principle. The concept of isomorphism means the search for a mathematical model similar to the original one, for which the result is known. In this case, the authors chose QS $M|G|1$ as a basis. As a result of the transformations, the following formula was obtained

$$Var[R_1] \approx \left[\frac{C_A^2 b^2 + \sigma_B^2}{2a(1-\rho)} \right]^2 + \frac{\dot{b}_3 + (3\sigma_A^4/a^4 - \dot{a}_3/a^3)b^3 + 3C_A^2\sigma_B^2 b}{3a(1-\rho)}, \tag{7}$$

where $\dot{a}_3$ and $\dot{b}_3$ are the third central moments of the corresponding random variables. According to numerical experiments from [22], this expression for various values of the coefficients of variation shows an acceptable approximation quality.

Next, we analyze the quality of this approximation for various types of distributions of the input flow, and, accordingly, for various values of its coefficient of variation.

4   Numerical Experiment

Let us consider the results of estimating the upper bound of the average response time in the case of analyzing the mathematical model as a "black box" and "white box". In this case, we will analyze various options for the number of sub-tasks $K = 16, 32, 64, 100$ and system load levels $\rho = 0.1, 0.2, ..., 0.9$.

For the objectivity of the experiment, we will study the results for different values of the coefficient of variation of the time between adjacent arrivals of tasks in the system. Namely, to check the obtained values calculated by formula (4), for $C_A < 1$ we will use the Erlang distribution with probability density function $A_1(x)$, for $C_A = 1$—the exponential distribution, and , finally, for $C_A > 1$—Gamma distribution with probability density function $A_2(x)$. To assess the quality of the approximation, we will use the relative error expressed as a percentage

$$error = \frac{\widehat{R}_K - E[R_K]}{E[R_K]} \cdot 100\%,$$

where $\widehat{R}_K$ is an approximation of the average response time. The $E[R_K]$ values were obtained by simulation in the Python programming environment.

In the first case, i.e. when it is possible to estimate the average characteristics of subsystems (at the level of sub-tasks) by technical means without determining

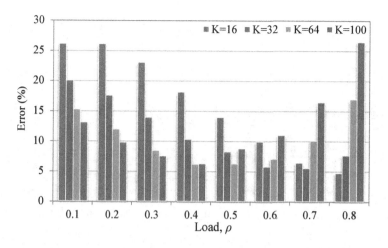

Fig. 2. Approximation errors for the average system response time $E[R_K]$, "black box" model, $C_A = 0.5$. (Color figure online)

a specific type of distribution for the input flow, we directly use the (4) formula, substituting the obtained values into it. As a result, we obtain the following results. If the coefficient of variation of the input flow has a value equal to 0.5, then the maximum approximation error for the load factor lies in the range from 0.1 to 0.8. and for all considered values K is approximately 25%. At the same time, judging by the Fig. 2, the approximation error decreases as the load increases to average values, after which it starts to increase again. In this case, the best performance is in the case of $K = 64$, since the maximum approximation error does not exceed 17%.

With $C_A = 0.7$ in the low load area, the situation is similar to the previous case (Fig. 3). However, when loading above 0.5, the values of the upper bound are slightly higher for all values of K. In particular, for $K = 100$ the maximum approximation error is about 31% for $\rho = 0.9$, which is approximately 4% more than the similar value for $C_A = 0.5$.

For $C_A = 1$, a similar trend is observed. In this case, the approximation error increases by several percent for all values of ρ (Fig. 4), and the maximum error for $K = 100$ and $\rho = 0.8$ is 40%. However, for the values $\rho \in [0.1, 0.6]$ the maximum error does not exceed 25.5%.

For higher values of the coefficient of variation $C_A = 2$ and $C_A = 3$, the emerging trend is reversed and the approximation error definitely increases with the growth of the system load and the number of sub-tasks. At the same time, the upper limit significantly exceeds the true values of the average system response time, and its use in its pure form does not make much sense.

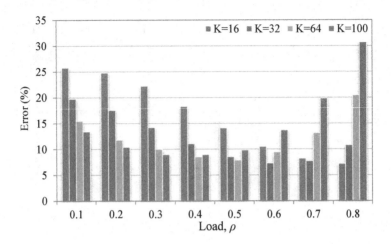

Fig. 3. Approximation errors for the average system response time $E[R_K]$, "black box" model, $C_A = 0.7$.

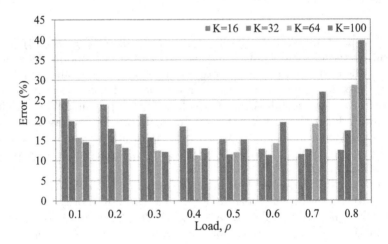

Fig. 4. Approximation errors for the average system response time $E[R_K]$, "black box" model, $C_A = 1$.

Next, we analyze the "white box" situation, i.e., when we do not know the specific values of the mathematical expectation and dispersion of the sojourn times in subsystems, but we have some idea of the distribution of not only the service time, but also the arrival flow. Then, knowing their average characteristics, it is necessary to estimate using the formulas (5) and (7) the average sojourn time by a sub-task in the corresponding subsystem.

In the case of $C_A < 1$, the behavior of the upper bound on the average response time is identical to the "black box" model (Figs. 5 and 6). However, for load factor values $\rho \in [0.1, 0.5]$ the error is higher. This is due to the following

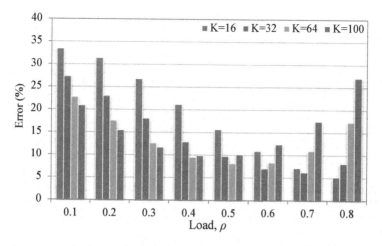

Fig. 5. Approximation errors for the average system response time $E[R_K]$, "white box" model, $C_A = 0.5$.

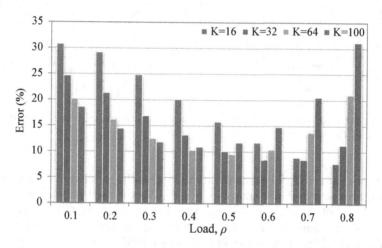

Fig. 6. Approximation errors for the average system response time $E[R_K]$, "white box" model, $C_A = 0.7$.

factors. Since we are approximating the average sojourn time of a sub-task in the subsystem, as well as its standard deviation, the quality of these estimates affects the final result. As for the average time $E[R_1]$, for $C_A = 0.5$ and $C_A = 0.7$ the approximation error is small and ranges from –4% to 1.5% (Fig. 7), and for small values of the load factor, the variance approximation error is somewhat overestimated (Fig. 8), which leads to an increase in the overall response time

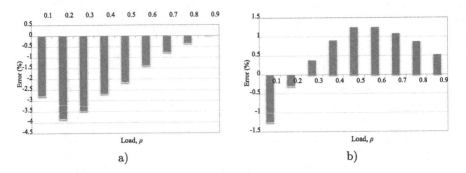

Fig. 7. Approximation error for $E[R_1]$, which is calculated using formula (5) a) $C_A = 0.5$; b) $C_A = 0.7$.

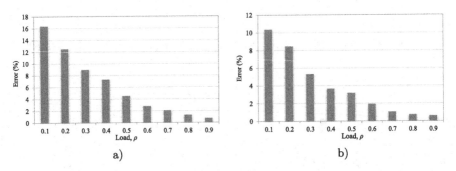

Fig. 8. Approximation error for $Var[R_1]$, which is calculated using formula (7) a) $C_A = 0.5$; b) $C_A = 0.7$.

estimate. For $C_A = 1$, formulas (5) and (7) give the exact result, since in this case they are converted into known expressions for the mathematical expectation and dispersion of the request's sojourn time in the system $M|G|1$. Therefore, the estimate of the upper bound coincides with the results of the estimate in the case of analyzing the system model as a "black box".

For the case of $C_A = 2$ and $C_A = 3$, the variance estimate turns out to be overestimated (Fig. 9), however, for $\rho > 0.3$ in the first $\rho > 0.4$ in the second case, the error of approximation of the variance of the $G|G|1$ system does not exceed 15%, which is a comparatively good result. However, the estimate of the mathematical expectation is seriously underestimated, due to the heavy tails of the distributions. So, for $\rho < 0.6$, the approximation error is below -20% and -30%, respectively, and decreases with decreasing load, but due to this, the overall response time is corrected and the approximation quality becomes better. This allows the use of these estimates in the area of low load factor values (Fig. 10).

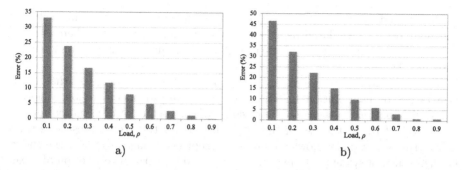

Fig. 9. Approximation error for $Var[R_1]$, which is calculated using the formula (7) a) $C_A = 2$; b) $C_A = 3$.

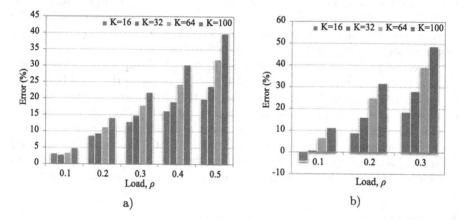

Fig. 10. Approximation errors for the average system response time $E[R_K]$, "white box" model a) $C_A = 2$; b) $C_A = 3$.

5   Discussion

Numerous publications are devoted to the study of mathematical models of various kinds of parallel structures of the fork-join type due to the breadth of their practical application. However, when analyzing systems with distributions that go beyond the Poisson input flow and the exponential distribution of service time, the developed approximate methods for estimating average characteristics are not distinguished by simplicity and breadth of their scope. As a rule, we are talking about the application of numerical algorithms to some specific special cases. Moreover, the approximation quality is limited by specific parameter values, such as high or low system load, the number of subsystems, as well as the complexity of the algorithms, which is expressed in significant time costs for their application.

The upper bound for the average response time of the system considered in the article is calculated using an analytical formula. The values of the parameters included in it can be practically determined at the level of sub-tasks, or they can also be calculated using simple analytical expressions if there is more information about the type of distributions of the flow of arrivals and service time in subsystems.

The truncated Pareto distribution for service time considered in the article is a rather difficult option for analysis, since it is considered to be a heavy-tailed distribution. In the case of low values of the coefficient of variation for the input flow of tasks, the use of the upper bound for the average response time as its primary estimate shows a relatively acceptable result for almost the entire studied range of system load factors ($0.1 \leqslant \rho \leqslant 0.8$), with the exception of areas of its high values, where the estimate is unnecessarily overestimated. Thus, the relative approximation error does not exceed 30%. The situation is similar for $C_A = 1$ for the considered values of K and $\rho \in [0.1, 0.7]$. For large values of the coefficient of variation of the input flow, the scope of the upper bound on the average response time is significantly narrowed down to the case of a low level of system load and its analysis as a "white box" model.

Nevertheless, despite the order of errors up to 30%, the upper bound can serve as an acceptable primary and fairly quick tool for assessing the main indicators of system performance in order to subsequently analyze the required amount of allocated resources. Since the main strategy of high-performance computing service providers is to overprovision the necessary resources: as practice shows, up to 50% of resources and even more are idle [23,24]. This is due to the need to fulfill obligations under the agreement on the quality of services provided to consumers and the lack of high-quality tools for making the necessary forecasts.

6   Conclusion

The paper analyzes the possibility of applying the upper bound formula for the maximum order statistics for the initial estimation of the average response time of a parallel structure of a high-performance computing environment. As a mathematical model of a parallel structure, we consider a fork-join queuing system with a truncated Pareto distribution of service time on servers and various distributions for the input flow.

According to the results of numerical experiments, the range of values of the system parameters for which the upper limit can be used as a first approximation is quite wide, although it requires correction and further research at high values of the coefficient of variation of the input flow. However, given the complexity and limited application of currently known algorithms for estimating the performance characteristics of parallel structures, the use of an analytical expression for the upper limit of extreme values as a primary estimate can serve as a simple and understandable forecasting tool when designing systems or improving their performance.

References

1. de Oliveira, D.C.M., Liu, J., Pacitti, E.: Data-intensive workflow management: for clouds and data-intensive and scalable computing environments. Synth. Lect. Data Manag. 14(4), 1–179 (2019)
2. Kumar, A., Shorey, R.: Performance analysis and scheduling of stochastic fork-join jobs in a multicomputer system. IEEE Trans. Parallel Distrib. Syst. 10(4), 1147–1164 (1993)
3. Varma, S., Makowski, A.M.: Interpolation approximations for symmetric fork-join queues. Perform. Eval. 20, 245–265 (1994)
4. Balsamo, S., Donatiello, L., Van Dijk, N.M.: Bound performance models of heterogeneous parallel processing systems. IEEE Trans. Parallel Distrib. Syst. 9(10), 1041–1056 (1998)
5. Takahashi, M., Osawa, H., Fujisawa, T.: On a synchronization queue with two finite buffers. Queueing Syst. 36, 107–123 (2000)
6. Squillante, M.S., Zhang, Y., Sivasubramaniam, A., Gautam, N.: Generalized parallel-server fork-join queues with dynamic task scheduling. Ann. Oper. Res. 160(1), 227–255 (2008)
7. Tsimashenka, I., Knottenbelt, W.J.: Reduction of subtask dispersion in fork-join systems. In: Balsamo, M.S., Knottenbelt, W.J., Marin, A. (eds.) EPEW 2013. LNCS, vol. 8168, pp. 325–336. Springer, Heidelberg (2013). https://doi.org/10.1007/978-3-642-40725-3_25
8. Nelson, R., Tantawi, A.N.: Approximate analysis of fork/join synchronization in parallel queues. IEEE Trans. Comput. 37, 739–743 (1988)
9. Thomasian, A.: Analysis of fork/join and related queueing systems. ACM Comput. Surv.(CSUR) 47(2), 17:1–17:71 (2014)
10. Qiu, Z., Pérez, J.F., Harrison, P.G.: Beyond the mean in fork-join queues: efficient approximation for response-time tails. Perform. Eval. 91, 99–116 (2015)
11. Wang, W., Harchol-Balter, M., Jiang, H., Scheller-Wolf, A., Srikant, R.: Delay asymptotics and bounds for multitask parallel jobs. Queueing Syst. 91(3), 207–239 (2019). https://doi.org/10.1007/s11134-018-09597-5
12. Gorbunova, A.V., Lebedev, A.V.: Bivariate distributions of maximum remaining service times in fork-join infinite-server queues. Prob. Inf. Trans. 56(1), 73–90 (2020)
13. Nguyen, M., Alesawi, S., Li, N., Che, H., Jiang, H.: A black-box fork-join latency prediction model for data-intensive applications. IEEE Trans. Parallel Distrib. Syst. 31(9), 1983–2000 (2020)
14. Gorbunova, A.V., Vishnevsky, V.M.: Estimating the response time of a cloud computing system with the help of neural networks. Adv. Syst. Sci. Appl. 20(3), 105–112 (2020)
15. Gorbunova, A.V., Lebedev, A.V.: Response time estimate for a fork-join system with Pareto distributed service time as a model of a cloud computing system using neural networks. In: Vishnevskiy, V.M., Samouylov, K.E., Kozyrev, D.V. (eds.) Distributed Computer and Communication Networks. DCCN 2021, CCIS, vol. 1552, pp. 318–332. Springer, Cham (2021). https://doi.org/10.1007/978-3-030-97110-6_25
16. Vishnevsky, V.M., Gorbunova, A.V.: Application of machine learning methods to solving problems of queuing theory. In: Dudin, A., Nazarov, A., Moiseev, A. (eds.) Information Technologies and Mathematical Modelling. Queueing Theory and Applications. ITMM 2021, CCIS, vol. 1605, pp. 304–316. Springer, Cham (2022). https://doi.org/10.1007/978-3-031-09331-9_24

17. Harchol-Balter, M.: Modeling and Design of Computer Systems: Queueing Theory in Action. Cambridge Univ. Press, U.K., Cambridge (2013)
18. Meisner, D., Junjie, W., Wenisch, T.F.: BigHouse: a simulation infrastructure for data center systems. In: Proceedings of IEEE International Symposium on Performance Analysis of Systems & Software, pp. 35–45. IEEE, New Brunswick, NJ, USA (2012)
19. David, H.A., Nagaraja, H.N.: Order Statistics, 3rd edn. John Wiley & Sons, New York (2003)
20. Bolch G.: Queueing Networks and Markov Chains: Modeling and Performance Evaluation with Computer Science Applications, 2nd edn, John Wiley & Sons (2006)
21. Krämer, W., Langenbach-Beiz, M.: Approximate formula for the delay in the queueing system $G/G/1$. In: Proceedings of 8th International Teletraffic Congress, pp. 235:1–235:8. Melbourne (1976)
22. Bhat, V.N.: Approximation for the variance of the waiting time in a $GI/G/1$ queue. Microelectron. Reliabil. **33**(13), 1997–2002 (1993)
23. Delimitrou, C., Kozyrakis, C.: Quasar: resource-efficient and QoS-aware cluster management. In: Proceedings of the 19th International Conference on Architectural Support for Programming Languages and Operating Systems, pp. 127–144. Association for Computing Machinery, New York, NY, USA (2014)
24. Reiss, C., Tumanov, A., Ganger, G.R., Katz, R.H., Kozuch, M.A.: Heterogeneity and dynamicity of clouds at scale: Google trace analysis. In: Proceedings of the Third ACM Symposium on Cloud Computing (SoCC 2012), pp. 7:1–7:13. Association for Computing Machinery, New York, NY, USA (2012)

Exponential Splitting Based Artificial Regeneration in Supercomputer Queueing Model

Alexander Rumyantsev[1,2,3(✉)] and Irina Peshkova[1,2]

[1] Institute of Applied Mathematical Research, Karelian Research Centre of RAS,
11 Pushkinskaya Str., Petrozavodsk, Russia
ar0@krc.karelia.ru
[2] Petrozavodsk State University, 33 Lenina Pr., Petrozavodsk, Russia
[3] Moscow Center for Fundamental and Applied Mathematics,
Moscow State University, Moscow 119991, Russia

Abstract. In this paper we suggest the method for performance estimation of a realistic size supercomputer model. To do so, we apply exponential splitting to the generalized semi-Markov process model of a supercomputer and obtain the so-called artificial regeneration points which allows to apply regenerative confidence estimation. Numerical results illustrate the method.

Keywords: Supercomputer model · Artificial regeneration · Regenerative estimation · Generalized semi-markov process

1 Introduction

Among the multiserver queueing models, supercomputer model has recently received significant attention [1,9,10] with key focus in the stability issues of this complicated model. The main source of the problem is the key attribute of a supercomputer, i.e. the possibility of simultaneous multiple servers seize/release by a single customer. Unlike classical multiserver model, in a supercomputer model each customer is served by a (random) number of servers simultaneously, for the same (random) time.

It is even more challenging to obtain steady-state performance of the supercomputer model. Specifically, even in Markovian case the queueing model is treated as a structured Markov chain known as the quasi-birth-death (QBD) process with block-tridiagonal infinitesimal generator [15]. However, the size of the blocks scales dramatically with the number of servers in the supercomputer model, and thus, explicit analysis is not possible for systems of realistic size even numerically.

On the contrast, simulation model does not suffer from the curse of dimensionality, and thus, is suitable for the analysis of relatively large systems. The main issue of the simulation model is to obtain steady-state performance estimates with given accuracy, which can be solved by using regenerative simulation [11]. Due to the independence of the regenerative cycles, in such a case parallel and distributed computing can be involved to speedup the numerical analysis [5]. However, the key challenge of regenerative simulation is to define regeneration epochs which may be rare or, say, do not exist in classical sense, i.e. the system may never become empty [3]. In the present paper we address this issue by applying the so-called exponential splitting [2] based artificial regeneration [16] to the discrete-event simulation model of a*supercomputer.

In this paper, we apply the generalized semi-Markov processes [8] as the discrete-event simulation tool. Taking the inspiration for the distributions of the large-scale model governing sequences in [7], we use the class-dependent discrete mixture of gamma distributions as the service time distribution, whereas the arrival process is governed by Weibull distributions. In such a case, due to the specific parameters of the gamma distributions, we need to impose a modification of the exponential splitting method suggested in [2]. Thus, the key contribution of the present paper is the novel modification of the exponential splitting based artificial regeneration in the generalized semi-Markov process model, which is demonstrated in the supercomputer model.

The structure of the paper is as follows. In Sect. 2 we introduce the general supercomputer model. The generalized semi-Markov process based model of a supercomputer and it's steady-state performance estimates are described in Sect. 3. In Sect. 4 we introduce our main result which is illustrated in Sect. 5 by a numerical experiment.

2   Supercomputer Model

In this section we briefly recall the properties of the supercomputer queueing model studied in this paper. Consider the c-server system with a single queue and customers of c classes, where class-i customer requires i servers simultaneously. The i servers dispatched to the customer are seized and released simultaneously, and thus the customer service time is the same (random) on these servers dispatched. However, upon insufficient number of servers available, the customer has to wait in a single queue operating on a first-come-first-served basis. Thus it may be seen that the system is non-work-conserving, i.e. non-empty queue may coincide with idle servers.

In what follows we assume that the customer interarrival times $\{\tau_i\}_{i\geq 1}$ are iid., service times $\{S_i\}_{i\geq 1}$ have general (class-dependent) distribution and the customer class $\{N_i\}_{i\geq 1}$ is sampled upon arrival from some (discrete) arbitrary distribution $\{p_1, \ldots, p_c\}$. However, the results below may be further extended to the case when the number of classes exceeds the number of servers (i.e. customer class defines the number of servers required, but not vice versa), which though results in unnecessary complication of the notation. In what follows, we address the system as the supercomputer model for brevity.

It is important to note that in general the stability criterion for the supercomputer model remains an open problem [10]. However, we may use a sufficient stability condition due to the following monotonicity result for the workload (residual work) vector W at each server proved in [14]:

Lemma 1. *Assume that in the systems Σ and $\hat{\Sigma}$ (with corresponding quantities designated accordingly with a hat symbol) the initial workload vectors are ordered as $\hat{W}_1 \leq W_1$ and the corresponding governing (generic) variables are (stochastically) ordered as $\hat{\tau} \geq \tau$, $\hat{S} \leq S$, $\hat{N} \leq N$. Then the workload vectors W_i and $\hat{W}_i$ just before the arrival epoch of i-th customer in the corresponding systems are ordered as $\hat{W}_i \leq W_i, i \geq 1$.*

Since the G/G/1 system with corresponding (generic) interarrival time τ and service time S can be treated as the system Σ with $N_i \equiv c$, it follows from Lemma 1 that the stability criterion of such a system is the (rough) sufficient stability condition for the supercomputer model, that is,

$$\mathrm{E}S < \mathrm{E}\tau. \tag{1}$$

At the same time, more precise stability conditions may be obtained numerically, e.g. by the method proposed in [13].

3 Generalized Semi-Markov Process Model of a Supercomputer

Generalized semi-Markov process (GSMP) is a specific method of a discrete-event simulation that allows regenerative estimation to be used in a convenient way. To define a GSMP model, the components and transitions of the following stochastic process are to be defined,

$$\Theta = \{X(t), T(t)\}_{t \geq 0}, \tag{2}$$

where $X(t)$ is the system *state* (vector) and $T(t)$ are the *timers*. It is generally assumed that the timers drift linearly (say, with unit speed) towards zero which is the timer expiration point. An expiration of the timer is interpreted as the corresponding *event* which may cause a transition in the system state vector. Apart from the event epochs $\{t_k\}_{k \geq 0}$, the system state remains constant. Upon system state transition, timers may be activated, which is, their initialization is performed in Markovian way by sampling from some distribution (possibly depending on the system states just before and right after the event). The GSMP method is described in more details in e.g. [4,8,11].

Hereinafter we define the necessary components for the supercomputer model. Let the system state vector have $c + 2$ components $X_1, \ldots, X_{c+2}$ defined as follows:

$X_1, \ldots, X_c$ – classes of the (not more than) c customers being served (or zero),
X_{c+1} – class of the customer in the head of the queue (or zero),
X_{c+2} – number of customers in the system.

Define the timers as follows:

$T_1, \ldots, T_c$ – remaining service time for the customers being served,

T_{c+1} – remaining interarrival time.

Note that the timer components correspond to the respective components in the system state vector, i.e. the timer T_i contains the residual service time of a customer being served whose class is X_i, $i = 1, \ldots, c$.

The transitions between the states are defined in relatively straightforward way, since the events are customer arrivals and departures. We summarize these transitions in the form of Algorithm 1 describing a single step of the GSMP supercomputer model.

Algorithm 1. Single step of the supercomputer GSMP model

Ensure: $U(0,1)$ – uniform random number sample from $(0,1)$, e – event that causes the transition, i.e. $T_e(t_{k+1}) = 0$

1: $X_i(t_{k+1}) \leftarrow X_i(t_k)$, $i = 1, \ldots, c + 2$ ▷ component copying
2: **if** $e = c + 1$, **then**   ▷ arrival
3:    **if** $X_{c+1}(t_k) > 0$, **then**   ▷ non-empty queue
4:       $X_{c+2}(t_{k+1}) \leftarrow X_{c+2}(t_k) + 1$
5:    **else**   ▷ empty queue
6:       sample the class K from the distribution $\{p_1, \ldots, p_c\}$
7:       **if** $\sum_{i=1}^{c} X_i(t_k) + K \leq c$, **then**   ▷ can be served
8:          $X_{\min\{i:X_i(t_k)==0\}}(t_{k+1}) \leftarrow K$
9:       **else**   ▷ first in the queue
10:          $X_{c+1}(t_{k+1}) = K$
11:       **end if**
12:    **end if**
13: **else**   ▷ service completion
14:    $X_e(t_{k+1}) = 0$
15:    **if** $X_{c+1}(t_k) > 0$ and $\sum_{i \in \{1, \ldots, M\} \setminus \{e\}} X_i(t_k) \leq c$, **then**
16:       $X_{\min\{i:X_i(t_{k+1})=0\}}(t_{k+1}) \leftarrow X_{c+1}(t_k)$
17:       $X_{c+1}(t_{k+1}) \leftarrow 0$
18:       **while** $X_{c+2}(t_{k+1}) > 0, \sum_{i=1}^{c} X_i(t_{k+1}) \leq c, X_{c+1}(t_{k+1}) = 0$ **do**
19:          sample the class K from the distribution $\{p_1, \ldots, p_c\}$
20:          **if** $\sum_{i=1}^{c} X_i(t_k) + K \leq c$, **then**   ▷ can be served
21:            $X_{\min\{i:X_i(t_k)==0\}}(t_{k+1}) \leftarrow K$
22:          **else**   ▷ first in the queue
23:            $X_{c+1}(t_{k+1}) \leftarrow K$
24:          **end if**
25:          $X_{c+2}(t_{k+1}) \leftarrow X_{c+2}(t_{k+1}) - 1$
26:       **end while**
27:    **end if**
28: **end if**

To consider the model of realistic size, we set $c = 1000$, i.e. the supercomputer has 1000 servers. It remains to define the distributions used for timer initialization. To do so, we get the inspiration from [7], where the properties of real-world workloads of supercomputers are studied. We configure the following setup for the supercomputer model. Let the generic interarrival time τ have the two-parameter Weibull distribution with shape parameter a and scale parameter b having distribution function

$$P\{\tau \le x\} = 1 - e^{-(x/b)^a}, \quad a, b > 0, \ x \ge 0. \tag{3}$$

The expected interarrival time is then obtained as $E\tau = b\Gamma(1 + 1/a)$, where Γ is the gamma function. We fix some a and take

$$b = 10/\Gamma(1 + 1/a) \tag{4}$$

to obtain $E\tau = 10$.

A generic customer requires i servers with probability (w.p.) p_i which is defined as the so-called two-phase log-uniform [7]. Following this model, the class N of a generic customer is obtained by the following procedure. It is taken to be 1 w.p. 0.24 or with complementary probability the logarithm $u = \log_2 N$ is taken from uniform distribution over the intervals $[l, m)$ or $[m, h]$, where the cumulative distribution at point m is taken as 0.86 (i.e. the density on the former interval is $0.86/(m - l)$, while on the latter is $0.14/(h - m)$). Finally, the u is rounded to nearest integer w.p. 0.75 and exponentiated i.e. $N = 2^{[u]}$, otherwise the number of servers required are obtained as $N = [2^u]$ (where $[\cdot]$ is the rounding operation). The constants l, m, h are configured in such a way to force $N \in [1, 1000]$.

It remains to define the service time distribution. Following suggestions in [7], we take the generic service time S to have the density f being the discrete mixture

$$f(x) = p(k)f_1(x) + (1 - p(k))f_2(x), \tag{5}$$

where the densities f_1 and f_2 are the gamma distributions given as

$$f_i(x) = \frac{x^{\alpha_i - 1}}{\beta_i^{\alpha_i}\Gamma(\alpha_i)}e^{-x/\beta_i}, \quad x \ge 0, \alpha_i, \beta_i > 0, i = 1, 2, \tag{6}$$

having the following parameters

$$\alpha_1 = 30, \ \beta_1 = 0.3, \ \alpha_2 = 330, \ \beta_2 = 0.03.$$

The mixing probability $p(k)$ depends on the customer class k in the following way

$$p(k) = 0.95 - 0.2k/1000.$$

As such, the service time distribution is class-dependent and thus the system is heterogeneous. Note that the average service time does not exceed 9.9 for any class, and thus, due to (1), the system is stable.

We are interested in obtaining the estimate of some performance measure $\chi(X_e)$ of the stationary system state $X_e = \lim_{t \to \infty} X(t)$ (where the limit is treated in distribution). Assume that a subsequence of event numbers β_j (starting from $\beta_0 = 0$) called discrete *regeneration points* may be defined such that the process Θ probabilistically restarts in each such a timepoint, and the regenerative cycles $\{X(t), T(t)\}_{t \in [t_{\beta_i}, t_{\beta_{i+1}})}$ are iid. for $i \ge 0$. Then the estimate for steady-state performance $\chi(X_e)$ can be obtained as the regenerative limit by measuring the required quantity over a regenerative cycle [3]. To do so, define the cumulative performance over the j-th regenerative cycle as

$$Y_j = \sum_{k=\beta_{j-1}}^{\beta_j-1} \chi(X(t_k))(t_{k+1} - t_k), \quad j \geq 1, \tag{7}$$

and let $R_j = t_{\beta_j} - t_{\beta_{j-1}}$ be the j-th regenerative cycle length. As such, the sequence $\{Y_j, R_j\}_{j\geq 1}$ is iid. and the required performance measure may be obtained as

$$\chi(X_e) = \frac{EY_1}{ER_1}.$$

Thus the estimate over n regenerative cycles can be obtained as

$$\bar{r}_n = \frac{Y_1 + \cdots + Y_n}{t_{\beta_n}} = \frac{S_{Y_n}}{S_{R_n}}, \tag{8}$$

where the corresponding partial sums are given as

$$S_{Y_n} = Y_1 + \cdots + Y_n, \quad S_{R_n} = R_1 + \cdots + R_n = t_{\beta_n}, \quad n \geq 1. \tag{9}$$

Assuming $E(Y_1 + R_1)^2 < \infty$, the central limit theorem for regenerative processes can be used to obtain the $(1 - 2\gamma)\%$ confidence interval for $\chi(X_e)$ as follows [6]:

$$\bar{r}_n \pm \frac{h_\gamma \sqrt{\mathrm{Var}(n)}}{\sqrt{n} S_{R_n}}. \tag{10}$$

Here the quantity $\overline{\mathrm{Var}}(n)$ is the unbiased estimate for the value of $\mathrm{Var}(Y_1) - 2r\mathrm{cov}(Y_1, R_1) + r^2\mathrm{Var}(R_1)$, which can be computed as

$$\overline{\mathrm{Var}}(n) = \frac{nS_{Y_n^2} - (S_{Y_n})^2 - 2\bar{r}_n(nS_{YR_n} - S_{Y_n}S_{R_n}) + \bar{r}_n^2(nS_{R_n^2} - (S_{R_n})^2)}{n(n-1)}, \tag{11}$$

where the corresponding parital sums are given in (9), and

$$S_{Y_n^2} = Y_1^2 + \cdots + Y_n^2, \quad S_{YR_n} = Y_1R_1 + \cdots + Y_nR_n, \quad S_{R_n^2} = R_1^2 + \cdots + R_n^2. \tag{12}$$

The constant $h_\gamma = \Phi^{-1}((1-\gamma)/2)$ is the corresponding quantile of the normal distribution, and the function $\Phi(x)$ is the Laplace function.

The choice of sequence of regeneration points can be done in multiple ways. Condition (1) guarantees the existence of classical regeneration (at zero) in the supercomputer model. However, in general the multiserver queue may be stable but never become empty [3]. Moreover, in some cases the regeneration at zero may be relatively rare. As such, alternative regeneration points can be considered, which is the issue studied in the next section.

4   Artificial Regeneration and Exponential Splitting

It can be seen that in a specific case $N_i \equiv c$ the stability condition (1) is a criterion. However, the same condition may be rather rough, which is observed in the boundary case $N_i \equiv 1$. In the latter case, the model essentially becomes the classical G/G/c model with a known stability criterion $ES < c E\tau$. However, such a stability condition alone does not guarantee that the process Θ is positive recurrent regenerative at zero (this issue is discussed in [3]).

One of the ways to obtain a classical regeneration in a non-empty system is to force the system to have memoryless property. It is clear that the timers $T(t)$ for any t in a GSMP model keep track of the prehistory of the process Θ in a specific way. Such a dependence becomes Markovian only on the timer initialization epochs. The system states evolution is performed with Markovian dependence type, and thus the timers need to be made memoryless. This can be done by the so-called exponential splitting of the density, applied at the time epochs corresponding to the timer initialization. In such a procedure, upon the corresponding timer initialization, the original density of the timer is replaced with the (possibly shifted) exponential density with some probability guaranteeing that the so-called *minorization* condition holds good [17]. The regeneration point is then defined as an event epoch in which all the timers are sampled from exponential distributions, that is, the component $T(t)$ becomes memoryless. To keep track of such an event, the system state is enriched with a discrete component vector $Z(t)$ having the same size as the component $T(t)$. Below we briefly give some details of this procedure given in more details in [2,16].

Consider the exponential splitting of a random variable (r.v.) X having density f. This may be performed if there exists such $0 < p < 1$, that the minorization condition holds good [17],

$$f(x) \geqslant p f^{(0)}(x), \tag{13}$$

where $f^{(0)}$ is the required density, say, shifted exponential

$$f^{(0)} = \theta e^{-\theta(x-x_0)}, x \geq x_0, \tag{14}$$

for some x_0 and θ constant. Define now the residual density

$$f^{(1)}(x) = \frac{f(x) - p f^{(0)}(x)}{1 - p}. \tag{15}$$

The original r.v. X is then stochastically equivalent to the combination of the two r.v. and splitting indicator,

$$X =_d \hat{X} = \xi X_0 + (1 - \xi) X_1, \tag{16}$$

where r.v. X_0 has the density $f^{(0)}$, X_1 has density $f^{(1)}$, and *splitting indicator* ξ is a Bernoulli r.v. with success probability $p = P\{\xi = 1\}$. The r.v. $\hat{X}$ is then called the splitting representation of X and has the required properties on the event $\{\xi = 1\}$. It follows from (14) and (15) that

$$f^{(1)} = \begin{cases} \dfrac{f(x)}{1-p}, & x \leqslant x_0; \\ \dfrac{f(x) - p\theta e^{-\theta(x-x_0)}}{1-p}, & x > x_0. \end{cases} \quad (17)$$

Note that the r.v. $\hat{X}$ has the memoryless property conditionally on event $\{\xi = 1\}$ and only when $\hat{X} > x_0$, since

$$P\{\hat{X} - t > x + y | \hat{X} - t > x\} = \frac{e^{-\theta(x+y+t-x_0)}}{e^{-\theta(x+t-x_0)}} = e^{-\theta y}, \quad x, y > 0, t > x_0. \quad (18)$$

Thus, to keep track of the memoryless phases of the timers, the system state is enriched with a component $Z(t)$ which plays the role of the vector governed by the (corresponding) splitting indicators of the timers. That is, the original process Θ is replaced with a stochastically equivalent $\hat{\Theta} = \{X(t), Z(t), \hat{T}(t)\}_{t \geq 0}$ provided that there exists such a subset of system states $\mathcal{X}^* \subseteq \mathcal{X}$ in which the timers initialization density f satisfies (13) (possibly with timer-specific constants $p_i, x_{0,i}, \theta_i$ for each timer). The *timer phases* $Z_i(t)$ are then defined in such a way to track memoryless property occurred (18). This is done by the so-called pre-exponential phase of the timer. Namely, the phase Z_i of the timer i takes values in the set $\{0, 1, 2\}$, which are interpreted as follows: $Z_i(t) = 0$ (exponential phase), if the splitting is successful (splitting indicator equals 1) and the time $x_{0,i}$ has passed since the timer initialization, $Z_i(t) = 1$ (pre-exponential phase) is set immediately upon successful splitting but before the time $x_{0,i}$ has passed since the timer initialization, finally $Z_i(t) = 2$ (non-exponential phase) is reserved for all other cases. Formally, the timer phase Z_i is defined upon the i-th timer initialization as follows:

$$Z_i(t) = \begin{cases} 1, & \text{if } X(t) \in \mathcal{X}^* \text{ and } \xi_i(t) = 1, \\ 2, & \text{otherwise.} \end{cases} \quad (19)$$

Then, given $Z_i(t) = 1$, at time $t + x_{0,i}$ the tranzition of Z_i to zero occurs,

$$Z_i(u) = \begin{cases} 1, & t < u < t + x_{0,i}, \\ 0, & u \geq t + x_{0,i}. \end{cases} \quad (20)$$

As such, given $Z_k(t) = 0$, the timer $\hat{T}_k(t)$ has (pure) exponential distribution with rate θ_i and hence $\hat{T}_k(t)$ can be re-sampled independently. We stress that splitting of the timer $\hat{T}_i$ occurs only at such t that the timer is initialized (i.e. the corresponding timer is/becomes active), and the system state is in the set $\mathcal{X}^*$. If $X(t) \notin \mathcal{X}^*$, the timer is initialized from the original density, say f, whereas the phase is set as $Z_i(t) = 2$. Finally, the regeneration of the process $\hat{\Theta}$ is defined as such an event epoch t^* that all the active timers are either in exponential phase or initialized anew, whereas the system state is fixed $X(t^*) = x^*$ for some fixed $x^* \in \mathcal{X}^*$. This allows to define a sequence of regeneration points and use regenerative confidence estimation described in Sect. 3.

In the supercomputer model the timers $1, \ldots, c$ are initialized from class-dependent mixture of gamma distributions, whereas the timer $c + 1$ is initialized from Weibull distribution. It is natural to consider the regeneration point as the arrival event epoch. As such, the value of the timer $c + 1$ is sampled anew and is independent of the prehistory. To split the timers $1, \ldots, c$, we need to define the splitting probability and the parameters of shifted exponential distributions for the components of the class-dependent mixture (5). Specifically, to split the gamma distribution components (6) and satisfy (13), where the density $f^{(0)}$ is given in (14), the following parameters can be used [2]

$$x_0 = \beta, \ \theta = \frac{1}{\beta}, \ p = \frac{1}{e\Gamma(\alpha)}.$$

However, this configuration is of low practical interest for large α due to an extremely low splitting probability p. Indeed, for, say, $\alpha = 30$, the value $p = (e \cdot 29!)^{-1} \approx 4 \cdot 10^{-32}$. To address this issue, we recommend the following alternative configuration to be used

$$x_0 = (\alpha - 1)\beta, \ \theta = \frac{1}{\beta}, \ p = \frac{(\alpha - 1)^{\alpha-1}}{e^{\alpha-1}\Gamma(\alpha)}. \tag{21}$$

To demonstrate the benefits of the proposed configuration (21), we use the asymptotic formula for $\Gamma(x), x \geq 0$, (in a slightly modified form) proven in [12] that goes back to Srinivasa Ramanujan,

$$\left(8x^3 + 4x^2 + x + \frac{1}{100}\right)^{\frac{1}{6}} < \frac{1}{\sqrt{\pi}}\Gamma(1 + x)\frac{e^x}{x^x} < \left(8x^3 + 4x^2 + x + \frac{1}{30}\right)^{\frac{1}{6}}. \tag{22}$$

Thus, taking $x = \alpha - 1$ and using (22) in (21), it follows that for $\alpha > 1$,

$$p > \frac{1}{\sqrt{\pi}}\left(8\alpha^3 - 20\alpha^2 + 17\alpha - 5 + \frac{1}{30}\right)^{-\frac{1}{6}}. \tag{23}$$

Moreover, asymptotically the r.h.s. of (23) is equivalent to $1/\sqrt{2\pi\alpha}$ as $\alpha \to \infty$, and thus the splitting occurs reasonably often even for large α values. Specifically, for $\alpha = 30$ it can be shown from (23) that $p > 0.073869$.

To perform the splitting of the mixture (5), the parameters defined in (21) can be used for the first component with parameters α_1, β_1, however, the splitting probability should be multiplied by the class-dependent mixture parameter $p(k)$ as follows:

$$p = p(k)\frac{(\alpha_1 - 1)^{\alpha_1-1}}{e^{\alpha_1-1}\Gamma(\alpha_1)}. \tag{24}$$

The splitting needs to be performed in such a case when the system state is in $\mathcal{X}^*$, and the corresponding timer needs to be initialized. It is important to note that since the splitting is done with the same parameters of the exponential distribution, the service times become class-independent after a successful splitting. As such, it is natural to assume the regeneration point as such an epoch when the system state becomes x^* such that a single customer of arbitrary class is being served. However, to guarantee identical distribution of the regeneration cycles, the class of such a customer can also be fixed. In the next section we illustrate this approach by performing a numerical experiment.

5   Numerical Experiment

Let us construct the regenerative estimate and 95% confidence interval for the steady-state average number of customers in the 1000-server system with distributions defined in Sect. 3. We take the parameter a of the interarrival time (Weibull) distribution (3) from the interval $[0.5, 2]$. The parameter b is then selected using (4), and thus the mean interarrival time equals 10. However, the properties (say, the coefficient of variation) of the interarrival distribution depend on the value of a. For each fixed a the trajectory of length 10^7 is constructed, and the regenerative epochs are selected by performing the exponential splitting with parameters defined in (21) and (24). The experimental results are shown on Fig. 1 and demonstrate the non-linear dependence of the performance estimate on a.

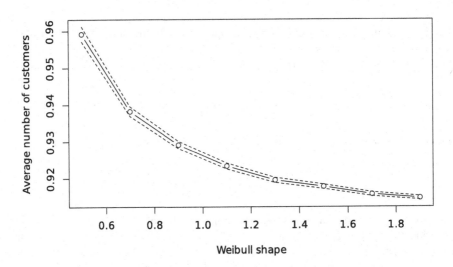

Fig. 1. Estimate and 95% confidence interval of the steady-state average number of customers in 1000-server supercomputer model vs. the parameter a of the interarrival times (Weibull) distribution (3) having parameter b given by (4).

6   Conclusion

In this paper we demonstrated the modified exponential splitting procedure to construct the artificial regeneration points in a GSMP supercomputer model of realistic size. It may be interesting to further study the more sophisticated types of dependency in a GSMP that nevertheless allow the artificial regeneration points to be constructed, however, we leave this as a possibility for future research.

Acknowledgements. The publication has been prepared with the support of Russian Science Foundation according to the research project No.21-71-10135, https://rscf.ru/en/project/21-71-10135/.

References

1. Afanaseva, L., Bashtova, E., Grishunina, S.: Stability analysis of a multi-server model with simultaneous service and a regenerative input flow. Methodol. Comput. Appl. Probab. **22**, 1439–1455 (2019)
2. Andronov, A.: Artificial regeneration points for stochastic simulation of complex systems. In: Simulation Technology: Science and Art. 10th European Simulation Symposium ESS 1998, pp. 34–40. SCS, Delft, The Netherlands (1998)
3. Asmussen, S.: Applied Probability and Queues. Springer, New York (2003)
4. Asmussen, S., Glynn, P.W.: Stochastic Simulation: Algorithms and Analysis. No. 57 in Stochastic Modelling and Applied Probability, Springer, New York (2007). oCLC: ocn123113652
5. Astafiev, S.N., Rumyantsev, A.S.: Distributed computing of R applications using RBOINC package with applications to parallel discrete event simulation. In: Vishnevskiy, V.M., Samouylov, K.E., Kozyrev, D.V. (eds.) DCCN. CCIS, vol. 1552, pp. 396–407. Springer, Cham (2022)
6. Crane, M.A., Iglehart, D.L.: Simulating stable stochastic systems, I: General multi-server queues. J. ACM (JACM) **21**(1), 103–113 (1974). http://dl.acm.org/citation.cfm?id=321805
7. Feitelson, D.G.: Workload Modeling for Computer Systems Performance Evaluation. Cambridge University Press, Cambridge (2015). https://doi.org/10.1017/CBO9781139939690
8. Glynn, P.W.: A GSMP formalism for discrete event systems. Proc. IEEE **77**(1), 14–23 (1989). https://doi.org/10.1109/5.21067
9. Grosof, I., Harchol-Balter, M., Scheller-Wolf, A.: Stability for Two-class Multiserver-job Systems. arXiv:2010.00631 [cs] (2020). http://arxiv.org/abs/2010.00631
10. Harchol-Balter, M.: The multiserver job queueing model. Queueing Syst. **100**, 201–203 (2022). https://doi.org/10.1007/s11134-022-09762-x
11. Henderson, S.G., Glynn, P.W.: Regenerative steady-state simulation of discrete-event systems. ACM Trans. Model. Comput. Simul. **11**(4), 313–345 (2001). https://doi.org/10.1145/508366.508367
12. Karatsuba, E.A.: On the asymptotic representation of the Euler gamma function by Ramanujan. J. Comput. Appl. Math. **135**(2), 225–240 (2001). https://doi.org/10.1016/S0377-0427(00)00586-0, https://linkinghub.elsevier.com/retrieve/pii/S0377042700005860
13. Leahu, H., Mandjes, M., Oprescu, A.M.: A numerical approach to stability of multiclass queueing networks. IEEE Trans. Autom. Control **62**(10), 5478–5484 (2017). https://doi.org/10.1109/TAC.2017.2699126, http://arxiv.org/abs/1606.07294, arXiv: 1606.07294

14. Morozov, E., Rumyantsev, A., Peshkova, I.: Monotonicity and stochastic bounds for simultaneous service multiserver systems. In: 2016 8th International Congress on Ultra Modern Telecommunications and Control Systems and Workshops (ICUMT), pp. 294–297. IEEE (2016). https://doi.org/10.1109/ICUMT.2016.7765374, http://ieeexplore.ieee.org/abstract/document/7765374/

15. Rumyantsev, A., Basmadjian, R., Golovin, A., Astafiev, S.: A three-level modelling approach for asynchronous speed scaling in high-performance data centres. In: Proceedings of the Twelfth ACM International Conference on Future Energy Systems, pp. 417–423. e-Energy 2021, Association for Computing Machinery, New York, NY, USA (2021). https://doi.org/10.1145/3447555.3466580

16. Rumyantsev, A., Peshkova, I.: Artificial regeneration based regenerative estimation of multiserver system with multiple vacations policy. In: Dudin, A., Nazarov, A., Moiseev, A. (eds.) ITMM 2019. CCIS, vol. 1109, pp. 38–50. Springer, Cham (2019). https://doi.org/10.1007/978-3-030-33388-1_4

17. Thorrison, H.: Coupling, Stationarity, and Regeneration. Springer, New York (2000)

Modeling and Analysis of Multi-channel Queuing System Transient Behavior for Piecewise-Constant Rates

K. A. Vytovtov$^{(\boxtimes)}$(iD), E. A. Barabanova(iD), and V. M. Vishnevsky(iD)

V. A. Trapeznikov Institute of Control Sciences of RAS,
Profsoyuznaya 65 Street, Moscow, Russia
vytovtov_konstan@mail.ru

Abstract. A transient behaviour of multi-channel queuing system is investigated in this paper. For analyses the analytical method of Kolmogorov equations system solution based on using so-called translation matrix is used. The cases when arrival and service rates are constant and piecewise-constant functions are considered. The analytical expressions for non-stationary probabilities of states, an average number of packets in the system, an average number of occupied channels, and an average queuing delay in transient mode are presented. The transient time and throughput of queuing system in transient mode have been analyzed. The numerical simulation of the multi-channel queuing system is carried out by using the example of $M/M/2/3$ system. The dependencies of probabilities of states, throughput and an average number of packets in the system for different initial conditions in transient mode are obtained. The case of periodic change of arrival rates corresponding to the periodic transmitting the control signals within the payload has been considered.

Keywords: Multi-channel queuing system · Piecewise-constant functions · Performance metrics · Transient behavior

1 Introduction

Nowadays queuing systems are used in various spheres. They are used in service sector, everyday life, economics, commerce, different organizations and enterprises. This mathematical approach is widely used in telecommunication systems and networks also [1,2]. Today there are a lot of various types of queuing systems. One of the important class of queuing systems is multi-channel queuing system with a finite buffer which is known as $M/M/n/m$ systems. The systems with several servicing devices are used for describing distributed computer networks with several servers or multi-channel switching systems [1,2]. Various types of multi-channel queuing systems have been considered in many domestic and foreign scientific works [1–6]. And in the most of them $M/M/n/m$ systems are

---

The reported study was funded by Russian Science Foundation, project number 22-49-02023.

V. M. Vishnevskiy et al. (Eds.): DCCN 2022, LNCS 13766, pp. 397–409, 2022.
https://doi.org/10.1007/978-3-031-23207-7_31

investigated in stationary mode. In paper [3] multi-server Feedback Markovian queuing model with encouraged arrivals, customer impatience, and retention of impatient customers have been performed. The authors presented expressions for steady-state probabilities and performance metrics calculating which had been derived using classical queuing theory approach. The $M/M/N/N$ queue where two types of users compete for the N resources has been considered in paper [4]. The authors have presented the two-dimensional queue model for this case and calculated steady-state probabilities of the system. In paper [5] the authors have analyzed a multi-server queue with customers' impatience and Bernoulli feedback under a variant of multiple vacations. They have investigated the mathematical model of the system and have developed the differential equations for the probability generating functions of the steady-state probabilities. A multi-server queuing system with reverse balking and impatient customers has been considered in [6]. The steady-state probabilities of system size are obtained using iterative method.

The most accurate analysis of multi-channel queuing system functioning can be obtained if the system is to be examined not only in stationary mode but in transient one too. The stationary mode occurs some time later after beginning system functioning and continue for $t \to \infty$. In contrast to the stationary mode the transient mode occurs right after beginning of the system functioning before the system will be in the steady state [7,8]. The system goes in the transient mode in result of rebooting or jumps of arrival and service rates caused by change in the information routes. There are only several works where two-channel queuing systems transient behavior has been described. In [9] authors have studied $M/M/2/N$ queuing system with two-heterogeneous servers and retention of reneging customers and obtained its transient solution by employing matrix method. Transient analysis of two-heterogeneous servers queue with impatient behavior has been carried out in [10]. Additionally the steady-state probabilities of the system size are studied. The paper [11] presents an expression for finding the transient probabilities of system states with an infinite number of servers.

In all listed above works the transient behaviour of queuing systems is considered for the case of constant arrival and serviced rates. Nevertheless, the problem of studying the behavior of a queuing system in the case of arrival or service rates are piecewise-constant functions is very important in a number of telecommunication applications. For example, arrival rate is changed abruptly when control signals with a payload are transmitted to the tethered UAV through determinate time intervals [12]. In [13] time-varying rate multi-channel queueing systems $M_t/G/L/L$ and $M_t/M/L/L$ describing telecommunication system intended for transmission of realtime communication services like voice calls and video on demand are investigating. The dependencies of loss probability for the case of change in the arrival rate according to a sinusoidal law are obtained. The other performance metrics are not studied.

The authors of this paper proposed the analytical approach of $M/M/1/n$ and $MAP/M/1/n$ queuing systems transient behavior investigation for cases of constant and piecewise-constant arrival and service rates on based of method of translation matrix [7, 8, 13]. In this paper we continue to developed our approach considering multi-channel queuing system $M|M|n|m$ with jumps in the arrival and service rates and investigating dependencies of system states transient probabilities in different initial conditions.

The paper is organized as follows. In Sect. 2 the statement of the problem is given. Section 3 presents the analytical approach of multi-channel queuing system investigation for cases of constant and piecewise-constant arrival or service rates in transient mode. Numerical simulation results are presented in Sect. 4. The paper is concluded in Sect. 5.

2   Statement of the Problem

Let us consider the system for transmitting information from a ground control station to a tethered UAV. The payload and control signals are transmitted over n wireless Wi-Fi channels with the same bandwidth. If all channels are busy, the information is stored in a buffer of the size m. So the considered telecommunication system can be described by a $M/M/n/m$ queuing system.

The state graph of the $M/M/n/m$ system is presented in the Fig. 1. The first state S_0 corresponds to the state when the system is empty. The second state S_1 corresponds to the state when the first channel is busy but the buffer is empty. The third state S_2 corresponds to the state when two channels are busy but the buffer is empty. The n state S_n corresponds to the state when n channels are busy but the buffer is empty. The state B_1 corresponds to the state when all n channels are busy and one packet is in the buffer. The m state B_m corresponds to the state when all n channels are busy, m packets are in the buffer and the next arriving packet is lost. Here we consider the case when $\lambda(t)$ or $\mu(t)$ (the arrival or service rates correspondingly) are the piecewise-constant functions (Fig. 2).

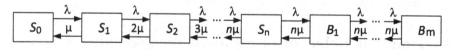

Fig. 1. The state graph of the $M/M/n/m$ system.

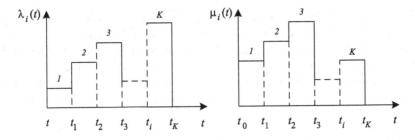

Fig. 2. The arrival and service rates time dependence graphs.

The Kolmogorov differential equations system for the multi-channel queuing system within the interval with constant parameters λ and μ has the form

$$
\begin{cases}
\dfrac{dP_0(t)}{dt} = -\lambda P_0(t) + \mu P_1(t) \\[2mm]
\dfrac{dP_i(t)}{dt} = \lambda P_{i-1}(t) - (\lambda + i\mu)P_i(t) + (i+1)\mu P_{i+1}(t), \quad 1 \le i \le n-1 \\[2mm]
\dfrac{dP_i(t)}{dt} = \lambda P_{i-1}(t) - (\lambda + n\mu)P_i(t) + n\mu P_{i+1}(t), \qquad n \le i \le n+m-1 \\[2mm]
\dfrac{dP_{n+m}(t)}{dt} = \lambda P_{n+m-1}(t) - n\mu P_{n+m}(t)
\end{cases}
\tag{1}
$$

Here $P_i(t)$ is the probability that the system is in i state at the time t. $P_0(t)$ is the probability that the system is empty, P_{n+m} is the probability of loss.

The main aim of this paper is investigation of the multi-channel queuing system transient behavior for cases of constant and piecewise-constant information flow parameters. It is the problem to obtain the analytic expressions for main transient performance metrics of this system such as transient time, throughput, an average number of packets in the system, an average queuing delay, and probability of packet loss. The novel aspects of queueing systems transient behavior study considered in this paper is investigation probabilities of states and main performance metrics for different initial conditions.

3   Transient Behavior of Multi-channel Queuing System with Constant and Piecewise-Constant Rates

To analyze of the multi-channel queuing system $M/M/n/m$ functioning in transient mode, the new approach based on the translation matrix method analogous to [7,8,14] is applied. According to this method the solutions of the system (1) are found as

$$
P_i(t) = \sum_{j=0}^{n+m} m_{ij} P_j(0),
\tag{2}
$$

where m_{ij} are elements of the translation matrix [7], which can be found using following expression ($i = 0..n + m$):

$$m_{ij} = \sum_{j=0}^{n+m} \xi_{ij} A_j \exp^{\gamma_j t}, \qquad (3)$$

where A_j are the integration constants determined by initial conditions, γ_j are the roots of the characteristic equation of the system (1), and ξ_{ij} are coefficients expressing probabilities $P_i(t)$ as functions of $P_0(t)$ ($i = \overline{1, n + m}$).

Thus the probabilities of the states and the performance metrics of the multi-channel queueing system for time-constant parameters λ and μ can be calculated by the expression (2). To determine analogous parameters of the system in case of piecewise-constant rates it is proposed to use the method of translation matrix [7,8].

Now, the translation matrix describing the behavior of the system with piecewise-constant parameters on the K-th interval (Fig. 2) can be write in the form [7,8]

$$\mathbf{M}_K(t) = \mathbf{M}_K(t - t_{K-1}) \prod_{i=K-1}^{1} \mathbf{M}_i(\Delta t_i) \qquad (4)$$

The matrix (4) relates the probabilities of states on the K-th interval to the probabilities of states in the initial time t_0:

$$\mathbf{U}(t) = \left(\mathbf{M}_K(t - t_{K-1}) \prod_{i=K-1}^{1} \mathbf{M}_i(\Delta t_i) \right) \mathbf{U}(t_0). \qquad (5)$$

Here $\mathbf{U}(t) = (P_i)^T$ ($i = \overline{0, n + m}$) is the $1 \times (n + m + 1)$-vector of the system states, T is the transpose operator.

The performance metrics of $M/M/n/m$ system can be found by using (2) and (4). According to the [7] the so-called transient instantaneous throughput is calculated as

$$A(t) = [1 - P_{n+m}(t)]\lambda(t), \qquad (6)$$

Here $P_{n+m}(t)$ is the instantaneous probability of loss:

$$P_{loss}(t) = \sum_{j=0}^{n+m} P_j(0) \sum_{j=0}^{n+m} \xi_{n+m,j} A_j exp(\gamma_j t) \qquad (7)$$

An average number of packets in the system $E(m(t))$ and an average number of busy channels $E(n(t))$ can be determine using following formulae:

$$E(m(t)) = \sum_{i=1}^{n+m} i P_i(t) \qquad (8)$$

$$E(n(t)) = \sum_{i=1}^{n} i P_i(t) \qquad (9)$$

Therefore an average number of packets in the buffer can be calculated using expression:

$$E(N(t)) = \sum_{i=1}^{n+m} iP_i(t) - \sum_{i=1}^{n} iP_i(t) \tag{10}$$

Given (2) and (3), the final expressions for an average number of packets in the system and an average number of busy channels can be written as follows

$$E(m(t)) = \sum_{i=1}^{n+m} i \sum_{j=0}^{n+m} P_j(0) \sum_{j=0}^{n+m} \xi_{ij} A_j exp^{\gamma_j t} \tag{11}$$

$$E(n(t)) = \sum_{i=1}^{n} i \sum_{j=0}^{n+m} P_j(0) \sum_{j=0}^{n+m} \xi_{ij} A_j exp^{\gamma_j t} \tag{12}$$

According to the [1] an average waiting time in a buffer (queuing delay) can be given as:

$$W(t) = E(N(t))/\lambda(t) \tag{13}$$

The other important characteristic of transient mode is transient time τ_{tr}. The detail theory of the transient time determination is presented in [7,8]. Here we use that method to the considered problem and obtain:

$$\tau_{tr} = (3 \div 5)\tau \tag{14}$$

where τ is the time constant:

$$\tau = \max\{\tau_i, i = \overline{0, n+m}\} = \max\{\frac{1}{|\gamma_i|}, i = \overline{0, n+m}\} \tag{15}$$

4   Numerical Results

In this section, we consider the $M/M/2/3$ system describing a wireless access point functioning installed on a tethered UAV and connected with a ground control station through two Wi-Fi channels. The size of wireless access point buffer is three packets.

At first, let us consider the case when the parameters of the information flow are constant ($\lambda = 200$ pps, $\mu = 300$ pps) and investigate the transient behavior of the system under different initial conditions. In Fig. 3, Fig. 4 and Fig. 5 we presented the dependencies of transient states probabilities for following initial conditions:

1. $[P_0(0), P_1(0), P_2(0), P_3(0), P_4(0), P_5(0)]^T = (1, 0, 0, 0, 0, 0)^T$.
2. $[P_0(0), P_1(0), P_2(0), P_3(0), P_4(0), P_5(0)]^T = (0, 1, 0, 0, 0, 0)^T$.
3. $[P_0(0), P_1(0), P_2(0), P_3(0), P_4(0), P_5(0)]^T = (0, 0, 0, 0, 1, 0)^T$.

The first case corresponds to a free state of both channels (Fig. 3), the second case corresponds to a working state only of the first channel (Fig. 4) and the third case corresponds to the state when both channels suddenly became unavailable but two packets remained in the buffer (Fig. 5). Figure 3 – Fig. 5 show that transient probabilities are differ for three cases, and the steady-state probabilities of the system states do not depend on the initial conditions: $\pi_0 = 0.501$, $\pi_1 = 0.334$, $\pi_2 = 0.112$, $\pi_3 = 0.037$, $\pi_4 = 0.012$, $\pi_5 = 0.004$. These values coincide with the values calculated by the well-known formula for steady-state probabilities applied to the $M/M/2/3$ system [1]:

$$
\pi_i =
\begin{cases}
\dfrac{\rho^i}{i!}\pi_0, & 0 \le i \le n-1 \\[2mm]
\dfrac{\rho^i}{n!n^{i-n}}\pi_0, & n \le i \le n+m
\end{cases}
\tag{16}
$$

Here $\pi_0 = \left[\displaystyle\sum_{i=0}^{n-1}\dfrac{\rho^i}{i!} + \dfrac{\rho^n}{n!}\dfrac{\left(1-\left(\frac{\rho}{n}\right)^{m+1}\right)}{1-\frac{\rho}{n}}\right]^{-1}$ is the empty state of the system and $\rho = \lambda/\mu$ is the system load factor.

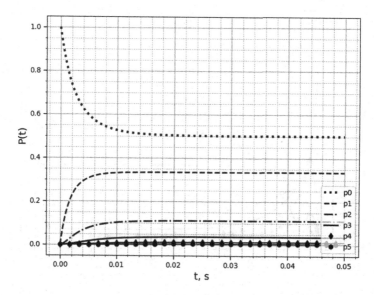

Fig. 3. Dependencies of the state probabilities on time for the first initial conditions.

It can be conclusion from Fig. 3 – Fig. 5 that there are differences between the speeds of establishing the transient mode for different initial conditions, but the transient time is the same $\tau_{tr} = 0.02$s. However, when calculating specific cases by using the classical theory, their own peculiarities may arise. Indeed, as it has been shown in [7] from a physical point of view the transient mode can be

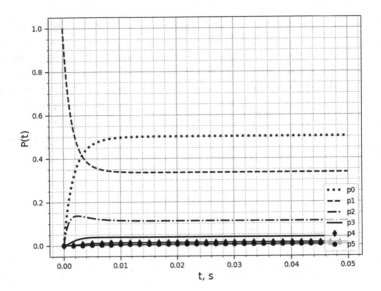

Fig. 4. Dependencies of the state probabilities on time for the second initial conditions.

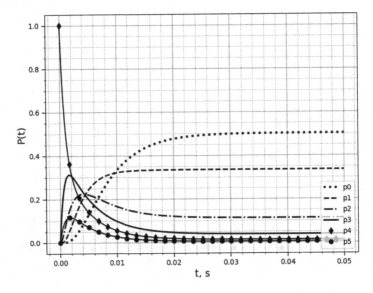

Fig. 5. Dependencies of the state probabilities on time for the third initial conditions.

considered as finite if $\mid P_j(t) - \pi_j \mid \leq \varepsilon$, where ε is some infinitesimal value, $P_j(t)$ is the transient probability of the j-th state at the time t, π_j is the stationary probability of the j- th state. Let determined the first state probability $P_1(t)$ at the time $t = 0.01$ s supposed that ε is constant for all cases. For the first case (Fig. 3) $P_1(0.01) = 0.335$ therefore $\mid P_1(t) - \pi_1 \mid = 0.001$; for the second case

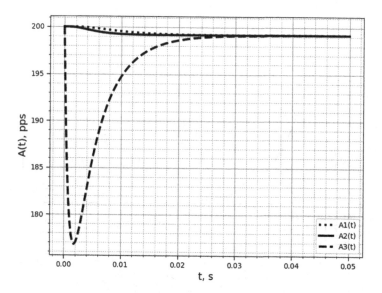

Fig. 6. Dependencies of throughput on time taking into account different initial conditions.

(Fig. 4) $P_1(0.01) = 0.34$ therefore $| P_1(t) - \pi_1 | = 0.006$; for the third case (Fig. 5) $P_1(0.01) = 0.325$ therefore $| P_1(t) - \pi_1 | = 0.009$. If suppose that $\varepsilon = 0.001$ we can prove that the transient mode is finished for the first initial conditions and has not been finished yet for the second and third initial conditions. In general case the choice of ε value is determined by the practical requirements.

As it can be seen from Fig. 6 the throughput of the system in transient mode depends on the initial conditions and in stationary mode $A = 199$ pps. Here $A_1(t)$, $A_2(t)$, $A_3(t)$ are values of throughput corresponding to the first, the second and the third cases of the initial conditions considered above and for the parameters of an information flow: $\lambda = 200$ pps, $\mu = 300$ pps. The Fig. 6 shows that a sudden unavailability of both channels leads to the most significant decrease of the system throughput. In the point of time $t = 0.0018$ s the system throughput has a minimum value $A_3(0.018) = 176.8$ pps. It corresponds throughput decreasing of 11.6% in comparison of initial moment of time and 11.2% in comparison of stationary mode.

The dependencies of an average number of packets in the system on time for different initial conditions described below are presented in Fig. 7. The value of steady-state probabilities is equal to 0.74 and coincides with the value obtained using existing approaches [1,9]:

$$L = \sum_{i=1}^{n+m} i\pi_i. \tag{17}$$

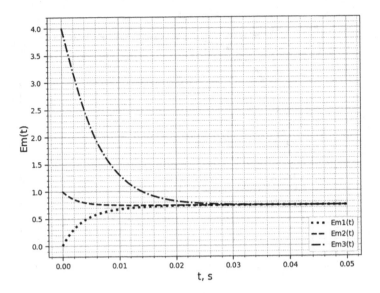

Fig. 7. An number of packets in the system size in transient mode.

Next, consider the case of piecewise-constant information flow for periodic jumps of arrival rate λ and find the transient probabilities of states (Fig. 8, Fig. 9). In practical point of view, it corresponds of periodic transmitting control signals parallel of payload transmitting. In the first case jumps of λ occur when the network is working normally ($\lambda < \mu$) (Fig. 8). The arrival rate is changed periodically from $\lambda_0 = 200$ pps to $\lambda_1 = 250$ pps, the service rate is constant $\mu_0 = \mu_1 = 300$ pps. In the second case jumps of λ occur when the network is overloaded ($\lambda > \mu$) (Fig. 9). Arrival rate is changed periodically from $\lambda_0 = 400$ pps to $\lambda_1 = 450$ pps, the service rate is constant $\mu_0 = \mu_1 = 300$ pps. Comparing the probabilities of states in Fig. 8 and Fig. 9 it can be conclude that the changing of state probabilities in the second case (Fig. 9) has smoother character. The maximum value of loss probability is 0.001 ($p_{loss1} = 0.001$) in the first case and it is in 55 times more in the second case ($p_{loss2} = 0.055$).

Dependencies of the average waiting time for a packet in the buffer on time (queuing delay) for cases of periodic piecewise-constant arrival rate are presented in Fig. 10. Here $W_1(t)$ is the average queuing delay for the case of $\lambda < \mu$ (λ is changed periodically from $\lambda_0 = 200$ pps to $\lambda_1 = 250$ pps, $\mu_0 = \mu_1 = 300$ pps), and $W_2(t)$ is the average queuing delay for the second case of $\lambda > \mu$ ($\lambda_0 = 400$ pps, $\lambda_1 = 450$ pps, $\mu_0 = \mu_1 = 300$ pps). Analysing the obtained results it can be conclude that the difference between maximum and minimum values of the average queuing delay $\mod W_{1max}(t) - W_{1min}(t) = 800\,\mu s$ in the first case, and $\mod W_{2max}(t) - W_{2min}(t) = 950\,\mu s$ in the second case.

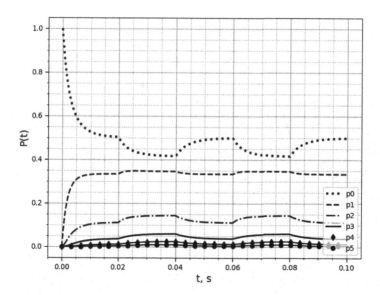

Fig. 8. Dependencies of the state probabilities on time ($\lambda < \mu$).

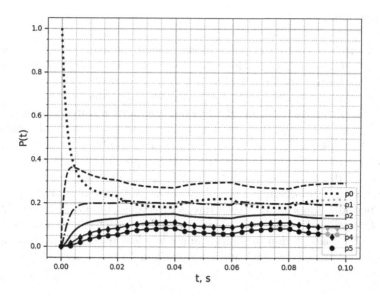

Fig. 9. Dependencies of the state probabilities on time ($\lambda > \mu$).

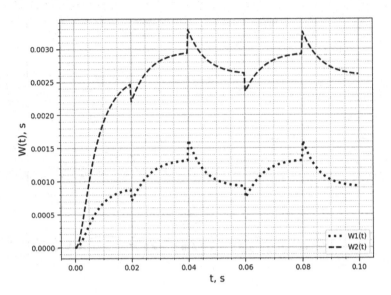

Fig. 10. An average waiting time for a packet in the buffer on time.

5 Conclusion

The transient behavior of the multi-channel queuing system $M/M/n/m$ for cases of constant and piecewise-constant rates has been studied for the first time. The analysis has been carried out by the accurate analytical method based on the so-called translation matrix. The method made it possible to study the change in the probabilities of the multi-channel queuing system states, throughput and an average number of packets in the system in transient mode depending on the initial conditions. The numerical simulation of $M/M/2/3$ system confirmed the analytical results. The correctness of the proposed approach also confirmed by the coincidence of the obtained steady-state probabilities and performance metrics with the results have been obtained by well-known methods [1,9]. The case of periodic jumps of an arrival rate is investigated numerically for different ratios of arrival and service rates. The proposed approach can be used for performance metrics calculation of multi-channel wireless telecommunication system designed to transmit payload and control signals from a ground station to a tethered UAV.

References

1. Dudin, A.N., Klimenok, V.I., Vishnevsky, V.M.: The Theory of Queuing Systems with Correlated Flows. Springer, Heidelberg (2020)
2. Lakatos, L., Szeidl, L., Telek, M.: Introduction to Queueing Systems with Telecommunication Applications. Springer, Heidelberg (2013)

3. Kumar, S.B., Sunny, S.: $M/M/c/N$ queuing systems with encouraged arrivals, reneging, retention and Feedback customers. Yugoslav J. Oper. Res. **28**(3), 333–344 (2018)
4. Smith, P.J., Firag, A., Dmochowski, P. A., Shafi, M.: Analysis of the $M/M/N/N$ queue with two types of arrival process: applications to future mobile radio systems. J. Appl. Math. **2012**, 123808 (2012)
5. Bouchentouf, A.A., Medjahri, L., Boualem, M., Kumar, A.: Mathematical analysis of a Markovian multi-server feedback queue with a variant of multiple vacations, balking and reneging. Discrete Continuous Models Appl. Comput. Sci. **30**(1), 21–38 (2022)
6. Kumar, R., Som, B.K.: A multi-server queue with reverse balking and impatient customers. Pak. J. Statist **36**(2), 91–101 (2022)
7. Vishnevsky, V..., Vytotvov, K.A., Barabanova, E.A., Semenova, O.V.: Transient behavior of the $MAP/M/1/N$ queuing system. Mathematics **9**(20), 2559 (2021)
8. Vytotvov, K., Barabanova, E., Vishnevsky, V.: The Analytical Method of Transient Behavior of the $M—M—1—n$ Queuing System for Piece-Wise Constant Information Flows. In: Vishnevskiy, V.M., Samouylov, K.E., Kozyrev, D.V. (eds.) DCCN 2021. LNCS, vol. 13144, pp. 167–181. Springer, Cham (2021). https://doi.org/10.1007/978-3-030-92507-9_15
9. Sampath, S. M. I. G., Kumar, R., Soodan, B. S., Liu, J., Sharma. S.: A matrix method for transient solution of an M/M/2/N queuing system with heterogeneous servers and retention of reneging customers. RT&A, **19**(4), 128–137 (2020)
10. Ammar, S.I.: Transient analysis of a two-heterogeneous servers queue with impatient behavior. J. Egypt. Math. Soc. **22**(1), 90–95 (2014)
11. Rubino, G.: Transient analysis of markovian queueing systems: a survey with focus on closed-forms and uniformization. In: Queueing Theory 2. Wiley, Hoboken, New Jersey, U.S (2021)
12. Perelomov, V.N., Myrova, L.O., Aminev, D.A., Kozyrev, D.V.: Efficiency enhancement of tethered high altitude communication platforms based on their hardware-software unification. In: Vishnevskiy, V.M., Kozyrev, D.V. (eds.) DCCN 2018. CCIS, vol. 919, pp. 184–200. Springer, Cham (2018). https://doi.org/10.1007/978-3-319-99447-5_16
13. Massey, W.A.: The analysis of queues with time-varying rates for telecommunication models. Telecommun. Syst. **21**(2–4), 173–204 (2002)
14. Vishnevsky, V., Vytotvov, K.A., Barabanova, E.A., Semenova, O.V.: Analysis of a $MAP/M/1/N$ queue with periodic and non-periodic piecewise constant input rate. Mathematics **10**(10), 1684 (2022)

On the Reliability Estimation of the Gaussian Multi-phase Degradation System

Oleg Lukashenko[1,2]([envelope]) [iD]

[1] Institute of Applied Mathematical Research of the Karelian Research Centre
of RAS, Petrozavodsk, Russia
lukashenko@krc.karelia.ru
[2] Petrozavodsk State University, Petrozavodsk, Russia

Abstract. The estimation of the reliability is an important and hard problem, arising in the performance analysis of degradation systems. The required performance measure is usually not analytically available. Hence, one has to rely on simulation methods. In this paper, we consider the multi-phase degradation model with a stepwise constant failure threshold based on the general Gaussian process with stationary increments and piecewise linear drift which seems to be more flexible alternative to the previously studied in the literature the Wiener degradation model. The variance reduction technique based on the special variant of the Conditional Monte Carlo method have been developed to estimate the reliability of this system. Numerical experiments are conducted to evaluate the performance of the proposed estimator in terms of relative error.

Keywords: Reliability · Degradation process · Gaussian process · Conditional Monte Carlo · Bridge process

1 Introduction

Degradation is the main cause of infrastructure failure in technical and engineering systems. Investigation of the degradation model performance characteristics is a very important problem in reliability analysis aimed at the prediction of fault tolerance. Thus, the development and evaluation of the models describing the degradation process is an actual research area.

It seems natural to describe the dynamic of the degradation process in terms of the stochastic process with an appropriate correlation structure. One class of such models is based on the stochastic process with a finite state space representing the stages of the degradation, where durations of stages are assumed to be independent (but not necessarily identical) random variables with a given distribution [7,13,18]. Such type of model describes the deterioration of the anti-corrosion coating [3,7].

The study was supported by the Russian Science Foundation, project 21-71-10135.

Other models are based on the continuous state space stochastic processes. In this framework, the Wiener process is one of the most popular degradation models (see [5,11,16,19,22] and references therein) thanks to its analytical tractability. There are some extensions on the general Gaussian process [21]. The system fails when the degradation process reaches the given failure threshold. So the performance analysis of such reliable systems is reduced to the evaluation of the so-called hitting time distribution. These models are widely used for different practical needs, such as fitting the fatigue data of metal [15] or modeling of the wear process of hard disk drives [20].

The standard models assume the fixed failure threshold which can be not realistic in practice since the deterioration system operates in a random environment and the thresholds and other parameters can change dynamically. That is why for practical application purposes a multi-phase degradation model is preferable [6]. A multi-phase degradation model based on the Wiener process with changing drift and threshold parameters was developed in [8]. The main goal of this research is to consider the more flexible model of the degradation process based on the general Gaussian process to capture different dependence structures including long-range dependence case. We mainly focus on the fractional Brownian motion (FBM) for which analytical results are not available and the required performance measure is estimated via simulation. Note that FBM was considered as a degradation model in [23,24] but they didn't consider the multi-phase case.

By definition, failure occurs when the degradation process reaches the predefined thresholds. For systems with high reliability, the failures may not occur during a short time under normal conditions. Since the event of interest becomes rare, the standard simulation technique can not be applied to estimate the target quantity with suitable accuracy. Thus, it is natural to use methods for variance reduction [1,12,17]. In this paper, a special case of Conditional Monte Carlo, based on the bridge process (BMC – Bridge Monte Carlo estimator) is proposed. The conditional MC approach always gives variance reduction but its performance should be investigated numerically. This research represents the development of the analysis started in [14] where the problem of the tail hitting time distribution estimation was considered.

The rest of this paper is organized as follows. Section 2 describes the degradation model in detail including the definition of the required performance measure, namely the reliability of the system. Additionally available analytical results are briefly discussed. In Sect. 3, the Conditional Monte Carlo method based on the Bridge process is proposed to estimate the reliability of the considered system. A simulation study is carried out in Sect. 4 to demonstrate the performance of the proposed estimator. Finally, some conclusions are stated in Sect. 5.

2   Model Description and Performance Measures

Let consider the so-called multi-phase degradation model with a deterministic sequence of the change points $\tau_1 < \cdots < \tau_n$. Each time period (τ_{i-1}, τ_i),

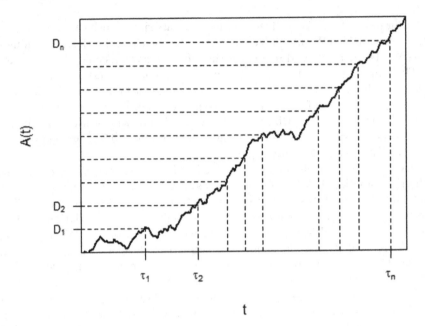

Fig. 1. The typical sample path of the degradation process.

$i = 1, ..., n$ is characterized by the failure threshold level D_i and the mean degradation rate m_i. Assume that the degradation dynamic of the considered deterioration system satisfies the following model

$$A(t) = \Lambda(t) + X(t), \tag{1}$$

where random fluctuations are described in terms of the Gaussian process $\{X(t),\ t \geq 0\}$ with stationary increments and covariance function

$$\Gamma(t, s) := \mathbb{E}\left[X(t)X(s)\right] = \frac{1}{2}\left(v(t) + v(s) - v(|t - s|)\right),$$

where the variance function v assumed to be strictly monotonically increasing.

The deterministic drift $\Lambda(t)$ is a piecewise linear function, namely

$$\Lambda(t) = \sum_{i=1}^{n} m_i t \cdot I(\tau_{i-1} < t < \tau_i),$$

where I denotes the indicator function.

Let denote the failure threshold set $\mathcal{D} = (D_1, .., D_n)$. Then, the lifetime of the deterioration system is

$$T_{\mathcal{D}} := \min\{t : A(t) \geq D(t)\}, \qquad (2)$$

i.e., the first time the process $\{X(t)\}$ hits the threshold piecewise constant curve

$$D(t) = \sum_{i=1}^{n} D_i I(\tau_{i-1} < t < \tau_i).$$

The reliability of the system is defined as the tail distribution of the lifetime $T_{\mathcal{D}}$:

$$R(u \mid \mathcal{D}) := \mathbb{P}(T_{\mathcal{D}} \geq u). \qquad (3)$$

The typical sample path of the degradation process of a deterioration system under changing thresholds is shown in Fig. 1.

Note that in case of two degradation phases

$$D(t) = D_1 \cdot I(t \leq \tau_1) + D_2 \cdot I(t > \tau_1),$$
$$A(t) = m_1 \cdot t \cdot I(t \leq \tau_1) + m_2 \cdot t \cdot I(t > \tau_1).$$

2.1 Available Analytic Results

Let start from the particular single-phase case when $D_1 = D_2 = \cdots = D_n = D$, $m_1 = m_2 = \cdots = m_n = m$ (which corresponds to the hitting time distribution evaluation). Although the *hitting time* of Gaussian processes is well investigated in the literature, closed form analytic results are available only when X is a Wiener process (i.e., $v(t) = t$). The probability density function of $T_{\mathcal{D}}$ has the following form [2]:

$$f_T(t|D) = \frac{D}{\sqrt{2\pi}t^{3/2}} \exp\left(-\frac{(D - mt)^2}{2t}\right) \qquad (4)$$

and the corresponding reliability

$$R(u \mid \mathcal{D}) = \rho(u|m, D),$$

where the function ρ is expressed through the standard normal distribution function Ψ:

$$\rho(u|m, V) := \Psi\left(\frac{V - mu}{\sqrt{u}}\right) - \exp\left(2\,mV\right) \Psi\left(\frac{-V - mu}{\sqrt{u}}\right).$$

Now consider the case when X is the fractional Brownian motion (FBM) with the variance

$$v(t) = t^{2H},$$

where $H \in (0, 1)$ is the Hurst parameter. In case of FBM there are no closed-form expressions for the distribution of $T_{\mathcal{D}}$ (there are only asymptotic results and some bounds). But it is interesting to consider the boundary case $H = 1$ when the process $X = t \cdot \eta$, η is the standard normal random variable $N(0,1)$, thus X is a straight line with a random slope. Then, obviously

$$R(u \mid \mathcal{D}) = \mathbb{P}\left(\eta < \frac{D}{u} - m\right)$$
$$= \Psi\left(\frac{D}{u} - m\right), \tag{5}$$

where Ψ denotes the distribution function of the standard normal random variable $N(0,1)$.

Finally, for a general Gaussian process with stationary increments and strictly monotonically increasing and convex variance such that $\lim_{t\to 0} v(t)/t = 0$, the following asymptotic holds [4]:

$$\mathbb{P}(T_{\mathcal{D}} \leq u) \sim \Phi\left(\frac{D - mu}{\sqrt{v(u)}}\right) \text{ as } D \to \infty, \tag{6}$$

where $\Phi = 1 - \Psi$. Thus, for sufficiently large values of D

$$R(u \mid \mathcal{D}) \approx \Psi\left(\frac{D - mu}{\sqrt{v(u)}}\right). \tag{7}$$

In case of the multi-phase degradation model, there are no closed-form expressions for system reliability except the Wiener process model with two failure thresholds [8]. For the multiple failure threshold, a recursive formula is developed. In general Gaussian case, the analytical solution is not available in closed form.

3   Monte Carlo Estimation

In this paper, our task is to estimate the system reliability for the degradation system with changing failure thresholds and mean rates via Monte Carlo (MC) simulation.

The standard MC approach is based on the estimation of the mean indicator of the event of interest:

$$R(u \mid \mathcal{D}) = \mathbb{E}I(T_{\mathcal{D}} \geq u), \tag{8}$$

The main problem is that for large values u the target probability is extremely small, hence a standard MC estimate fails to evaluate the quantity of interest with given accuracy (usually expressed in terms of the relative error). Thus, a special rare-event simulation technique is required. In this paper, we apply a special variant of the Conditional MC (CMC) method [12,17]. Under this approach, a conditional expectation of the target quantity with respect to some auxiliary random variable is estimated, provided this expectation is available in closed form. In general, it is not easy to select an appropriate auxiliary random variable, however in the current setting it can be expressed in terms if the so-called bridge process [9,10]

$$Y(t) = X(t) - \phi(t)X(s), \tag{9}$$

where s is a some prefixed instant, ϕ is expressed in terms of the covariance function of the process X:

$$\phi(t) := \frac{\Gamma(t,s)}{\Gamma(s,s)}.$$

Note that the function ϕ is strictly positive since the variance function is assumed to be strictly monotonically increasing.

Conditional MC based on the properties of the bridge process was earlier successfully applied to the hitting time distribution estimation [14] which corresponds to the single-phase degradation case.

The target probability can be rewritten in terms of the corresponding bridge process as follows:

$$
\begin{aligned}
R(u\,|\,\mathcal{D}) &= \mathbb{P}(T_{\mathcal{D}} \geq u) \\
&= \mathbb{P}\big(\forall t \in [0,u] : \Lambda(t) + X(t) \leq D(t)\big) \\
&= \mathbb{P}\left(\forall t \in [0,u]\; X(s) \leq \frac{D(t) - Y(t) - \Lambda(t)}{\phi(t)}\right) \\
&= \mathbb{P}\left(X(s) \leq \overline{Y}\right),
\end{aligned}
$$

where

$$\overline{Y} := \inf_{t \in [0,u]} \frac{D(t) - Y(t) - \Lambda(t)}{\phi(t)}, \tag{10}$$

and, due to the independence between $\overline{Y}$ and $X(s)$ and properties of the conditional expectation:

$$
\begin{aligned}
R(u\,|\,\mathcal{D}) &= \mathbb{P}\left(X(s) \leq \overline{Y}\right) \\
&= \mathbb{E}\left[\mathbb{P}\left(X(s) \leq \overline{Y}\,|\,\overline{Y}\right)\right] \\
&= \mathbb{E}\left[\Psi\left(\frac{\overline{Y}}{\sqrt{\Gamma(s,s)}}\right)\right].
\end{aligned}
$$

Hence, given N independent samples $\{\overline{Y}^{(n)},\ n = 1,...,N\}$ of $\overline{Y}$, the BMC estimator of $R(u\,|\,\mathcal{D})$ is

$$\widehat{R}_N^{\mathrm{BMC}} := \frac{1}{N} \sum_{n=1}^{N} \Psi\left(\frac{\overline{Y}^{(n)}}{\sqrt{\Gamma(s,s)}}\right). \tag{11}$$

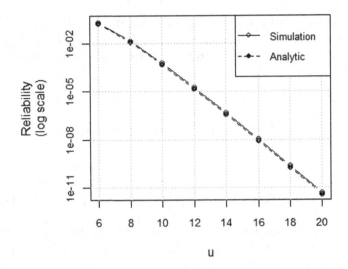

Fig. 2. Reliability of the single phase Wiener model: simulation vs. analytic formula.

4 Simulation Results

In this section, a few numerical examples are presented to illustrate the results given in the previous sections. We start from the description of the general simulation procedure. To estimate the required performance measures one has to generate the sample paths of the Gaussian process X, i.e.

$$X_i^h := X(ih), \quad i = 1, ..., L,$$

where $h > 0$ is the so-called sampling step; L is the simulation horizon. Note that for the considered problem it is enough to take $L = u/h$.

To demonstrate the quality of the proposed BMC estimator $\widehat{R}_N^{\mathrm{BMC}}$ (see Eq. (11)) we compare it with the standard MC estimator $\widehat{R}_N^{\mathrm{MC}}$ (sample mean of the indicator of the target event, see (8)) both calculated from the $N = 10^4$ independent realizations of the Gaussian process X. The typical performance measure is the relative error (RE) defined as the coefficient of variation of the estimator

$$\mathrm{RE}(\widehat{R}_N) := \frac{\sqrt{\mathbb{V}\mathrm{ar}\left[\widehat{R}_N\right]}}{\mathbb{E}\widehat{R}_N},$$

which is calculated for both estimators.

Table 1. Performance of the estimators for the Wiener single phase model.

u	$\widehat{R}^{\mathrm{MC}}$	$\widehat{R}^{\mathrm{BMC}}$	RE($\widehat{R}^{\mathrm{MC}}$)	RE($\widehat{R}^{\mathrm{BMC}}$)
6	1.9e−01	1.91e−01	2.06e−02	1.33e−03
8	1.47e−02	1.42e−02	8.18e−02	2.49e−03
10	8e−04	5.98e−04	3.53e−01	3.46e−03
12	−	1.84e−05	−	4.4e−03
14	−	4.76e−07		5.23e−03
16	−	1.11e−08	−	5.88e−03
18	−	2.41e−10	−	6.59e−03
20	−	4.99e−12	−	7.13e−03

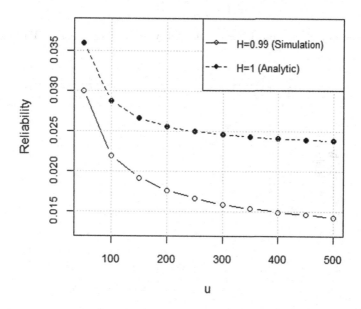

Fig. 3. Reliability of the single-phase FBM model with $H = 0.99$: simulation vs. analytic formula.

4.1 Single Phase

We start from the single phase model when $D_1 = D_2 = \cdots = D_n = D$, $m_1 = m_2 = \cdots = m_n = m$ in order to validate simulations results comparing them with available analytic results.

The first experiment was conducted for the Wiener case with the following values of the parameters: $m = 2$; $D = 10$; $h = 0.1$. Motivated by previous research [14] the conditioning point $s = u$ (the same value was used also in all experiments bellow). The simulation results compared with the available analytic results are shown in Fig. 2 and the corresponding values of the RE are presented in Table 1. Recall that BMC approach being special case of the conditional MC

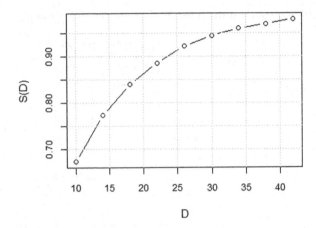

Fig. 4. Accuracy of asymptotic for the reliability of the single-phase FBM model with $H = 0.7$.

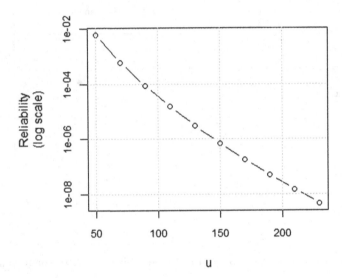

Fig. 5. Reliability of the two-phase FBM model with the Hurst parameter $H = 0.7$.

always leads to the variance reduction and presented results illustrate this property numerically. Note that RE of the BMS estimator increases comparatively slow.

The second experiment (Fig. 3) related to the FBM degradation model with the following parameters: $H = 0.99$, $m = 2$, $D = 10$, $h = 1$. Since the Hurst parameter was chosen close to one, the obtained numerical results are compared with the corresponding formula (5).

The final experiment in this subsection was held of FBM with the Hurst parameter $H = 0.7$ to check the accuracy of the asymptotic (7). To do this the following ratio between estimator and asymptotic is calculated as a function of the threshold level D

$$S(D) := \frac{\widehat{R}_N(u|D)}{\Psi\left(\frac{D-mu}{u^H}\right)}$$

for some fixed value u. The results are shown in Fig. 4, the following values of the parameters were used: $m = 1$; $u = 50$; $H = 0.7$; $h = 1$.

Table 2. Performance of the estimators for the two-phase FBM model with $H = 0.7$.

u	$\widehat{R}^{MC}$	$\widehat{R}^{BMC}$	RE($\widehat{R}^{MC}$)	RE($\widehat{R}^{BMC}$)
50	6.1e−03	6.16e−03	1.27e−01	2.28e−03
70	6e−04	6.17e−04	4.08e−01	3.29e−03
90	2e−04	8.56e−05	7.07e−01	4.03e−03
110	–	1.48e−05	–	4.62e−03
130	–	3.01e−06		5.23e−03
150	–	6.98e−07	–	5.64e−03
170	–	1.79e−07	–	5.99e−03
190	–	4.97e−08	–	6.31e−03
210	–	1.46e−08	–	6.69e−03
230	–	4.57e−09	–	7.07e−03

4.2 Multiple Phases

In this subsection, we provide simulation analysis of the accuracy of the BMC estimator for the two-phase degradation with the mean degradation rates $m_1 = 1, m_2 = 1.2$ and the failure thresholds $D_1 = 20, D_2 = 22$. $N = 10000$ replications of the FBM with Hurst parameter $H = 0.7$, the sampling step $h = 1$. To verify the effectiveness of the proposed estimator, we considered the dependence of the RE on the rarity parameter u in comparison with the standard MC estimator. The numerical results presented in Table 2 illustrate that a limited number of replications permits to estimate probabilities of the order of 10^{-9} with comparatively moderate value of the relative error. The corresponding reliability is shown in Fig. 5.

5 Conclusion

In this paper, the problem of the reliability estimation of the degradation system with multiple phases with different mean degradation rates and failure thresholds was studied. The required performance measure of the considered system is reliability defined as the tail distribution of the lifetime. Since the target probability is small, special rare-event simulation methods should be applied. To tackle this problem, the BMC estimator has been derived and several numerical experiments were carried out indicating the efficiency of the presented estimator compared with the standard MC approach. In future research, it is reasonable to consider the more general degradation model with the random thresholds. Also from the theoretical point of view, it seems interesting to study analytically the asymptotic properties of the proposed estimator expressed in terms of the relative error increasing rate.

References

1. Asmussen, S., Glynn, P.W.: Stochastic Simulation: Algorithms and Analysis. Springer, New York (2007). https://doi.org/10.1007/978-0-387-69033-9
2. Borodin, A.N., Salminen, P.: Handbook of Brownian Motion - Facts and Formulae. Birkhauser, Basel (2002)
3. Borodina, A., Efrosinin, D., Morozov, E.: Application of splitting to failure estimation in controllable degradation system. In: Vishnevskiy, V.M., Samouylov, K.E., Kozyrev, D.V. (eds.) DCCN 2017. CCIS, vol. 700, pp. 217–230. Springer, Cham (2017). https://doi.org/10.1007/978-3-319-66836-9_18
4. Caglar, M., Vardar, C.: Distribution of maximum loss of fractional Brownian motion with drift. Statist. Probab. Lett. **83**, 2729–2734 (2013)
5. Chetvertakova, E.S., Chimitova, E.V.: The Wiener degradation model in reliability analysis. In: 2016 11th International Forum on Strategic Technology (IFOST), pp. 488–490 (2016)
6. Dong, Q., Cui, L.: A study on stochastic degradation process models under different types of failure thresholds. Reliab. Eng. Syst. Saf. **181**, 202–212 (2019)
7. Efrosinin, D.V., Farhadov, M.P.: Optimal management of the system with the gradual and instantaneous failures. Dependability **28**(1), 27–42 (2009). (in Russian)
8. Gao, H., Cui, L., Kong, D.: Reliability analysis for a Wiener degradation process model under changing failure thresholds. Reliab. Eng. Syst. Saf. **171**, 1–8 (2018). https://doi.org/10.1016/j.ress.2017.11.006
9. Giordano, S., Gubinelli, M., Pagano, M.: Bridge Monte-Carlo: a novel approach to rare events of Gaussian processes. In: Proceedings of the 5th St. Petersburg Workshop on Simulation, St. Petersburg, Russia, pp. 281–286 (2005)
10. Giordano, S., Gubinelli, M., Pagano, M.: Rare events of Gaussian processes: a performance comparison between bridge Monte-Carlo and importance sampling. In: Next Generation Teletraffic and Wired/Wireless Advanced Networking, St. Petersburg, Russia, pp. 269–280 (2007)
11. Kahle, W., Lehmann, A.: The Wiener process as a degradation model: modeling and parameter estimation, pp. 127–146. Birkhäuser, Boston (2010). https://doi.org/10.1007/978-0-8176-4924-1_9
12. Kroese, D.P., Taimre, T., Botev, Z.I.: Handbook of Monte Carlo Methods. Wiley, New York (2011)
13. Lisnuansky, A., Levitin, G.: Multi-state System Reliability, Assessment, Optimization and Application. Series on Quality, Reliability and Engineering Statistics, vol. 6 (Book 6). World Scientific Pub Co Inc., New Jersey (2003)
14. Lukashenko, O., Pagano, M.: Rare-event simulation for the hitting time of Gaussian processes. In: Vishnevskiy, V.M., Samouylov, K.E., Kozyrev, D.V. (eds.) DCCN 2020. LNCS, vol. 12563, pp. 589–603. Springer, Cham (2020). https://doi.org/10.1007/978-3-030-66471-8_45

15. Park, C., Padgett, W.: New cumulative damage models for failure using stochastic processes as initial damage. IEEE Trans. Reliab. **54**(3), 530–540 (2005). https://doi.org/10.1109/TR.2005.853278
16. Prakash, G., Kaushik, A.: A change-point-based wiener process degradation model for remaining useful life estimation. Safety and Reliability **39**(3–4), 253–279 (2020). https://doi.org/10.1080/09617353.2020.1801165
17. Ross, S.M.: Simulation. Elsevier, Amsterdam (2006)
18. Rykov, V., Dimitrov, B.: On multi-state reliability systems. In: Proceedings of Seminar Applied Stochastic Models and Information Processes, pp. 128–135 (2002)
19. Si, X.S., Wang, W., Hu, C.H., Chen, M.Y., Zhou, D.H.: A wiener-process-based degradation model with a recursive filter algorithm for remaining useful life estimation. Mech. Syst. Signal Process. **35**(1), 219–237 (2013). https://doi.org/10.1016/j.ymssp.2012.08.016
20. Wang, Y., Ye, Z.S., Tsui, K.L.: Stochastic evaluation of magnetic head wears in hard disk drives. IEEE Trans. Magn. **50**(5), 1–7 (2014). https://doi.org/10.1109/TMAG.2013.2293636
21. Wang, Z., et al.: A generalized degradation model based on Gaussian process. Microelectron. Reliab. **85**, 207–214 (2018). https://doi.org/10.1016/j.microrel.2018.05.001
22. Xiao, M., Zhang, Y., Li, Y., Wang, W.: Degradation modeling based on wiener process considering multi-source heterogeneity. IEEE Access **8**, 160982–160994 (2020). https://doi.org/10.1109/ACCESS.2020.3020723
23. Zhang, H., Chen, M., Shang, J., Yang, C., Sun, Y.: Stochastic process-based degradation modeling and RUL prediction: from Brownian motion to fractional Brownian motion. SCIENCE CHINA Inf. Sci. **64**(7), 1–20 (2021). https://doi.org/10.1007/s11432-020-3134-8
24. Zhang, H., Zhou, D., Chen, M., Xi, X.: Predicting remaining useful life based on a generalized degradation with fractional Brownian motion. Mech. Syst. Signal Process. **115**, 736–752 (2019). https://doi.org/10.1016/j.ymssp.2018.06.029

On Steady State Reliability and Sensitivity Analysis of a k-out-of-n System Under Full Repair Scenario

N. M. Ivanova[1,2]([✉]) [iD]

[1] Peoples' Friendship University of Russia (RUDN University),
6 Miklukho-Maklaya Street, Moscow 117198, Russian Federation
nm_ivanova@bk.ru

[2] V. A. Trapeznikov Institute of Control Sciences of Russian Academy of Sciences,
65 Profsoyuznaya Street, Moscow 117997, Russia

Abstract. The paper is devoted to the investigation of a k-out-of-n system under full repair scenario after its failure. This regime implies the restoration of all failed components for some random time. As a result, the system runs like a new one. A repairable system is one that is repaired not only after a component failure (partial repair), but also after the failure of the entire system (full repair). It is supposed that these repair times have both arbitrary and different distributions, while the components' lifetime is exponentially distributed.

In some previous works, time-dependent reliability characteristics have been obtained with the theory of decomposable semi-regenerative processes [1] and method of characteristics [2] in the same assumptions about life and repair time distributions. Markovization method in particular method of supplementary variables [3] has been applied for some special cases of parameters k and n for calculation of the stationary characteristics. In the current paper, the closed-form representations for the steady state system reliability characteristics for arbitrary k and n under similar assumptions about life and repair time distributions are presented. The obtained expressions are demonstrated in terms of Laplace transform of components' partial repair time. The results are validated on an example of a 3-out-of-6 system by substituting the exponential distribution of repair time. In addition, the probabilistic characteristics of this system in the case of rare failures, as well as some numerical example to show their insensitivity are considered.

Keywords: k-out-of-n system · System reliability · Markovization method · Steady state probabilities · Sensitivity analysis · Rare failures

1 Introduction

Nowadays, the issue of improving system reliability becomes more and more significant and relevant in connection with the complex mechanization and automation of processes occurring in numerous areas of human activity [4]. The importance of this problem is due to the fact that the unsatisfactory reliability of the system generates large costs for its maintenance and, in some cases, can lead to serious consequences.

The k-out-of-n systems have been a popular object of research for many years. These systems are a simple example of redundancy, which is considered as a method of increasing reliability. Such a system consists of n components and fails when at least k of them fail. This type of system is commonly referred to as a k-out-of-n: F system [5]. So, this definition is implied throughout the paper without the symbol "F".

Due to the wide applications of such a system in many spheres of human activity and industries (telecommunication [6], engineering [7], maintenance optimization [8] and others), many papers are devoted to its investigation. There is an extensive literature on the study of such systems (for example, see [5] and the bibliography within).

The development of technology poses new challenges for researchers in engineering and reliability fields. At present, there is an extensive bibliography on the study of various types of k-out-of-n systems. Such systems are considered under various assumptions about the shape of life and repair time distributions [9], the dependency and several types of failures [10], load sharing of failed components [11] and others. There is also an extensive list of papers devoted to the investigation of k-out-of-n systems under several assumptions about repair policies [12–14]. Moreover, among the tasks set in the framework of such investigations, the calculation of stationary and non-stationary characteristics is particularly distinguished. For example, the book by Kuo and Zuo [15] is devoted to the study of time and probabilistic characteristics of various models of k-out-of-n systems.

In addition, in the context of reliability research, it is also of interest to analyze the sensitivity of a system's objective characteristics to the shapes of its input distributions. Many authors have dealt with this problem, for example, B. Sevastyanov, I. Kovalenko, B. Gnedenko, A. Soloviev (a brief overview of their works is presented in [3]). A number of recent studies have also addressed this issue. For example, the investigation of some reliability models in case of rare failures [16], the sensitivity of reliability measures of k-out-of-n systems to the shape of repair time distribution as well as its coefficient of variation [17]. For such purposes, various analytical methods on the basis of multidimensional Markov processes as well as simulation ones are used. Note that the term "sensitivity" can be defined differently in various spheres like engineering and basic sciences [18].

In order to continue research in the reliability field, the current paper considers a k-out-of-n system using one of the markovization method consists in the introduction of supplementary variables. This method was firstly introduced by Cox [19] and further found its application in many studies (see, for example, [20]). This method allows describing the system's behavior by a two-dimensional Markov process that enables one to find analytical expressions for the system's steady state probabilities (s.s.p.) that will be considered. Moreover, sensitivity analysis including the case of rare failures will be conducted.

In the current paper, a modified Kendall's notation $\langle M_{k<n}|GI|1\rangle$ to describe the system under consideration is used. Symbols "$\langle\ \rangle$" denote a closed system. M in the first position correspond to the exponential distribution of component's lifetime, GI means (general independent) arbitrary distribution of component's repair time. Number 1 in the last position corresponds to the number of repair units.

The paper is organized as follows. In the next section, some notations, assumptions as well as problem setting will be done. Section 3 deals with analytical results for steady state system reliability characteristics of $\langle M_{k<n}|GI|1\rangle$ model. The problem is solved with the method of supplementary variables. On the example of a 3-out-of-6 system, the obtained results are validated, and the case of rare failures are also investigated. Some numerical example is also presented. The paper ends with the conclusion and some future research directions.

2   Notations, Assumptions, and Problem Setting

Consider a reparable k-out-of-n $(k < n)$ system according to the definition from the Introduction. If $k = 1$, the model looks like a series system, and with $k = n$ becomes to be a system in parallel. The case of $k = 2$ is the simplest one and slightly different from the general one, so through the paper it is supposed that $k > 2$.

A repairable k-out-of-n system is a system that is repaired not only after a component failure (partial repair), but also after the failure of the entire system (full repair). *Partial repair* means a component repair during some random time, after which the restored component works again and another failed components are repaired one-by-one. *Full repair* occurs, when any k components are failed which leads to their one-time restoration during some random time, after which the entire system works like a new one.

Introduce some assumptions about the shape of components life and repair time distributions. Suppose that

- the lifetimes of system components are independent and exponentially distributed with parameter α and mean time $a = \alpha^{-1}$;
- according to the system structure all n components work simultaneously, that is the system is in a hot redundancy, therefore, the system failure intensity when i components of n fail is $\lambda_i = (n - i)\alpha$, $(i = \overline{0, k - 1})$;
- the repair times for any failed components (the case of partial repair) are independent identically distributed (i.i.d.) random variables (r.v.'s) B_i ($i = 1, 2, \ldots$) with common cumulative distribution function (c.d.f.) $B(x) = \mathbf{P}\{B_i \leq x\}$ which is absolute continuous with its probability density function (p.d.f.) $b(x)$;
- the repair time for failed system (the case of full repair) is also i.i.d. r.v. F_i ($i = 1, 2, \ldots$) with corresponding c.d.f. $F(x) = \mathbf{P}\{F_i \leq x\}$, its p.d.f. is $f(x)$;

- the instantaneous repairs are impossible, their mean times are finite,

$$B(0) = F(0) = 0, \quad b = \int_0^\infty (1 - B(x))dx < \infty, \quad f = \int_0^\infty (1 - F(x))dx < \infty;$$

- corresponding Laplace transforms (LTs) of p.d.f.'s $b(t)$ and $f(t)$ are

$$\tilde{b}(s) = \int_0^\infty e^{-st} b(t)dt, \quad \tilde{f}(s) = \int_0^\infty e^{-st} f(t)dt.$$

Note that repair time distribution for partial and full repair can be equal. Such a regime was considered in [3] on the example of a 3-out-of-6 system.

Denote i as the number of failed components out of n, at that, according to the repair rule, one of such components is repaired. Thus, $(n - i)$ is the number of working components. This system state space can be represented as $E = \{0, 1, 2, ..., k - 1, k\}$. Here the state 0 shows that all the components are working, no one is repaired. The state k means the full system failure and its repair with r.v. F, after which the system becomes as a new one.

To perform reliability analysis, we describe the system behavior as a random process $J = \{J(t), \ t \geq 0\}$ on the space set E:

$$J(t) = j, \ j \in E, \text{ if the system is in state } j \text{ in time } t.$$

Suppose that $J(0) = 0$.

In this paper we deal with the calculation of time-dependent system state probabilities (t.d.s.s.p.'s)

$$\pi_j(t) = \mathbf{P}\{J(t) = j\}, j \in E,$$

steady state probabilities (s.s.p.'s)

$$\pi_j = \lim_{t \to \infty} \mathbf{P}\{J(t) = j\}, j \in E,$$

as well as the availability coefficient

$$K_{av} = \sum_{0 \leq i \leq k-1} \pi_i = 1 - \pi_k.$$

Moreover, the properties of their asymptotic insensitivity to the shapes of system components repair time distributions, including the case of rare failures, are investigated.

3 Main Results

3.1 Markovization Method and T.d.s.s.p.'s

Present analytical results of the s.s.p.'s calculation of $\langle M_{k<n} | GI | 1 \rangle$ system. For this, the method of supplementary variables (one of the markovization methods)

is used [20]. In our case, as a supplementary variable, consider the elapsed repair time of the failed component. Thus, denote by

$$Z(t) = \{J(t), X(t)\}_{t \geq 0}$$

a two-dimensional process, where $J(t)$ is defined as above, and $X(t)$ means the elapsed repair time of the failed component or the whole system. Assuming elapsed repair time x, the conditional intensities of partial and full repairs are, respectively

$$\beta(x) = \frac{b(x)}{1 - B(x)}, \quad \phi(x) = \frac{f(x)}{1 - F(x)}.$$

Due to the method used, the process $Z(t)$ is a Markov one with finitely continuous states space $\bar{E} = \{0, (i, x) \mid i = \overline{1, k}\}$. The corresponding transition graph is shown in Fig. 1.

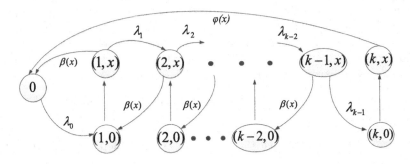

Fig. 1. Transition graph of the process $Z(t)$

According to the new process $Z(t)$, denote by

- $\pi_0(t) = \mathbf{P}\{J(t) = 0\}$ – the probability of a working state of all system components at time t;
- $\pi_i(t; x)dx = \mathbf{P}\{J(t) = i; x < X(t) \leq x + dx\}$ – the joint probability that at time t there are i failed components, among which one is repaired with the elapsed repair time in the interval x and $x + dx$, $i = \overline{1, k}$.

Thus, from Fig. 1 as well as comparing the process $Z(t)$ in the closed interval t and $t + \Delta t$, the following Kolmogorov forward system of partial differential equations for the t.d.s.s.p.'s calculation are obtained.

Theorem 1. *T.d.s.s.p.'s of the process $Z(t)$ are followed from the system of Kolmogorov forward partial differential equations,*

$$\frac{d}{dt}\pi_0(t) = -\lambda_0\pi_0(t) + \int_0^t \beta(x)\pi_1(t,x)dx + \int_0^t \phi(x)\pi_k(t,x)dx, \tag{1}$$

$$\left(\frac{\partial}{\partial t} + \frac{\partial}{\partial x}\right)\pi_1(t,x) = (\lambda_1 + \beta(x))\pi_1(t,x),$$

$$\left(\frac{\partial}{\partial t} + \frac{\partial}{\partial x}\right)\pi_i(t,x) = -(\lambda_i + \beta(x))\pi_i(t,x) + \lambda_{i-1}\pi_{i-1}(t,x), \; i = \overline{2, k-1},$$

$$\left(\frac{\partial}{\partial t} + \frac{\partial}{\partial x}\right)\pi_k(t,x) = -\phi(x)\pi_k(t,x),$$

jointly with the initial

$$\pi_0(0) = 1, \quad \pi_i(0;x) = 0, \; i = \overline{1,k} \quad \forall x \geq 0 \tag{2}$$

and boundary conditions

$$\pi_1(t,0) = \lambda_0\pi_0(t) + \int_0^t \beta(x)\pi_2(t,x)dx, \tag{3}$$

$$\pi_i(t,0) = \int_0^t \beta(x)\pi_{i+1}(t,x)dx, \; i = \overline{2, k-2},$$

$$\pi_{k-1}(t,0) = 0,$$

$$\pi_k(t,0) = \lambda_{k-1}\int_0^t \pi_{k-1}(t,x)dx.$$

Proof. To construct the system of finite difference equations by the usual method of comparison of the corresponding process state probabilities changes on an infinitesimal small-time epochs t and $t + \Delta t$. Then, passing to the limit $\Delta t \to 0$, we obtain the system of differential equations (1).

The initial conditions (2) follows from the assumption that in the very beginning the system is in the state with all components in the good state.

Also by comparing of the process state probabilities at close time epochs t and $t + \Delta t$, when the supplementary variable take values close to zero, we obtain the first two boundary conditions (3). The next boundary condition follows from the fact that the process $Z(t)$ never occurs in the state $k - 1$ with the elapsed repair time equal to zero (see Fig. 1). The last boundary condition follows from the fact that after failing a component from the state $k - 1$, the system occurs in the state k and the residual repair time will be equal to zero. □

The solution of the systems (1)–(3) can be found in [1,2], where the theory of decomposable semi-regenerative processes as well as the method of characteristics were proposed to use.

3.2 Steady State Probabilities of $\langle M_{k<n}|GI|1 \rangle$ System

A further step in reliability analysis is calculation of s.s.p.'s. For this, there are at least two possibilities. From the first one, s.s.p.'s can be obtained by passing to $\pi_i = \lim_{s \to 0} s\tilde{\pi}_i(s)$ from t.d.s.s.p.'s (see, for example, [2]). However, in this paper another way is applied.

The process $Z(t)$ is a Harris one, so according to the Harris Markov processes theory, it has a stationary regime, and, therefore, for $t \longrightarrow \infty$ its t.d.s.s.p.'s tends to corresponding s.s.p.'s. It means that the process $Z(t)$ has a stationary distribution for which the stationary regime differential equations (balance equations) hold with corresponding transition in the systems (1)–(3). Such a transition and the application of the method of constants variation give the following result.

Theorem 2. *The s.s.p.'s of the process $Z(t)$ in terms of LT of partial repair time distribution $\tilde{b}(\lambda_i)$, $i = 1, 2$, and mean full repair time f have the form*

$$\pi_0 = \lambda_0^{-1} \left[C_1 \left(1 + \frac{\lambda_1}{\lambda_1 - \lambda_2} \tilde{b}(\lambda_1) \right) - C_2 \tilde{b}(\lambda_2) \right], \tag{4}$$

$$\pi_1 = C_1 \frac{1 - \tilde{b}(\lambda_1)}{\lambda_1},$$

$$\pi_i = C_i \frac{1 - \tilde{b}(\lambda_i)}{\lambda_i} + S(i-1) \frac{1 - \tilde{b}(\lambda_{i-1})}{\lambda_{i-1}}, \ i = \overline{2, k-1},$$

$$\pi_k = C_k \cdot f,$$

where

$$C_i = C_{i+1} \tilde{b}(\lambda_{i+1}) + S(i)\tilde{b}(\lambda_i) - S(i-1), \ i = \overline{2, k-2}, \tag{5}$$

$$C_{k-1} = -S(k-2),$$

$$C_k = \lambda_{k-1} C_{k-1} \left(\frac{1 - \tilde{b}(\lambda_{k-1})}{\lambda_{k-1}} - \frac{1 - \tilde{b}(\lambda_{k-2})}{\lambda_{k-2}} \right),$$

$$S(i) = \sum_{j=1}^{i} (-1)^{i-j+1} \left(\prod_{m=j}^{i} \frac{\lambda_m}{\lambda_j - \lambda_{m+1}} \right) C_j,$$

and C_1 is calculated according to the normalization condition $\sum_{i \in E} \pi_i = 1$.

Proof. From the systems (1)–(3) with $t \to \infty$ one can write down the following system of balance equations for s.s.p.'s of the process

$$\lambda_0 \pi_0 = \int_0^\infty \beta(x)\pi_1(x)dx + \int_0^\infty \phi(x)\pi_k(x)dx, \tag{6}$$

$$\dot{\pi}_1(x) = -(\lambda_1 + \beta(x))\pi_1(x),$$

$$\dot{\pi}_i(x) = -(\lambda_i + \beta(x))\pi_i(x) + \lambda_{i-1}\pi_{i-1}(x), \ i = \overline{2, k-1},$$

$$\dot{\pi}_k(x) = -\phi(x)\pi_k(x),$$

jointly with boundary conditions

$$\pi_1(0) = \lambda_0\pi_0 + \int_0^\infty \beta(x)\pi_2(x)dx, \tag{7}$$

$$\pi_i(0) = \int_0^\infty \beta(x)\pi_{i+1}(x)dx, \ i = \overline{2, k-2},$$

$$\pi_{k-1}(0) = 0,$$

$$\pi_k(0) = \lambda_{k-1}\int_0^\infty \pi_{k-1}(x)dx.$$

For the calculation of s.s.p.'s the method of constant variation is proposed to use. The solution of the second and the last equations of system (6) gives the probabilities $\pi_1(x)$ and $\pi_k(x)$ in the form

$$\pi_1(x) = C_1 e^{-\lambda_1 x}(1 - B(x)),$$

$$\pi_k(x) = C_k(1 - F(x)).$$

The solution of the third equation is found using the method of constants variation

$$\pi_i(x) = C_i(x)e^{-\lambda_i x}(1 - B(x)), \ i = \overline{2, k-1},$$

where

$$C_i(x) = C_i + \sum_{j=1}^{i-1}(-1)^{i-j+1}\left(\prod_{m=j}^{i}\frac{\lambda_m}{\lambda_j - \lambda_{m+1}}\right)C_j e^{-(\lambda_{i-1}-\lambda_i)x}.$$

The boundary conditions (7) allow finding constants C_i, $(i = \overline{1, k})$. The first boundary condition gives the representation of probability π_0 via C_1 and C_2,

$$C_1 = \lambda_0\pi_0 + C_2\tilde{b}(\lambda_2) - \frac{\lambda_1}{\lambda_1 - \lambda_2}C_1\tilde{b}(\lambda_1),$$

$$\lambda_0\pi_0 = C_1\left(1 + \frac{\lambda_1}{\lambda_1 - \lambda_2}\tilde{b}(\lambda_1)\right) - C_2\tilde{b}(\lambda_2).$$

The second boundary condition gives recursive expression for C_i via C_{i+1} for $i = \overline{2, k-2}$

$$C_i = C_{i+1}\tilde{b}(\lambda_{i+1}) + S(i)\tilde{b}(\lambda_i) - S(i-1),$$

where

$$S(i) = \sum_{j=1}^{i}(-1)^{i-j+1}\left(\prod_{m=j}^{i}\frac{\lambda_m}{\lambda_j - \lambda_{m+1}}\right)C_j.$$

The additional condition for probability $\pi_{k-1}(0)$ gives the representation of C_{k-1},

$$C_{k-1} = -S(k-2).$$

The final constant C_k is found from the last equation of (7),

$$C_k = \lambda_{k-1}\pi_{k-1},$$

$$C_k = \lambda_{k-1}C_{k-1}\left(\frac{1 - \tilde{b}(\lambda_{k-1})}{\lambda_{k-1}} - \frac{1 - \tilde{b}(\lambda_{k-2})}{\lambda_{k-2}}\right).$$

Thus, all the s.s.p.'s are expressed in terms of constants C_i, $(i = \overline{1,k})$, which are recursively computed in terms of themselves. The last one C_1 is found from the normalization condition,

$$\sum_{0 \le i \le k} \int_0^\infty \pi_i(x)dx = 1$$

The simple calculation $\pi_i = \int_0^\infty \pi_i(x)dx$, $i = \overline{1,k}$, and the substitution by $S(i)$, where it is possible, end the proof of the theorem.   □

Remark 1. Since the process $Z(t)$ describe the behavior of the $\langle M_{k<n}|GI|1\rangle$ system, the obtained s.s.p.'s should be used for the investigation of a k-out-of-n system with exponential lifetime and arbitrary distributed repair time.

Consider the application of the Theorem 2 on the example of a 3-out-of-6 system.

Example 1. Present the case $\langle M_{3<6}|GI|1\rangle$. From the Theorem 2 it follows,

$$\pi_1 = \frac{6}{5} \cdot \frac{1 - \tilde{b}(5\alpha)}{1 + 5\tilde{b}(5\alpha) - 5\tilde{b}(4\alpha)}\pi_0, \qquad \pi_2 = \frac{3}{2} \cdot \frac{1 + 4\tilde{b}(5\alpha) - 5\tilde{b}(4\alpha)}{1 + 5\tilde{b}(5\alpha) - 5\tilde{b}(4\alpha)}\pi_0, \quad (8)$$

$$\pi_3 = \frac{6\alpha f(1 + 4\tilde{b}(5\alpha) - 5\tilde{b}(4\alpha))}{1 + 5\tilde{b}(5\alpha) - 5\tilde{b}(4\alpha)}\pi_0, \qquad \pi_0 = 1 - \sum_{1 \le i \le 3} \pi_i,$$

and the availability coefficient

$$K_{av} = \frac{37 + 58\tilde{b}(5\alpha) - 75\tilde{b}(4\alpha)}{60\alpha f(1 + 4\tilde{b}(5\alpha) - 5\tilde{b}(4\alpha)) + 37 + 58\tilde{b}(5\alpha) - 75\tilde{b}(4\alpha)}. \quad (9)$$

It should be noted that for exponential distribution of partial repair time $B(x) = 1 - e^{-\beta x}$ with LT $\tilde{b}(\alpha) = \beta(\alpha + \beta)^{-1}$ and mean time $b = \beta^{-1}$ as well as distribution-independent mean full repair time f, these probabilities match with those, obtained with a simple birth and death process,

$$\pi_0 = \frac{20\alpha^2 + 4\alpha\beta + \beta^2}{120\alpha^3 f + 74\alpha^2 + 10\alpha\beta + \beta^2}, \qquad \pi_1 = \frac{6\alpha(4\alpha + \beta)}{20\alpha^2 + 4\alpha\beta + \beta^2}\pi_0, \quad (10)$$

$$\pi_2 = \frac{30\alpha^2}{20\alpha^2 + 4\alpha\beta + \beta^2}\pi_0, \qquad \pi_3 = \frac{120\alpha^3 f}{20\alpha^2 + 4\alpha\beta + \beta^2}\pi_0.$$

3.3   Rare Failures

The s.s.p.'s from the Theorem 2, as well as the formulas (8)–(9) derived from it, are presented in terms of LT of repair time distribution of the system components. Thence, the obvious dependence of these probabilities on the shape of repair time distribution is observed. On the other hand, papers [16,21] show that with "rare" failures, the shape of such a distribution does not affect the reliability measures. In stochastic systems, this property is called insensitivity. In this section, we consider the behavior of the s.s.p.'s under the rare failures condition.

For the considered model, the rare failures should be understood as the low failures' intensity with respect to the fixed mean repair time. Thus, suppose that $\alpha \to 0$, that consequently leads to the following result.

Theorem 3. *Under the rare components' failures, when $\alpha \to 0$, the s.s.p.'s and the availability coefficient of the 3-out-of-6 system take the form,*

$$\pi_1 \approx \frac{6\rho_1(2 - 5\rho_1)}{2 - 5\rho_1(2 - 9\rho_1)}\pi_0, \quad \pi_2 \approx \frac{30\rho^2}{2 - 5\rho_1(2 - 9\rho_1)}\pi_0, \tag{11}$$

$$\pi_3 \approx \frac{120\rho_1^2\rho_2}{2 - 5\rho_1(2 - 9\rho_1)}\pi_0, \quad \pi_0 \approx \frac{2 - 5\rho_1(2 - 9\rho_1)}{2 + \rho_1(2 + 15\rho_1(3 + 8\rho_2))},$$

$$K_{av} \approx \frac{2 + 2\rho_1 + 45\rho_1^2}{2 + \rho_1(2 + 15\rho_1(3 + 8\rho_2))},$$

where $\rho_1 = \alpha b$ and $\rho_2 = \alpha f$.

Proof. Applying Taylor series up to the second order of α,

$$\tilde{b}(\lambda_i) \approx 1 - b\lambda_i + \frac{b^2\lambda_i^2}{2}, \ i = 1, 2,$$

from (8)–(9) the following ones can find,

$$\pi_1 \approx \frac{6\alpha b(2 - 5\alpha b)}{2 - 5\alpha b(2 - 9\alpha b)}\pi_0, \quad \pi_2 \approx \frac{30(\alpha b)^2}{2 - 5\alpha b(2 - 9\alpha b)}\pi_0,$$

$$\pi_3 \approx \frac{120\alpha^3 b^2 f}{2 - 5\alpha b(2 - 9\alpha b)}\pi_0, \quad \pi_0 \approx \frac{2 - 5\alpha b(2 - 9\alpha b)}{2 + \alpha b(2 + 15\alpha b(3 + 8\alpha f))},$$

$$K_{av} \approx \frac{2 + 2\alpha b + 45(\alpha b)^2}{2 + 2\alpha b + 45(\alpha b)^2 + 120\alpha^3 b^2 f}.$$

Substitutions of $\rho_1 = \alpha b$ and $\rho_2 = \alpha f$ end the proof.

Consider further some numerical example to show the rate of convergence of the availability coefficient K_{av} in case of rare failures, different distributions of repair time as well as different values of its coefficient of variation.

Example 2. Consider again the 3-out-of-6 system. Suppose that the repair time has Gamma (Γ) distribution, $\Gamma(v^{-2}, bv^2)$, where b is the mean repair time, v is the coefficient of variation, that is the ratio of the standard deviation to the mean. According to a modified Kendall's notation, such a system is defined as $\langle M_{3<6}|\Gamma|1\rangle$. If $v = 1$, Γ distribution pass to the exponential one, $\langle M_{3<6}|M|1\rangle$, with the mean time b. Then, the s.s.p.'s coincide with the formulas (10) from the Example 1.

Define $b = 1$, $v = 2$ and 0.5, mean full repair time $f = 2$. Since the time f does not depend on the shape of distribution, the corresponding coefficient of variation is not fixed. Mean lifetime of system elements $a = \overline{0.1, 20}$, then the failure intensity $\alpha = a^{-1}$. Figure 2 shows the dependence of the coefficient of availability K_{av} from the mean lifetime of system elements for different repair time distributions, as well as the case of rare failures.

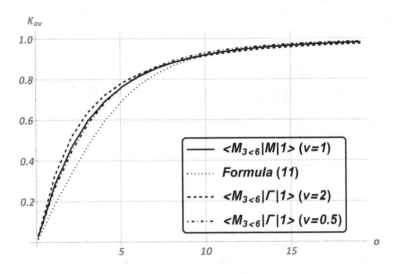

Fig. 2. K_{av} of a 3-out-of-6 system

According to the figure, over the entire interval a under consideration all curves become very close to each other despite the different values of v. As $a \approx 10$, the asymptotic expression (11) from the Theorem 3 shows the absolute accuracy in comparison with the obtained expressions from the Theorem 2.

4   Conclusion

An explicit form of the steady state probabilities of a k-out-of-n system with exponential lifetime and arbitrary partial repair time distribution is presented. These expressions are performed in terms of Laplace transform of repair time and show their dependence on the shape of the corresponding distribution. The produced analytical results reveal asymptotic insensitivity of the steady state reliability measures of a k-out-of-n system under the "rare" failures of its components. Presented numerical examples confirm the obtained analytical results. Further research will continue studies of a k-out-of-n system in direction of reliability and sensitivity analysis.

Acknowledgments. This publication has been supported by the RUDN University Scientific Projects Grant System, project No. P02 IF, theme No. 021930-2-000 (review and analytic results). The publication has been partially funded by RFBR according to the research project No. 20-01-00575A (problem setting) and RSF project No. 22-49-02023 (formal analysis, validation).

References

1. Rykov, V., Ivanova, N., Kozyrev, D.: Application of decomposable semi-regenerative processes to the study of k-out-of-n systems. Mathematics 9(16), 1933 (2021). https://doi.org/10.3390/math9161933
2. Dimitrov, B., Rykov, V.: On k-out-of-n system under full repair and arbitrary distributed repair time. In: Vishnevskiy, V.M., Samouylov, K.E., Kozyrev, D.V. (eds.) DCCN 2021. LNCS, vol. 13144, pp. 323–335. Springer, Cham (2021). https://doi.org/10.1007/978-3-030-92507-9_26
3. Rykov, V.V., Ivanova, N.M., Kozyrev, D.V.: Sensitivity analysis of a k-out-of-n:F system characteristics to shapes of input distribution. In: Vishnevskiy, V.M., Samouylov, K.E., Kozyrev, D.V. (eds.) DCCN 2020. LNCS, vol. 12563, pp. 485–496. Springer, Cham (2020). https://doi.org/10.1007/978-3-030-66471-8_37
4. Trivedi, K.S., Bobbio, A.: Reliability and Availability Engineering - Modeling, Analysis, and Applications. Cambridge University Press, Cambridge (2017). https://doi.org/10.1017/9781316163047
5. Zuo, M.J., Tian, Z.: k-out-of-n Systems. Wiley Encyclopedia of Operations Research and Management Science. Ed. by James J. Cochran (2010)
6. Wang, Y., Hu, L., Yang, L., Li, J.: Reliability modeling and analysis for linear consecutive-k-out-of-n: F retrial systems with two maintenance activities. Reliab. Eng. Syst. Saf. **226**, 108665 (2022). https://doi.org/10.1016/j.ress.2022.108665
7. Zhang, T., Zhang, Y., Du, X.: Reliability analysis for k-out-of-n systems with shared load and dependent components. Struct. Multidiscip. Optim. **57**(3), 913–923 (2018). https://doi.org/10.1007/s00158-017-1893-z
8. Eruguz, A.S., Tan, T., van Houtum, G.-J.: Optimizing usage and maintenance decisions for k-out-of-n systems of moving assets. Nav. Res. Logist. **64**, 418–434 (2017). https://doi.org/10.1002/nav.21764
9. Rykov, V.V., Ivanova, N.M.: Reliability and sensitivity analysis of a repairable k-out-of-n: F system with general life- and repair times distributions. In: Baraldi, P., Di Maio, F., Zio, E. (eds.) Proceedings of the 30th European Safety and Reliability Conference and the 15th Probabilistic Safety Assessment and Management Conference. Research Publishing Services, Singapore (2020). https://doi.org/10.3850/978-981-14-8593-05750-cd
10. Hu, L., Liu, S., Peng, R., Liu, Z.: Reliability and sensitivity analysis of a repairable k-out-of-n: G system with two failure modes and retrial feature. Commun. Stat. Theory Methods (2020). https://doi.org/10.1080/03610926.2020.1788083

11. Amari, S., Bergman, R.: Reliability analysis of k-out-of-n load-sharing systems. Ann. Reliab. Maintainab. Sympos. 440–445 (2008). https://doi.org/10.1109/RAMS.2008.4925836

12. Krishnamoorthy, A., Ushakumari, P., Lakshmi, B.: K-out-of-n system with repair: the N-policy. Asia Pacific J. Oper. Res. **19**, 47–61 (2002)

13. Ushakumari, P., Krishnamoorthy, A.: k-out-of-n system with repair: the $max(N,T)$ policy. Perform. Eval. **57**(2), 221–234 (2004). https://doi.org/10.1016/j.peva.2003.10.006

14. Wu, W., Tang, Y., Yu, M., Jiang, Y.: Computation and profit analysis of a k-out-of-n: G repairable system under N-policy with multiple vacations and one replaceable repair facility. RAIRO-Oper. Res. **49**(4), 717–734 (2015). https://doi.org/10.1051/ro/2015001

15. Kuo, W., Zuo, M.: Optimal Reliability Modeling: Principles and Applications. Wiley, New York (2003)

16. Houankpo, H.G.K., Kozyrev, D.: Mathematical and simulation model for reliability analysis of a heterogeneous redundant data transmission system. Mathematics **9**, 2884 (2021). https://doi.org/10.3390/math9222884

17. Ivanova, N.: Modeling and simulation of reliability function of a k-out-of-n: F system. Commun. Comput. Inf. Sci. **1337** (2020). https://doi.org/10.1007/978-3-030-66242-4_22

18. Kala, Z.: New importance measures based on failure probability in global sensitivity analysis of reliability. Mathematics **9**, 2425 (2021). https://doi.org/10.3390/math9192425

19. Cox, D.R.: The analysis of non-Markovian stochastic processes by the inclusion of supplementary variables. Math. Proc. Cambridge Philos. Soc. **51**(3), 433–441 (1955). https://doi.org/10.1017/S0305004100030437

20. Kalashnikov, V.: Method of supplementary variables. In: Mathematical Methods in Queuing Theory. Mathematics and Its Applications, vol. 271. Springer, Dordrecht (1994). https://doi.org/10.1007/978-94-017-2197-4_10

21. Rykov, V.: On steady state probabilities of renewable system with Marshal–Olkin failure model. Stat. Pap. **59**(4), 1577–1588 (2018). https://doi.org/10.1007/s00362-018-1037-6

Author Index

Adou, Y.   41
Ageeva, Anastasia   176
Ahmed, Riyadh Khlf   3
Alwan, Mohammed Hasan   17
Aminev, Dmitry   144

Bács, Zoltán   231
Barabanova, E. A.   338, 397
Berezkin, A.   90, 103
Bobrikova, Ekaterina   132
Bogdanova, Evgenia   144
Botvinko, A. Yu.   190

Dudin, Alexander   243, 257
Dudin, Sergey   243
Dudina, Olga   243
Dvorkovich, A. V.   338

Efrosinin, Dmitry   284, 297

Gaidamaka, Yu. V.   29, 41, 132
Glushkova, Anastasia G.   117
Gorbunova, A. V.   371
Gorshenin, Andrey   176
Granin, S.   361

Hammadi, Yousif I.   3, 17
Hilquias, Viana C. C.   309

Ivanova, N. M.   422

Jain, Vidyottama   205, 218

Keba, Anastasia   349
Keyela, P.   29
Khayrov, E. M.   161
Kim, Chesoong   243
Kirichek, R.   90, 103
Klimenok, Valentina   257
Kochetkova, Irina   176
Kozyrev, Dmitry   144
Kroshin, Fedor S.   324
Kukunin, D.   90, 103

Laptin, V.   361
Lukashenko, Oleg   410

Mahmood, Omar Abdulkareem   3, 17
Makeeva, Elena   176
Mandel, A.   361
Markova, E.   41
Markovich, Natalia M.   67, 75
Matyushenko, S. I.   309
Medvedeva, Ekaterina   132
Milovanova, T. A.   309
Morozov, Evsey   297
Morozova, Mariya   270
Muhi, Mamoon A.   17
Muthanna, Ammar   3
Muthanna, Mohammed Saleh Ali   17

Nazarov, Anatoly   270
Nekrasova, Ruslana   297
Nezhel'skaya, Ludmila   349

Paul, Svetlana   270
Peshkova, Irina   385
Phung-Duc, Tuan   270
Pintér, Ákos   231
Platonova, Anna   132, 161
Prosvirov, V. A.   161

Raj, Raina   205, 218
Rumyantsev, Alexander   385
Ryzhov, Maksim S.   67, 75

Samouylov, K. E.   190
Selvamuthu, Dharmaraja   205, 218
Sevastianov, Leonid A.   117
Shahoud, Ayham   54
Shashev, Dmitriy   54
Shchetinin, Eugene Yu.   117
Shidlovskiy, Stanislav   54
Shorgin, Sergey   132
Stepanov, Mikhail S.   324
Stepanov, Sergey N.   324

Stepanova, Natalia   284
Sztrik, János   231

Tóth, Ádám   231
Tselykh, Alexey   17

Vishnevsky, V. M.   218, 338, 371, 397
Vytovtov, G. K.   338
Vytovtov, K. A.   338, 397

Yartseva, I. S.   29

Zaryadov, I. S.   309

Printed in the United States
by Baker & Taylor Publisher Services